2^3-8
3^3-27

product - MULIPLY
Sum - add

ELEMENTARY ALGEBRA

Second Edition

VIVIAN SHAW GROZA
Sacramento City College

W. B. SAUNDERS COMPANY
Philadelphia • London • Toronto

W. B. Saunders Company: West Washington Square
Philadelphia, PA 19105

1 St. Anne's Road
Eastbourne, East Sussex BN21 3UN, England

1 Goldthorne Avenue
Toronto, Ontario M8Z 5T9, Canada

Library of Congress Cataloging in Publication Data

Groza, Vivian Shaw

Elementary Algebra

1. Algebra. I. Title.

QA152.2.G75 1978 512.9'042 77-78571

ISBN 0-7216-4322-1

Elementary Algebra ISBN 0-7216-4322-1

Last digit is the print number: 9 8 7 6 5 4 3 2

PREFACE
TO THE SECOND EDITION

This second edition of *Elementary Algebra* contains revisions and modifications resulting mainly from direct and indirect suggestions of students during three years of classroom usage.

The Exercises and Sample Problems have been carefully examined; wording has been clarified, explanations have been given in greater detail when it was deemed appropriate, problems proving to be too difficult have been deleted, and problems more suitable to the needs of the students have been added.

The unit on the fractions of arithmetic has been incorporated with the unit on algebraic fractions, since this has proved to be a more effective method of instruction. The sections on stated problems have been altered, rearranged, and improved to better suit the needs and abilities of the students.

The intention remains to provide an individualized course of instruction in beginning algebra, serving the needs of many types of students.

Words and development of theory have been minimized, since it is felt that comprehension and performance are best developed by actively doing the mathematics. For this purpose, the text contains many illustrative examples and an abundance of drill exercises. Each unit of the text contains a pretest and a posttest keyed by number to the subsections of the unit. In this way, any difficulty can be readily identified, and the student can be directed to the appropriate subsections for further study. All answers are provided in the back of the text.

A student who has not previously taken a course in algebra might use the text as follows:

1. Read the pretest (without actually doing the problems) to know what is expected.
2. Study the subunits in order and do the "A" exercises.
3. Take the pretest.
4. If test results are satisfactory, proceed to the next unit.
5. If test results are not satisfactory, restudy the subunit (or subunits) corresponding to the problems worked incorrectly and

do the "B" exercises. Since the test problems are keyed by number to match the subunits, this can readily be done.

6. Take the posttest.

A student who is reviewing algebra or who is using the text as supplementary material might use the text in the following way:

1. Take the pretest of a unit of interest.
2. If results of the test are satisfactory, proceed to another unit.
3. If results are not satisfactory, proceed to the subunits corresponding to the problems worked incorrectly and do these problems.
4. Take the posttest.

In addition to informing the student of exactly what is expected of him and directing him to the basic source of his difficulty, this text also has the advantages of allowing the student to progress at his own rate and of providing him latitude in selecting those topics which he wants to study. In essence, *Elementary Algebra* is intended to meet the needs of each individual student.

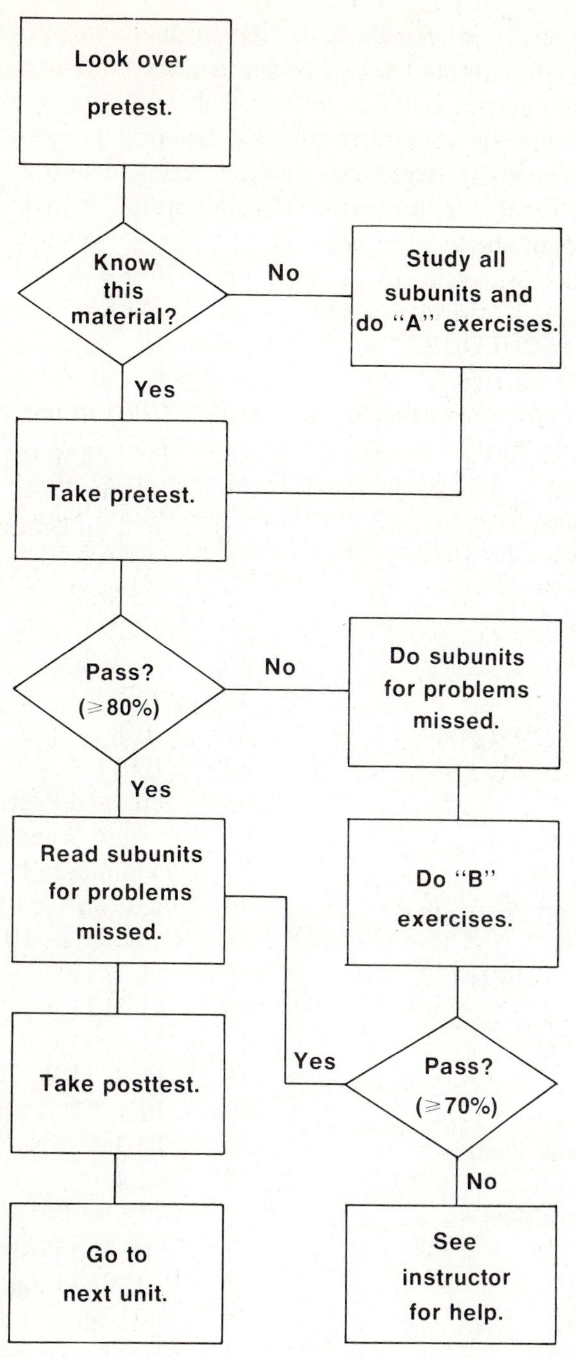
Look over
pretest.
Know
this
material?
No
Study all
subunits and
do "A" exercises.
Yes
Take pretest.
Pass?
(≥80%)
No
Do subunits
for problems
missed.
Yes
Read subunits
for problems
missed.
Do "B"
exercises.
Take posttest.
Yes
Pass?
(≥70%)
No
Go to
next unit.
See
instructor
for help.

This book is also adaptable to the traditional course of instruction in elementary algebra. It can be used either as a textbook or as a supplement to the text. A schedule that has proved to be effective is shown below. It is designed for a semester course of 80 instructional days, 50 minutes per day. However, alternatives for shorter terms are indicated by the insertion of "Final Comprehensive Examination," which signals a convenient stopping point.

SUGGESTED SCHEDULE

Day	*Units to be covered*
1	1.1 and 1.2
2	1.3, 1.4, 1.5
3	1.6, 1.7
4	2.1, 2.2
5	2.3
6	3.1, 3.2
7	3.3, 3.4
8	3.5
9	4.1, 4.2, 4.3, 4.4
10	4.5, 4.6
11	4.7
12	4.8, 4.9
13	4.10, 4.11
14	Pre- and Posttests, Units 1, 2, 3, 4
15	Test on Units 1, 2, 3, 4
16	5.1, 5.2, 5.3, 5.4
17	5.5, 5.6
18	5.7
19	5.8
20	Pre- and Posttests, Unit 5
21	Test on Unit 5
22	7.1, 7.2
23	7.3, 7.4
24	7.5, 7.6
25	7.7
26	7.8, 7.9, 7.10
27	8.1
28	8.2
29	8.3
30	8.4
31	8.5
32	8.6
33	8.7
34	Pre- and Posttests, Units 7, 8
35	Test on Units 7 and 8
36	9.1, 9.2
37	9.3, 9.4
38	9.5, 9.6
39	9.7
40	9.8
41	10.1
42	10.2
43	Pre- and Posttests, Unit 9 and Unit 10, numbers 1 – 10
44	Test on Unit 9 and Sections 10.1, 10.2
45	11.1, 11.2
46	11.3, 11.4
47	11.5, 11.6
48	11.7, 11.8
49	12.1, 12.2
50	12.3
51	12.4
52	12.5
53	Pre- and Posttests, Unit 11 and Unit 12, numbers 1 – 10
54	Test on Unit 11 and Sections 12.1 – 12.5

FINAL COMPREHENSIVE EXAMINATION or as follows:

55	6.1
56	6.2, 6.3

Day	*Units to be covered*
57	6.4
58	6.5, 6.6, 6.7
59	6.9, 6.10, 6.11
60	10.3
61	10.4
62	10.5
63	12.7
64	12.8
65	Review, Stated Problems on Pretests and Posttests of Units 6, 10, and 12
66	Test on Stated Problems

FINAL COMPREHENSIVE EXAMINATION or as follows:

Day	*Units to be covered*
67	13.1
68	13.2
69	13.3
70	13.4
71	13.5
72	13.6
73	13.7
74	13.8
75	Pre- and Posttests, Unit 13

Test on Unit 13 or FINAL COMPREHENSIVE EXAMINATION

V. S. G.

PREFACE
TO THE FIRST EDITION

Elementary Algebra has been designed as an individualized course of instruction in beginning algebra. This text is intended to serve the needs of many types of students: students enrolled in a regular semester or quarter course in elementary algebra; students enrolled in an elementary algebra course that involves independent study, such as that given in schools having a mathematics laboratory; students wishing to review elementary algebra by independent study; students (currently enrolled in an elementary algebra course) needing supplementary drill material; and students (enrolled in courses requiring an algebraic background) needing review or reinforcement in special algebraic topics.

Words and development of theory have been minimized, since it is felt that comprehension and performance are best developed by actively doing the mathematics. For this purpose, the text contains many illustrative examples and an abundance of drill exercises. Each unit of the text contains a list of the unit objectives, a pretest and a posttest – all keyed by number to the subsections of the unit. In this way, any difficulty can be readily identified, and the student can be directed to the appropriate subsections for further study. All answers are provided in the back of the text.

A student who has not previously taken a course in algebra might use the text as follows:

1. Read the unit objectives and read the pretest (without actually doing the problems) to know what is expected.
2. Study the subunits in order and the "A" exercises.
3. Take the pretest.
4. If test results are satisfactory, proceed to the next unit.
5. If test results are not satisfactory, restudy the subunit (or subunits) corresponding to the problems worked incorrectly and do the "B" exercises. Since the test problems are keyed by number to match the subunits, this can readily be done.
6. Take the posttest.

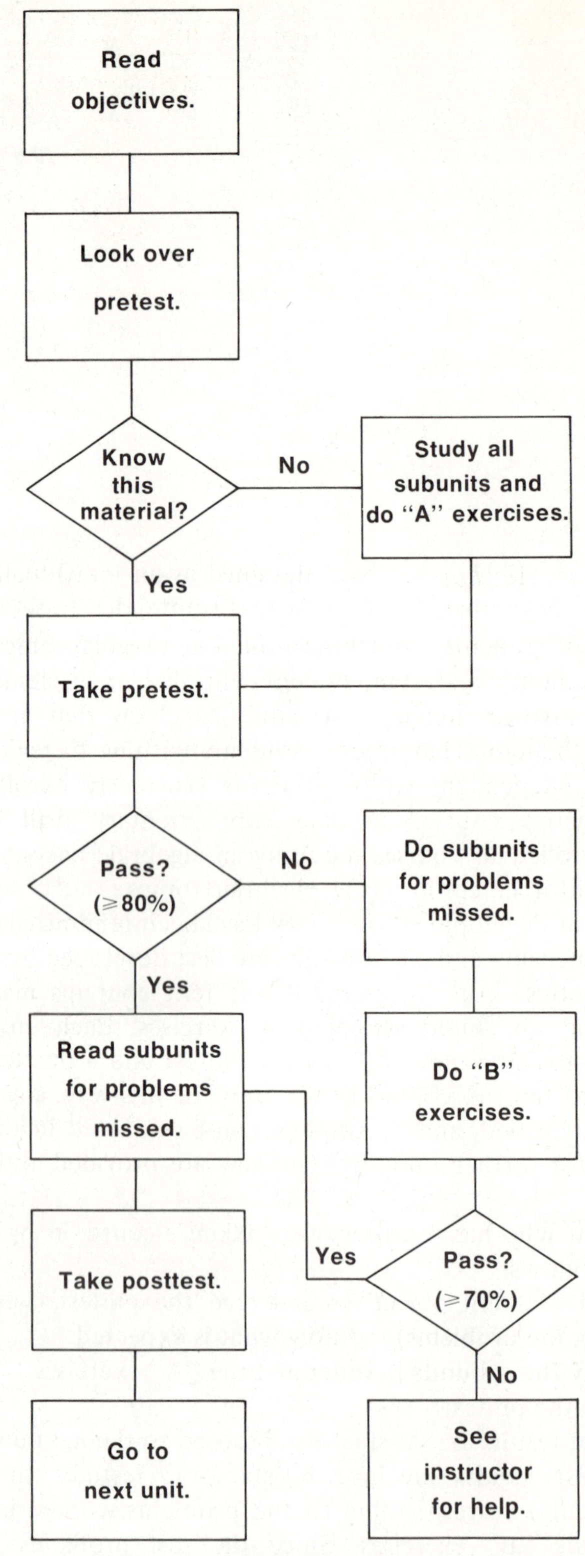
Read objectives.
Look over pretest.
Know this material?
No
Study all subunits and do "A" exercises.
Yes
Take pretest.
Pass? (≥80%)
No
Do subunits for problems missed.
Yes
Read subunits for problems missed.
Do "B" exercises.
Yes
Pass? (≥70%)
Take posttest.
No
Go to next unit.
See instructor for help.

This book is also adaptable to the traditional course of instruction in elementary algebra. It can be used either as the textbook or as a supplement to the text. Since an attempt has been made to isolate each basic idea involved in a unit of study and to present this as a single subunit, it will be necessary, in some cases, to cover two or more subunits in one instructional day. One possible procedure would be as follows:

1. The instructor presents the idea of a subunit and one or more sample problems, leaving a sample problem displayed.
2. Students work one or more sample problems. (The instructor has the opportunity to offer individual help at this time. Also, faster students may offer help to slower students.)
3. The instructor repeats steps 1 and 2 with the next subunit, continuing in this way as time permits.

The pretests and posttests may be used as sample tests for the students. The instructor's manual contains two additional tests for each unit, and these can be used as the regular examinations. The instructor's manual also contains explanatory comments and suggestions for using the text.

A student who is reviewing algebra or who is using the text as supplementary material might use the text in the following way:

1. Take the pretest of a unit of interest.
2. If results of the test are satisfactory, proceed to another unit.
3. If results are not satisfactory, proceed to the subunits corresponding to the problems worked incorrectly and do these problems.
4. Take the posttest.

In addition to informing the student of exactly what is expected of him and directing him to the basic source of his difficulty, this text also has the advantages of allowing the student to progress at his own rate and of providing him latitude in selecting those topics which he wants to study. In essence, *Elementary Algebra* is intended to meet the needs of each individual student.

I would like to express my gratitude to the following persons for their invaluable suggestions for improving the manuscript: Raymond F. Bryant of Florissant Valley Community College, St. Louis, Missouri; Peter C. Foltz of Harrisburg Area Community College, Harrisburg, Pennsylvania; Ken Seydel of Skyline College, San Bruno, California; Tsutomu Kanematsu, a mathematics student at Sacramento City College, Sacramento, California; and J. B. Curtis, President of Western Colorado University at Grand Junction, Colorado.

V. S. G.

CONTENTS

UNIT 1

BASIC OPERATIONS

UNIT 1
PRETEST

(1–4) Express each of the following in symbols. (1.1)

1. The sum of 4 and x. 1. ______________
2. The product of 3 and y. 2. ______________
3. The difference when x is subtracted from 9. 3. ______________
4. The quotient when 7 is divided by y. 4. ______________

(5–8) Express each of the following in symbols. Do not evaluate. (1.2)

5. The square of 6. 5. ______________
6. The square of n. 6. ______________
7. The cube of 5. 7. ______________
8. The cube of y. 8. ______________

(9–13) Express each of the following in symbols. (1.3)

9. Three less than the product of 4 and y. 9. ______________
10. Nine times the sum of x and 6. 10. ______________
11. The difference when the sum of n and 3 is subtracted from 10. 11. ______________
12. The square of the sum of y and 4. 12. ______________
13. One-half the cube of the sum of t and 1. 13. ______________

(14–16) Evaluate. (1.4)

14. $25 - (15 - 5)$ 14. ______________

15. $4(8 - 3)^3$ 15. ______________

16. $\dfrac{6^2 - 4^2}{6 + 4}$ 16. ______________

(17–19) Evaluate. (1.5)

17. $3 + 2[9 - 3(8 - 6)]$ 17. ______________

18. $9 - \{8 - (7 - 6)\}$ 18. ______________

19. $\dfrac{(5 + 2)^2 - (5 - 2)^2}{7 + 3}$ 19. ______________

(20–22) Evaluate. (1.6)

20. $x(x - 5)$ for $x = 7$ 20. ______________

21. $\dfrac{3n + n^2}{3n + 9}$ for $n = 6$ 21. ______________

22. $(4y - 2)(y - 3)$ for $y = 5$ 22. ______________

(23–25) Evaluate. (1.7)

23. $\dfrac{x^2 + xy - 6y^2}{x - 2y}$ for $x = 7$ and $y = 2$ 23. ______________

24. $(x + y - 5)(x - y + 5)$ for $x = 8$ and $y = 6$ 24. ______________

25. $\dfrac{x^3 + (y + 3)^3}{x + y + 3}$ for $x = 5$ and $y = 4$ 25. ______________

1.1 ONE OPERATION: ADDITION, SUBTRACTION, MULTIPLICATION, DIVISION

OBJECTIVE: To translate a verbal expression about the sum, difference, product, or quotient of two numbers into algebraic symbols.

Algebra is the theory of arithmetic. It is a study of numbers and their properties. The numbers of arithmetic are also some of the numbers of algebra. The term, **numbers of arithmetic**, includes the **counting numbers**,

$$1,\ 2,\ 3,\ 4,\ 5,\ 6,\ 7,\ 8,\ 9,\ 10,\ 11,\ 12,\ \ldots$$

(the three dots mean that this pattern continues without end), the number zero, 0, and the quotients of counting numbers, also called fractions, such as $\frac{1}{2}$, $\frac{3}{5}$, and $\frac{16}{7}$.

The other numbers of algebra will be introduced gradually as the study progresses. For now, "number" shall mean one of the numbers of arithmetic.

A **numeral** is a symbol that names a number according to a specified system of numeration, such as the decimal system. Examples of numerals are 3, 5, 12, $\frac{1}{2}$, and $10\frac{1}{4}$.

In algebra, a **letter** may also be used as the name of a number.

A **constant** is a numeral or letter that is the name of exactly one number. The letters a, b, and c are commonly used as constants.

A **variable** is a letter that names an unspecified number that belongs to a set of numbers. The letters x, y, and z are commonly used as variables. Later on we will work with expressions such as $2x^2 + 3x + 5$, $3x^2 - 2x - 1$, and $x^2 - 2x + 3$, where x may represent one of several numbers for each problem. To talk about these problems in a general way, we say that each has the form $ax^2 + bx + c$. Here, a, b, and c are constants, and each is given exactly one value for a special problem. The letter x is the variable because it may be replaced by any one of several numbers for each special problem.

The **sum** of two numbers is the result obtained by adding the two numbers. The numbers that are added are called the **terms** of the sum. For $3 + 4 = 7$, 3 and 4 are terms, and 7 is the sum.

The **difference, or remainder,** is the result obtained by subtracting one number from another. The numbers that form the difference are called the **terms** of the difference. For $10 - 4 = 6$, 10 and 4 are terms, and 6 is the difference.

The **product** is the result obtained by multiplying two numbers, and the **factors** are the numbers that are being multiplied. For $3 \times 4 = 12$, 3 and 4 are factors and 12 is the product.

The **quotient** is the result obtained when one number, the **dividend,** is divided by another number, the **divisor.** For $\frac{35}{5} = 7$, 7 is the quotient, 35 is the dividend, and 5 is the divisor.

The ways to write the result when two numbers are combined by the operations of addition, subtraction, multiplication, or division are shown in the sample problems.

SAMPLE PROBLEMS 1-20

Express each of the following in symbols.

Words	Symbols
1. the sum of 4 and 5	1. $4 + 5$
2. 4 added to 5	2. $5 + 4$
3. the sum of 4 and a number x	3. $4 + x$
4. a number x plus a number y	4. $x + y$
5. a number x minus a number y	5. $x - y$
6. 5 less than a number x	6. $x - 5$
7. 4 subtracted from 5	7. $5 - 4$
8. 5 subtracted from y	8. $y - 5$
9. 5 increased by x	9. $5 + x$
10. 5 decreased by y	10. $5 - y$
11. the product of 4 and 5	11. $4 \cdot 5$, or $4(5)$, or $(4)(5)$
12. 4 times a number x	12. $4x$
13. a number x multiplied by a number y	13. xy
14. 4 divided by 5	14. $\frac{4}{5}$
15. the quotient of 4 divided by x	15. $\frac{4}{x}$
16. x divided by y	16. $\frac{x}{y}$
17. the quotient when 8 is divided into x	17. $\frac{x}{8}$
18. twice x	18. $2x$
19. one-half of x	19. $\frac{x}{2}$
20. one-third of x	20. $\frac{x}{3}$

The product of 1 and x is written as x, as is the product of x and 1, since $1 \cdot x = x$ and $x \cdot 1 = x$ for all numbers x. For example, $1 \cdot 5 = 5$, $1 \cdot \frac{1}{2} = \frac{1}{2}$, $20 \cdot 1 = 20$, and so on. While it is correct to write the product of 1 and x as $1 \cdot x$ or as $1x$, the most commonly used form for the product is x.

It is correct to express a product involving a numeral and a letter, or two letters, by using a dot (such as $3 \cdot a$, $x \cdot 4$, and $x \cdot y$), although the most commonly used forms for these products are $3a$, $4x$, and xy. In such cases, the dot is omitted, the numeral precedes the letter, and literal factors are written in alphabetical order.

When a numeral and a letter, or when two letters, are written next to each other, multiplication is implied. For example, $4x$ means 4 times x, and xy means x times y.

A product involving two numerals, however, cannot be written without using a dot. For example, the product of 4 and 5 must be written as $4 \cdot 5$. Note that $4 \cdot 5 = 20$, whereas 45 means forty-five.

EXERCISES 1.1

Express each of the following in symbols.

A.

1. The product of 7 and a number y. 1. ____________
2. The sum of a number s and 11. 2. ____________
3. The sum of 3 and a number x. 3. ____________
4. The difference when a number x is subtracted from 5. 4. ____________
5. The difference when 5 is subtracted from a number x. 5. ____________
6. The quotient when a number n is divided by 4. 6. ____________
7. The quotient when a number x is divided by 8. 7. ____________
8. The quotient when 4 is divided by a number n. 8. ____________
9. The sum of a number q and 6. 9. ____________
10. The difference when q is subtracted from 6. 10. ____________

11. q less than 6. 11. ______

12. The quotient when q is divided into 6. 12. ______

13. The product of 1 and y. 13. ______

14. The product of y and 9. 14. ______

15. The product of 3 and 25. 15. ______

16. The difference when r is subtracted from 8. 16. ______

17. The quotient when 7 is divided by z. 17. ______

18. The sum of x and 9. 18. ______

19. The product of x and 9. 19. ______

20. The difference when 2 is subtracted from t. 20. ______

21. t decreased by z. 21. ______

22. The product of a number x and a number y. 22. ______

23. The product of 2 and 3. 23. ______

24. The product of 2 and x. 24. ______

25. The quotient when n is divided by 2. 25. ______

26. The quotient when 2 is divided by n. 26. ______

27. The difference when x is subtracted from y. 27. ______

28. The difference when y is subtracted from x. 28. ______

29. The sum of y and z. 29. ______

30. The quotient when r is divided by s. 30. ______

31. Twice y. 31. ______

32. One-half of y. 32. ______

33. One-third of t. 33. ____________

34. One-fourth of n. 34. ____________

35. One-fifth of x. 35. ____________

B.

1. The sum of a number n and 8. 1. ____________

2. The product of a number n and 8. 2. ____________

3. The difference when 12 is subtracted from t. 3. ____________

4. The difference when t is subtracted from 12. 4. ____________

5. The quotient when x is divided by 7. 5. ____________

6. The quotient when 7 is divided by x. 6. ____________

7. The product of a and b. 7. ____________

8. The sum of a and b. 8. ____________

9. The quotient when a is divided by b. 9. ____________

10. The difference when a is subtracted from b. 10. ____________

11. The product of 1 and n. 11. ____________

12. 4 less than x. 12. ____________

13. The quotient when 4 is divided into t. 13. ____________

14. The quotient when t is divided into 4. 14. ____________

15. 9 increased by y. 15. ____________

16. 9 decreased by y. 16. ____________

17. Twice t. 17. ____________

18. Three times r. 18. ____________

19. One-half of z. 19. ____________

20. One-third of n. 20. ____________

1.2 SQUARING, CUBING, RAISING TO A POWER

OBJECTIVE: To translate a verbal expression about the square, cube, or nth power of a number into algebraic symbols.

The **square** of a number is the product obtained by using the given number as a factor two times. The squaring operation is indicated in symbols by writing a 2 to the upper right of the number that is to be squared.

As examples,

3^2 means $3 \cdot 3$, or 9

5^2 means $5 \cdot 5$, or 25

x^2 means xx

The **cube** of a number is the product obtained by using the given number as a factor three times. The cubing operation is indicated in symbols by writing a 3 to the upper right of the number that is to be cubed.

As examples,

3^3 means $3 \cdot 3 \cdot 3$, or 27

5^3 means $5 \cdot 5 \cdot 5$, or 125

x^3 means xxx

In general, a counting number, called an **exponent**, written to the upper right of another number, called the **base**, shows how many times the base is to be used as a factor. This special type of multiplication is called **raising to a power**.

As an example, 2^5 (read "two to the fifth" or "the fifth power of two") means $2 \cdot 2 \cdot 2 \cdot 2 \cdot 2$, or 32.

Similarly,

x^4, x to the fourth, means $xxxx$

x^5, x to the fifth, means $xxxxx$

In general,

x^n means $xx \ldots x$ (n factors of x)

SAMPLE PROBLEMS 1-10

Express each of the following in symbols.

Words	*Symbols*
1. the square of 5 (5 times 5)	**1.** 5^2
2. the square of x **(**x **times** x**)**	**2.** x^2
3. y **squared**	**3.** y^2
4. the cube of 2 (2 times 2 times 2)	**4.** 2^3
5. the cube of x **(**x **times** x **times** x**)**	**5.** x^3
6. y **cubed**	**6.** y^3
7. the fourth power of 3	**7.** 3^4
8. the fourth power of x	**8.** x^4
9. the sixth power of y	**9.** y^6
10. z **to the fifth power**	**10.** z^5

Note that while x **times** x **can be written as** xx**, the preferred form for the product is** x^2**. Similarly,** x^3 **is preferred to** xxx**.**

EXERCISES 1.2

Express each of the following in symbols

A.

1. The product of y times itself.	1.	
2. The square of n.	2.	N^2
3. The product of 4 times 4 times 4.	3.	4^3
4. The product of n times n times n.	4.	
5. The cube of 6.	5.	
6. The cube of t.	6.	
7. The square of t.	7.	
8. The third power of x.	8.	
9. The square of 10.	9.	10^2
10. The square of 9.	10.	9^2
11. The cube of 9.	11.	9^3

12. The cube of y. 12. ______

13. The 4th power of y. 13. ______

14. The 4th power of 10. 14. ______

15. The 5th power of 8. 15. ______

16. The 5th power of x. 16. ______

17. 4 to the 5th power. 17. ______

18. 9 to the seventh power. 18. ______

19. The sixth power of n. 19. ______

20. y to the 8th power. 20. ______

B.

1. The square of b. 1. ______

2. The cube of z. 2. ______

3. The cube of 7. 3. ______

4. The square of 18. 4. ______

5. The third power of 5. 5. ______

6. The fourth power of r. 6. ______

7. The third power of s. 7. ______

8. The fifth power of t. 8. ______

9. The 23rd power of 10. 9. ______

10. The 12th power of x. 10. ______

11. 5 to the 6th power. 11. ______

12. z to the 3rd power. 12. ______

13. r to the 9th power. 13. ______

14. 6 factors of x. 14. ______

15. n factors of y. 15. ______

1.3 EVALUATION: ONE SET OF GROUPING SYMBOLS

OBJECTIVE: To evaluate a numerical expression involving two or more operations or one set of grouping symbols.

To evaluate an algebraic expression means to do the operations that are indicated, thereby finding a single number, the value of the expression.

Grouping symbols and conventions are used to determine the order in which the operations are to be done.

Grouping Symbols

The grouping symbols used most often are:

Parentheses	( )
Brackets	[ ]
Braces	{ }
Bar (also called vinculum)	——
as used in division:	$\frac{x + 3}{2}$

Conventions

1. The operation within the set of grouping symbols is performed first.

2. Unless grouping symbols indicate otherwise, the operations are done in the following order:

first, squaring and/or cubing, as read from left to right.
second, multiplication and/or division, as read from left to right.
third, addition and/or subtraction, as read from left to right.

3. The operations within grouping symbols are done in the order stated in 2.

SAMPLE PROBLEM 1

Evaluate $20 - 2\,(3 + 4)$.

SOLUTION

$20 - 2\,(3+4) = 20 - 2\,(7)$ **Do the operation inside the parentheses first.**

$= 20 - 14$ **Multiply before subtracting, following the convention.**

$= 6$

SAMPLE PROBLEM 2

Evaluate $15 - 8 - 3$.

SOLUTION

$15 - 8 - 3 = 7 - 3 = 4$ **Do subtraction as read from left to right.**

SAMPLE PROBLEM 3

Evaluate $15 - [8 - 3]$.

SOLUTION

$15 - [8 - 3] = 15 - 5 = 10$ **Do the operation inside the brackets first.**

SAMPLE PROBLEM 4

Evaluate $\frac{4^2 - 6}{5}$.

SOLUTION

The division bar is a grouping symbol, and therefore, the operations in the numerator must be done before the division by 5.

$\frac{4^2 - 6}{5} = \frac{16 - 6}{5}$ **Do squaring before subtraction, by convention.**

$= \frac{10}{5}$

$= 2$

SAMPLE PROBLEM 5

Evaluate $6 + 3\{7 - 5\}$

SOLUTION

$6 + 3\{2\} = 6 + 6 = 12$ — **The subtraction inside braces is done first, then the multiplication by 3, and *last* the addition of 6.**

Note that braces and brackets are also used to indicate multiplication; $5[3] = 5\{3\} = 5(3) = 15$.

Evaluate.

EXERCISES 1.3

A.

		Solution
1. $(14 + 3) - 4^2$	1.	1
2. $3^2 - (3 + 5) + 5^2$	2.	26
3. $15 + 3\,[7 - 2]$	3.	30
4. $4\,[3^2 + 3^3]$	4.	144
5. $6\left(\frac{12}{4}\right) - 2^3$	5.	2
6. $9 + \frac{45}{5} - 3^2$	6.	9
7. $\frac{6\,(3 + 4)}{2}$	7.	21
8. $\frac{1}{7}\,\{37 + 12\}$	8.	243
9. $56 - \{8 - 4\}$	9.	52
10. $3\,[8 + 12 - 14]$	10.	21
11. $5\,\{5 + 5^2\}$	11.	150
12. $\frac{9^2 - 11}{10}$	12.	7

13. $\frac{2(3+5)}{4}$ — 13. 4

14. $3\{3^2 - 7\}$ — 14. 6

15. $\frac{8+7-9}{3}$ — 15. 2

16. $\frac{5+7}{2}$ — 16. 6

17. $\frac{30-6}{4}$ — 17. 6

18. $6^2 + \frac{6+4}{5}$ — 18. 38

19. $15[3^2 - 2^2]$ — 19. 75

20. $4 - \{41 - 39\}$ — 20. 1

21. $3\{4+7\}$ — 21. 33

22. $3\{7+4\}$ — 22. 33

23. $(3+7)4$ — 23. 40

24. $3[7-4]$ — 24. 9

25. $(7-4)^2$ — 25. 9

26. $(7-4)^3$ — 26. 27

27. $3(7) - 4$ — 27. 17

28. $(3+7) - 4$ — 28. 6

29. $3 + (7-4)$ — 29. 6

30. $3 - [7-4]$ — 30. 0

B.

1. $10(5-2)$ — 1. 30

2. $(10)(5) - 2$ — 2. 48

3. $10(5 \cdot 2)$ — 3. 100

4. $(10 \cdot 5)(10 \cdot 2)$ 4. ______________

5. $(10 - 2)(5 - 2)$ 5. ______________

6. $10 - 2(5 - 2)$ 6. ______________

7. $15 - [10 - 5]$ 7. ______________

8. $[15 - 10] - 5$ 8. ______________

9. $18 - 3(2 + 3)$ 9. ______________

10. $18 - 3(2) + 3$ 10. ______________

11. $(18 - 3)(2 + 3)$ 11. ______________

12. $\dfrac{125 - 27}{5 - 3}$ 12. ______________

13. $\dfrac{125}{5} - \dfrac{27}{3}$ 13. ______________

14. $\dfrac{2(8 + 7)}{2 + 3}$ 14. ______________

15. $\dfrac{10^3 - 5^3}{10^2 + 5(10 + 5)}$ 15. ______________

1.4 COMBINED OPERATIONS

OBJECTIVE: To translate a verbal expression involving two or more algebraic operations into algebraic symbols.

When a second operation is performed on a sum or difference, there must be a way to indicate that the sum or difference is considered as a single number. Usually, parentheses ( ) are used, or the division bar is used when the operation of division is involved.

Note in each of the sample problems that the **first** word for an operation is the **last** operation that is done in the calculation.

SAMPLE PROBLEMS 1-11

Express each of the following in symbols.

Words	*Symbols*
1. Twice the sum of x and 5.	1. $2(x + 5)$
2. One-third the difference of x decreased by 5.	2. $\frac{x - 5}{3}$
3. Five less than the product of 2 and x.	3. $2x - 5$
4. Three times the quotient of x divided by 5.	4. $3\frac{x}{5}$ or $\frac{3x}{5}$
5. The sum of x and the difference when 2 is subtracted from y.	5. $x + (y - 2)$
6. The difference when the sum of y and 2 is subtracted from x.	6. $x - (y + 2)$
7. The square of the sum of x and 5.	7. $(x + 5)^2$
8. The cube of the difference of x less than y.	8. $(y - x)^3$
9. The cube of the sum of x and 2.	9. $(x + 2)^3$
10. The square of the difference when 2 is subtracted from y.	10. $(y - 2)^2$
11. The quotient obtained when the sum of a and 4 is divided by the product of 4 and a.	11. $\frac{a + 4}{4a}$

EXERCISES 1.4

Express each of the following in symbols.

Words	*Symbols*
A.	
1. Nine times the sum of x and 3.	1. ______
2. One-half the difference when 3 is subtracted from x.	2. ______
3. One-fifth the sum of y and 2.	3. ______
4. Four times the difference of y decreased by 6.	4. ______
5. Three times the quotient of y divided by 8.	5. ______
6. The sum of 4 and the product of 5 and x.	6. ______

7. Five times the sum of x and 4. — 7. ________

8. The difference when 1 is subtracted from the product of 6 and y. — 8. $6y-1$

9. Six times the difference when 1 is subtracted from y. — 9. ________

10. The difference when the sum of t and 6 is subtracted from s. — 10. $s-(t+6)$

11. The sum of x and the difference when y is subtracted from 8. — 11. ________

12. The sum of x and the quotient of 3 divided by x. — 12. ________

13. The difference when the product of 2 and r is subtracted from 6. — 13. $6-2r$

14. The sum of 5 and the product of 9 and z. — 14. ________

15. The product of x and the difference when 3 is subtracted from y. — 15. ________

16. The quotient when the sum of 4 and z is divided by 9. — 16. ________

17. The product of t multiplied by the sum of t and 8. — 17. ________

18. The square of the product of z and 7. — 18. $(7\cdot z)^2$

19. The cube of the sum of x and y and z. — 19. ________

20. The sum of the square of n and the square of 3. — 20. ________

21. The quotient when 5 is divided by the sum of n and 3. — 21. ________

22. The sum of q and the product of 7 and q. — 22. ________

23. One-ninth the product of 4 and t. — 23. ________

24. One-ninth the cube of t. — 24. ________

25. The quotient when the difference of 6 less than n is divided by 2. 25. ____________

B.

1. The difference when the product of x and 3 is subtracted from the cube of x. 1. ____________

2. The sum when the product of 5 and z is added to 7. 2. ____________

3. One-third the product of 8 and y. 3. ____________

4. The product of 9 times the difference when s is subtracted from 9. 4. ____________

5. The sum of x and the difference of x decreased by 6. 5. ____________

6. The difference when the sum of x and 6 is subtracted from y. 6. ____________

7. The difference when x less than 6 is subtracted from 12. 7. ____________

8. Twice the sum of t and 3. 8. ____________

9. The difference when the sum of 4 and n is subtracted from n. 9. ____________

10. The product of 2 and the product of x and y. 10. ____________

11. The quotient when the product of x and y is divided by 2. 11. ____________

12. The product of 5 and the square of y. 12. ____________

13. The product of 6 and the cube of x. 13. ____________

14. One less than the square of x. 14. ____________

15. The sum of one and the cube of y. 15. ____________

16. The difference when the square of y is subtracted from the square of x. 16. $x^2 - y^2$

17. The square of the difference when y is subtracted from x. 17. ~~$(y-x)^2$~~ $(x-y)^2$

18. The sum of the square of x and the square of 5. 18. ____________

19. The sum of the cube of y and the cube of 2. 19. ____________

20. The difference when y decreased by z is subtracted from x. 20. ____________

21. The sum when y decreased by z is added to x. 21. ____________

22. The product of x and the difference obtained by subtracting 3 from x. 22. ____________

23. The sum of x and the product of 3 times the difference when 3 is subtracted from x. 23. ____________

24. One-half the product of 5 and the sum of y and 3. 24. ____________

25. The quotient when the sum of x and y is divided by twice the difference obtained by subtracting y from x. 25. ____________

1.5 EVALUATION: SEVERAL SETS OF GROUPING SYMBOLS

OBJECTIVE: To evaluate a numerical expression involving several sets of grouping symbols.

When two or more sets of grouping symbols occur in a problem, the operations are performed as follows:

First, do the operation or operations within the innermost set of grouping symbols.

Then, do the indicated operation or operations within the remaining innermost set of grouping symbols.

Within each set of grouping symbols, follow the order of operations as stated in Section 1.3.

SAMPLE PROBLEM 1

Evaluate $8 - [30 - (2+3)^2]$.

SOLUTION

$$
\begin{aligned}
8 - [30 - (2+3)^2] &= 8 - [30 - (5)^2] \\
&= 8 - [30 - 25] \\
&= 8 - 5 \\
&= 3 \text{ (answer)}
\end{aligned}
$$

SAMPLE PROBLEM 2

Evaluate $2\,\{19 - 3\,[5^2 - 2\,(7+3)]\}$.

SOLUTION

$$
\begin{aligned}
2\{19 - 3[5^2 - 2(7+3)]\} &= 2\,\{19 - 3\,[5^2 - 2\,(10)]\} \\
&= 2\,\{19 - 3\,[25 - 20]\} \\
&= 2\,\{19 - 3\,[5]\} \\
&= 2\,\{19 - 15\} \\
&= 2\,\{4\} \\
&= 8 \text{ (answer)}
\end{aligned}
$$

SAMPLE PROBLEM 3

Evaluate $\frac{(8+7)(8-5)}{2 \cdot 7 - 5}$.

SOLUTION

$$
\begin{aligned}
\frac{(8+7)(8-5)}{2 \cdot 7 - 5} &= \frac{(15)(3)}{14 - 5} \\
&= \frac{45}{9} = 5 \text{ (answer)}
\end{aligned}
$$

When evaluating a numerical expression, some of the steps may be done mentally and may be omitted in the written solution, according to one's experience and the complexity of the problem. However, omissions should be done mentally, they should be reasonable in order to avoid mistakes, and clear solutions should be written, so that they can be understood by another reader.

When two or more sets of grouping symbols are needed, any type can be used, and the sets can be written in any order desired. Usually parentheses are used first because they are the easiest to write, then brackets, and finally braces, which are considered the most difficult to write.

EXERCISES 1.5

Evaluate.

	Solution
A.	
1. $6(2^2 - 3) + 5(6 - 2^2)$	1.
2. $\dfrac{7(7+3)}{5(9-2)}$	2.
3. $6\{(2^2 - 3) + 5\}$	3.
4. $8[(4+5)(3+2)]$	4.
5. $\dfrac{(5+1)(5+2)}{2^2 - 1}$	5.
6. $2[16 - 5(40 - 38)]$	6.
7. $\{6 + (3 - 1)\}\{5 + (3 - 1)\}$	7.
8. $\dfrac{5^2 - 4^2}{5 + 4}$	8.
9. $\left(\dfrac{5^2 - 4^2}{3^2}\right)(8 + 9 - 4)$	9.
10. $\dfrac{1}{4}[53 - (2^3 - 3)]$	10.
11. $7\{[5+6]\ [5-2]\}$	11.
12. $7\{[5+6] - [5-2]\}$	12.
13. $36 - [5(5^2) - (3+7)^2]$	13.
14. $2[(3^2 - 3)(3^2 + 3)] - 24$	14.
15. $\dfrac{(3^2 - 3)(3^2 + 3)}{24}$	15.

B.

1. $\dfrac{(3^2 - 3)(3^2 + 3)}{(3^3 - 3)}$ 1.

2. $8\left\{\dfrac{3(4^2 + 6)}{2^2 + 7}\right\}$ 2.

3. $3^2 (7 - 4)^2$ 3.

4. $(7 + 4)^2 - (7 + 3)^2$ 4.

5. $(4[7 - 3]^2)^2$ 5.

6. $(3\{7 - 3\})^2$ 6.

7. $5(6 - 3) - 6(5 - 3)$ 7.

8. $4^2 - [6 - (20 - 18)]$ 8.

9. $\dfrac{7^2 - 4^2}{7 - 4}$ 9.

10. $\dfrac{7^2 - 4^2}{7 + 4}$ 10.

11. $10 - \{8 - [6 - (4 - 2)]\}$ 11.

12. $10 - \{8 + [6 - (4 + 2)]\}$ 12.

13. $2(5 - 2)^2$ 13.

14. $[2(5 - 2)]^2$ 14.

15. $2(8 - 2[6 - 2\{5 - 3\}])$ 15.

1.6 SUBSTITUTION: ONE VARIABLE

OBJECTIVE: To evaluate an algebraic expression in one variable by using the substitution principle.

The Equality Relation

An equation is a statement having the form $A = B$, which is read "A equals B." The statement $A = B$ means that A and B are names of the same number.

The Substitution Principle

If $A = B$, then A may replace B (or B may replace A) in any expression naming a number without changing the number.

SAMPLE PROBLEM 1

Evaluate $2x - \frac{x+5}{3}$ for $x = 7$.

SOLUTION

(1) Rewrite the expression, removing the letter and holding its place with open parentheses.	$2x - \frac{x+5}{3}$ $2(\ ) - \frac{(\ )+5}{3}$
(2) Insert the given value of the letter within the parentheses.	$2(7) - \frac{(7)+5}{3}$
(3) Do the indicated operations and write the answer as a single number.	$14 - \frac{12}{3}$ $14 - 4$ **10 (answer)**

SAMPLE PROBLEM 2

Evaluate $(2t^2 - 9)(t + 1)$ for $t = 5$.

SOLUTION

(1) Rewrite expression.	$(2t^2 - 9)(t + 1)$
(2) Remove t, holding its place.	$(2[\ ]^2 - 9)([\ ] + 1)$
(3) Insert 5 within the brackets.	$(2[5]^2 - 9)([5] + 1)$
(4) Do the operations.	$= (2[25] - 9)(6)$
	$= (50 - 9)(6)$
	$= (41)(6) = 246$ **(answer)**

SAMPLE PROBLEM 3

Evaluate $\frac{2y^2 + 9y - 5}{2y - 1}$ for $y = 4$.

SOLUTION

(1) Rewrite expression. $\frac{2y^2 + 9y - 5}{2y - 1}$

(2) Remove y, holding its place. $\frac{2(\ \ )^2 + 9(\ \ ) - 5}{2(\ \ ) - 1}$

(3) Insert 4 within the parentheses. $\frac{2(4)^2 + 9(4) - 5}{2(4) - 1}$

(4) Do the operations.

$$= \frac{2(16) + 36 - 5}{8 - 1}$$

$$= \frac{32 + 36 - 5}{7}$$

$$= \frac{68 - 5}{7} = \frac{63}{7} = 9 \text{ (answer)}$$

EXERCISES 1.6

Evaluate.

A.

1. $5x - 2x - 2$ for $x = 3$

2. $6y + \frac{y^2}{3 + 2}$ for $y = 5$

3. $\frac{6n - 15}{2n - 5}$ for $n = 8$

4. $4(5x + 3) - x$ for $x = 4$

5. $4(5x) - 3(5 - x)$ for $x = 4$

6. $5z - \left[z - \frac{z}{3}\right]$ for $z = 9$

7. $7\ \{4 + 12q\}$ for $q = 2$

8. $q\ \{4 + 12q\}$ for $q = 3$

9. $\dfrac{6r - 16}{r}$ for $r = 4$

10. $t^2 - t + 3$ for $t = 7$

11. $2 + \dfrac{5x}{2}$ for $x = 6$

12. $3s + \dfrac{s^2 - 5}{2}$ for $s = 9$

13. $n\ (n - 3)$ for $n = 5$

14. $z^2 - 2\ [z - 3]$ for $z = 8$

15. $\dfrac{65 - y}{y + 7}$ for $y = 2$

16. $3n^2 + \dfrac{n}{n - 7}$ for $n = 14$

17. $5n^2\ \left(\dfrac{5n}{7 - n}\right)$ for $n = 2$

18. $4t + \{16 - t^3\}$ for $t = 2$

19. $4t - \{16 - t^3\}$ for $t = 2$

$4[2] - \{16 - 2^3\}$

$8 - \{16 - 8\}$

$8 - \{8\} = 0$

20. $7y + \dfrac{y^2 - 3}{y - 3}$ for $y = 5$

B.

1. $\frac{x+3}{2} - \frac{x^2+x}{5x}$ for $x = 9$

2. $[3m - 2] + m^2$ for $m = 6$

3. $7(r^2 - 4) - r(r^2 - 4)$ for $r = 3$

4. $\frac{8\{x+7\}}{2x}$ for $x = 4$

5. $30 - \frac{y^2 - y}{6}$ for $y = 7$

6. $5y - 2(y + 3)$ for $y = 4$

7. $12 - 3(x - 2)$ for $x = 6$

8. $6x + x - 3x$ for $x = 8$

9. $5y - y - 2y$ for $y = 10$

10. $(x + 3)(x + 5)$ for $x = 4$

11. $x + 3(x + 5)$ for $x = 4$

12. $(3y - 5)(y - 3)$ for $y = 5$

13. $3y - 5(y - 3)$ for $y = 5$

14. $[2(x - 7)]^2 - 2(x - 7)^2$ for $x = 10$

1.7 SUBSTITUTION: SEVERAL VARIABLES

OBJECTIVE: To evaluate an algebraic expression in two variables by using the substitution principle.

SAMPLE PROBLEM 1

Evaluate $\frac{x^2 - 2xy + y^2}{x - y}$ for $x = 7$ and $y = 2$.

SOLUTION

(1) Replace one variable, say x, by open parentheses. $\frac{(\ )^2 - 2(\ )y + y^2}{(\ ) - y}$

(2) Insert the value for x. $\frac{(7)^2 - 2(7)y + y^2}{(7) - y}$

(3) Replace the other variable, y, by open parentheses. $\frac{(7)^2 - 2(7)(\ ) + (\ )^2}{(7) - (\ )}$

(4) Insert the value for y. $\frac{(7)^2 - 2(7)(2) + (2)^2}{(7) - (2)}$

(5) Evaluate. $= \frac{49 - 28 + 4}{5}$

$= \frac{21 + 4}{5} = \frac{25}{5} = 5$

SAMPLE PROBLEM 2

Evaluate $\frac{3x^2 - 5y^2 - 1}{xy}$ for $x = 3$ and $y = 2$.

SOLUTION

(1) Replace one variable, say y, by open parentheses. $\frac{3x^2 - 5(\ )^2 - 1}{x(\ )}$

(2) Insert the value for y. $\frac{3x^2 - 5(2)^2 - 1}{x(2)}$

(3) Replace the other variable, x, by open parentheses. $\frac{3(\ )^2 - 5(2)^2 - 1}{(\ )(2)}$

(4) Insert the value for x. $\frac{3(3)^2 - 5(2)^2 - 1}{(3)(2)}$

(5) Evaluate.

$$= \frac{3(9) - 5(4) - 1}{(3)(2)}$$

$$= \frac{27 - 20 - 1}{6}$$

$$= \frac{6}{6} = 1$$

With practice, both substitutions may be made at the same time. Also, some steps may be done mentally and may be omitted from the written solution. These omissions should be reasonable, and they should be clearly understood by both the writer and any reader of the written work.

SAMPLE PROBLEM 3

Evaluate $(5x - y)(25x^2 - 10xy + y^2)$ for $x = 4$ and $y = 6$.

SOLUTION

$$\begin{aligned}(5x - y)(25x^2 - 10xy + y^2) &= (5 \cdot 4 - 6)(25 \cdot 4^2 - 10[4 \cdot 6] + 6^2) \\ &= (20 - 6)(25 \cdot 16 - 240 + 36) \\ &= (14)(400 - 240 + 36) \\ &= 14(196) = 2744\end{aligned}$$

EXERCISES 1.7

Evaluate each of the following.

A.

1. $x - [2y - x]$ for $x = 4$ and $y = 3$

2. $a^2 + 4ab + b^2$ for $a = 9$ and $b = 5$

3. $2m(n - m)$ for $m = 2$ and $n = 7$

4. $5xy^2 - 3x^2y$ for $x = 3$ and $y = 5$

5. $3(a - b)^2$ for $a = 10$ and $b = 7$

6. $\dfrac{m^2 - n^2}{m - n}$ for $m = 6$ and $n = 5$

7. $30 - [s^2 - \{3t - s\}]$ for $s = 3$ and $t = 2$

8. $\dfrac{10ab}{a + b}$ for $a = 7$ and $b = 3$

9. $\dfrac{a^2b + ab^2}{a + b}$ for $a = 7$ and $b = 3$

10. $y^2 + \dfrac{x^3 + y^3}{x + y}$ for $x = 5$ and $y = 4$

B.

1. $3\{y^2 - z^2\}$ for $y = 8$ and $z = 7$

2. $a^2b - b^2a - 3ab$ for $a = 6$ and $b = 2$

3. $r^3s + r^2s - rs$ for $r = 3$ and $s = 5$

4. $(x + y)(x^2 - xy + y^2)$ for $x = 6$ and $y = 4$

5. $(x - y)(x^2 + xy + y^2)$ for $x = 5$ and $y = 3$

6. $(x + 3y)(x - 3y)$ for $x = 8$ and $y = 2$

7. $(x + 3y)(x + 3y)$ for $x = 4$ and $y = 2$

8. $(x^2 + 2)(y^2 - 3)$ for $x = 3$ and $y = 4$

9. $\dfrac{xy - 4x + 5y - 20}{(x + 5)(y - 4)}$ for $x = 2$ and $y = 7$

10. $(r^2 + 2s + 2)(r^2 - 2s + 2)$ for $r = 6$ and $s = 4$

11. $a^2 + b^2 + c^2 - abc$ for $a = 2$, $b = 3$, and $c = 4$

12. $(x + y - 2z)^2$ for $x = 1$, $y = 4$, and $z = 2$

UNIT 1 POSTTEST

(1–4) Express each of the following in symbols. (1.1)

1. The product of 8 and x. 1. ______________
2. The sum of n and 6. 2. ______________
3. The difference when 4 is subtracted from y. 3. ______________
4. The quotient when x is divided by 3. 4. ______________

(5–8) Express each of the following in symbols. (1.2)

5. The square of 7. 5. ______________
6. The square of t. 6. ______________
7. The cube of 8. 7. ______________
8. The cube of n. 8. ______________

(9–13) Express each of the following in symbols. (1.3)

9. Seven times the sum of n and 8. 9. ______________
10. Five less than the product of 3 and x. 10. ______________
11. The square of the difference when 8 is subtracted from y. 11. ______________
12. The difference when the cube of the sum of x and 2 is subtracted from y. 12. ______________
13. The quotient when the product of x and 2 is divided by the sum of x and 2. 13. ______________

(14–16) Evaluate. (1.4)

14. $5\,(7 - 2)$ 14. ______________
15. $\dfrac{5\,[8 - 2]}{5 - 2}$ 15. ______________
16. $30 - \{3^2 - 2^3\}$ 16. ______________

(17–19) Evaluate. (1.5)

17. $10 - \{9 - (5 - 2)\}$ 17. ______________

18. $(7 + [6 - 4])(7 - [6 - 4])$ 18. ______

19. $\dfrac{(4 + 1)^2 - (3 + 1)^2}{(4 - 1) - (3 - 1)}$ 19. ______

(20–22) Evaluate. (1.6)

20. $\dfrac{x + 2}{x - 2}$ for $x = 4$ 20. ______

21. $(y - 6)(y + 2)$ for $y = 8$ 21. ______

22. $n^2 - 5(n - 2)$ for $n = 6$ 22. ______

(23–25) Evaluate. (1.7)

23. $\dfrac{x^2 - (y - 3)^2}{x + y - 3}$ for $x = 4$ and $y = 5$ 23. ______

24. $(x - y)(x^2 + xy + y^2)$ for $x = 7$ and $y = 4$ 24. ______

25. $\dfrac{(x^2 + y)^3 - 5^3}{x^2 + y - 5}$ for $x = 3$ and $y = 1$ 25. ______

UNIT 2

SETS

UNIT 2
PRETEST

(1–4) List the members of the set described. (2.1)

1. The set of digits greater than 4. 1. ______________

2. The natural numbers, n, for which $n - 10$ is a natural number. 2. ______________

3. The set of natural numbers less than 15. 3. ______________

4. The set of digits less than 7. 4. ______________

(5–8) List the element in each described subset of the set of natural numbers. (2.2)

5. The multiples of 6. 5. ______________

6. The divisors of 24. 6. ______________

7. The factors of 50. 7. ______________

8. The numbers divisible by 30. 8. ______________

(9–12) List the elements in each described subset of the set of natural numbers. (2.3)

9. The set of even numbers greater than 50. 9. ______________

10. The set of odd numbers greater than 50 but less than 60. 10. ______________

11. The prime divisors of 28. 11. ______________

12. The composite factors of 30. 12. ______________

2.1 THE SET CONCEPT, THE LISTING METHOD

OBJECTIVE: To represent a subset of the set of digits or the set of counting numbers by the listing method.

A **set** is a well-defined collection of objects which are called elements or members of the set.

"**Well-defined**" means that it is always possible to determine if a particular object is or is not a member of a given set. A capital letter is often used as the name of a set.

The listing method is a commonly used method for defining a particular set. When this method is used, the elements of the set are written inside braces and are separated by commas.

Example 1. Represent the **set of digits**, D, by the listing method.

Solution. $D = \{0, 1, 2, 3, 4, 5, 6, 7, 8, 9\}$

Example 2. Represent the **set of natural numbers**, N (another name for the counting numbers), by the listing method.

Solution. $N = \{1, 2, 3, 4, 5, \ldots\}$
The three dots, called an elision, mean that this pattern continues without end.

In Sample Problems 1–5, use the listing method to represent each set which is described.

SAMPLE PROBLEM 1

The set of digits greater than 4.

SOLUTION

$\{5, 6, 7, 8, 9\}$

SAMPLE PROBLEM 2

The set of natural numbers greater than 20.

SOLUTION

$\{21, 22, 23, 24, 25, \ldots\}$

SAMPLE PROBLEM 3

The set of digits x for which $x + 2$ is less than 7.

SOLUTION

Test each digit.

$x = 0$ Is $0 + 2$ less than 7? Yes
$x = 1$ Is $1 + 2$ less than 7? Yes
$x = 2$ Is $2 + 2$ less than 7? Yes
$x = 3$ Is $3 + 2$ less than 7? Yes
$x = 4$ Is $4 + 2$ less than 7? Yes
$x = 5$ Is $5 + 2$ less than 7? No
$x = 6$ Is $6 + 2$ less than 7? No

Digits larger than 4 do *not* meet the requirement. Thus the set is $\{0, 1, 2, 3, 4\}$.

SAMPLE PROBLEM 4

The set of natural numbers x for which $x + 2$ is less than 7.

SOLUTION

$\{1, 2, 3, 4\}$

SAMPLE PROBLEM 5

The set of natural numbers x for which $\frac{12}{x}$ is a natural number.

SOLUTION

Test each natural number.

$x = 1$ Is $\frac{12}{1}$ a natural number? Yes

$x = 2$ Is $\frac{12}{2}$ a natural number? Yes

$x = 3$ Is $\frac{12}{3}$ a natural number? Yes

$x = 4$ Is $\frac{12}{4}$ a natural number? Yes

$x = 5$ Is $\frac{12}{5}$ a natural number? No

$x = 6$ Is $\frac{12}{6}$ a natural number? Yes

Continue in this way. The set is $\{1, 2, 3, 4, 6, 12\}$.

EXERCISES 2.1 Use the listing method to represent each set which is described.

A.

1. The set of digits less than 5. 1. ______________
2. The set of digits greater than 6. 2. ______________
3. The set of natural numbers greater than 6. 3. ______________
4. The set of natural numbers x for which $x + 5$ is less than 10. 4. ______________
5. The set of digits x for which $x + 5$ is less than 10. 5. ______________
6. The set of digits d for which $d + 4$ is greater than 9. 6. ______________
7. The set of natural numbers n for which $n + 4$ is greater than 9. 7. ______________
8. The set of natural numbers y for which $\frac{45}{y}$ is a natural number. 8. ______________
9. The set of digits y for which $\frac{45}{y}$ is a natural number. 9. ______________
10. The set of digits d for which $\frac{d}{2}$ is a digit. 10. ______________
11. The set of natural numbers n for which $\frac{n}{2}$ is a natural number. 11. ______________

B.

1. The set of natural numbers less than 7. 1. ______________
2. The set of digits less than 7. 2. ______________
3. The set of digits greater than 4. 3. ______________
4. The set of natural numbers greater than 4. 4. ______________

5. The set of digits x for which $\frac{60}{x}$ is a natural number. 5. ____________

6. The set of natural numbers x for which $\frac{60}{x}$ is a natural number. 6. ____________

7. The set of natural numbers x for which $\frac{x}{60}$ is a natural number. 7. ____________

8. The set of digits d for which $\frac{d}{3}$ is a digit. 8. ____________

9. The set of natural numbers n for which $\frac{n}{3}$ is a natural number. 9. ____________

10. The set of digits d for which $\frac{d}{3}$ is *not* a digit. 10. ____________

11. The set of natural numbers x for which $x + 0 = x$. 11. ____________

2.2 EMPTY SET, UNIVERSAL SET, MULTIPLES, DIVISORS, FACTORS

OBJECTIVES: To use the symbol $\emptyset$ or { } to identify the empty set, described as a set having a certain property.

To list the multiples of a given natural number, using set notation.

To list the divisors or factors of a given natural number, using set notation.

Empty Set. Also called the null set, the empty set is the set that has no members. Either $\emptyset$ or { } is used to indicate the empty set. As an example, the set of digits greater than 10 is the empty set, $\emptyset$.

Universal Set. For a particular discussion, the universal set is a set to which all elements of the discussion must belong. The universal set may differ for different discussions. For now, the set of natural numbers N is selected as the universal set.

Multiple. A natural number m is a **multiple** of another natural number n if it can be obtained by multiplying n by a natural number. For example, 20 is a multiple of 5, since $20 = 5(4)$. We also say that 20 is **divisible** by 5.

Divisor. A natural number d is a **divisor** of another natural number n if it can be obtained by dividing n by a natural number. For example, 6 is a divisor of 18 because $\frac{18}{3} = 6$.

Factor. A **factor** of a natural number n is a divisor of n. For example, 6 is a factor of 18, since 6 is a divisor of 18.

SAMPLE PROBLEM 1

List the multiples of 4.

SOLUTION

A multiple of 4 is a number obtained by multiplying 4 by a natural number. Since $4(1) = 4$, $4(2) = 8$, $4(3) = 12$, $4(4) = 16$, $4(5) = 20$, and so on, the set of multiples of $4 = \{4, 8, 12, 16, 20, ...\}$

SAMPLE PROBLEM 2

List the divisors of 12.

SOLUTION

Divide 12 by each natural number. Accept only those quotients that are natural numbers. $\frac{12}{1} = 12, \frac{12}{2} = 6, \frac{12}{3} = 4, \frac{12}{4} = 3, \frac{12}{5}$ reject, ..., $\frac{12}{12} = 1$

$\{1, 2, 3, 4, 6, 12\}$

SAMPLE PROBLEM 3

List the factors of 15.

SOLUTION

The factors are the divisors. Test each natural number. $\frac{15}{1} = 15, \frac{15}{2}$ reject, $\frac{15}{3} = 5$, ..., $\frac{15}{15} = 1$

$\{1, 3, 5, 15\}$

SAMPLE PROBLEM 4

List the natural numbers divisible by 12.

SOLUTION

Numbers divisible by 12 are multiples of 12. Multiply 12 by each natural number. $12(1) = 12$, $12(2) = 24$, $12(3) = 36$, and so on.

$\{12, \; 24, \; 36, \; 48, \; 60, \; ...\}$

SAMPLE PROBLEM 5

List the divisors of 7 different from 1 and 7.

SOLUTION

$\emptyset$, the empty set

SAMPLE PROBLEM 6

List the divisors of 6 different from 1 and 6.

SOLUTION

$\{2, \; 3\}$

SAMPLE PROBLEM 7

List all natural numbers x for which $x \cdot 1 = x$.

SOLUTION

Test each natural number: $1 \cdot 1 = 1, 2 \cdot 1 = 2, 3 \cdot 1 = 3, 4 \cdot 1 = 4, 5 \cdot 1 = 5, ...$

From our knowledge of arithmetic, it should be clear that $x \cdot 1 = x$ for all natural numbers.

Either N or $\{1, 2, 3, 4, 5, ...\}$ is acceptable as the written solution.

EXERCISES 2.2

List the elements in each set described. The universal set is N, the set of natural numbers. This means that each indicated number must be a natural number.

A.

1. The divisors of 20. 1. ______
2. The multiples of 20. 2. ______
3. The factors of 21. 3. ______
4. The numbers divisible by 8. 4. ______
5. The factors of 35 different from 1 and 35. 5. ______
6. The factors of 17 different from 1 and 17. 6. ______
7. The multiples of 30. 7. ______
8. The factors of 30. 8. ______
9. The divisors of 35. 9. ______
10. All numbers x for which $5x = x5$. 10. ______

B.

1. The multiples of 45. 1. ______
2. The divisors of 45. 2. ______
3. The numbers divisible by 30. 3. ______
4. The factors of 30. 4. ______
5. The factors of 28 different from 1 and 28. 5. ______
6. The factors of 43 different from 1 and 43. 6. ______
7. The multiples of 42. 7. ______
8. The multiples of 10. 8. ______

9. All numbers x for which $x + 5 = 5 + x$. 9. ____________________

10. All numbers x for which $x - 5 = 5 - x$. 10. ____________________

2.3 EVEN, ODD, PRIME, AND COMPOSITE NUMBERS

OBJECTIVES: To list the members of the following:

A subset of the set of even numbers.

A subset of the set of odd numbers.

A subset of the set of prime numbers.

A subset of the set of composite numbers.

Definition of Even Number. A natural number n is **even** if and only if 2 is a divisor of n. As examples, 2, 4, 6, and 8 are even numbers.

Definition of Odd Number. A natural number n is **odd** if and only if 2 is not a divisor of n. As examples, 1, 3, 5, and 7 are odd numbers.

Definition of Prime Number. A natural number p is **prime** if and only if p has exactly 2 factors, namely p and 1. As examples, 2, 3, 5, and 7 are prime numbers.

Definition of Composite Number. A natural number n is **composite** if and only if n is not prime and n is not 1. As examples, 4, 6, 12, 15, and 21 are composites.

The set of even numbers, the set of odd numbers, the set of primes, and the set of composites are important *subsets* of the set of natural numbers.

A set A is said to be a **subset** of a set B if and only if each element of A is also an element of B. In symbols,

$$A \subset B \text{ means } A \text{ is a subset of } B.$$

Note from the definitions that the special number, 1, is neither prime nor composite.

For Sample Problems 1–4, list the members of each set described.

SAMPLE PROBLEM 1

The set of even numbers less than 15.

SOLUTION

$\{2, 4, 6, 8, 10, 12, 14\}$

SAMPLE PROBLEM 2

The set of odd numbers less than 15.

SOLUTION

$\{1, 3, 5, 7, 9, 11, 13\}$

SAMPLE PROBLEM 3

The set of prime numbers less than 20.

SOLUTION

$\{2, 3, 5, 7, 11, 13, 17, 19\}$

SAMPLE PROBLEM 4

The set of composite numbers less than 20.

SOLUTION

$\{4, 6, 8, 9, 10, 12, 14, 15, 16, 18\}$

EXERCISES 2.3

List the members of each set described.

A.

1. The set of even numbers greater than 10 and less than 20. 1. ________________

2. The set of odd numbers greater than 20 and less than 30. 2. ________________

3. The set of primes greater than 20 and less than 30. 3. ____________

4. The set of composites greater than 20 and less than 30. 4. ____________

5. The set of even primes. 5. ____________

6. The set of odd primes. 6. ____________

7. The prime divisors of 60. 7. ____________

8. The even divisors of 60. 8. ____________

9. Numbers neither prime nor composite. 9. ____________

10. The odd composites less than 20. 10. ____________

11. The even factors of 15. 11. ____________

12. The even multiples of 7. 12. ____________

13. The set of odd multiples of 7. 13. ____________

14. The set of odd multiples of 6. 14. ____________

15. The prime factors of 105. 15. ____________

B.

1. The set of primes greater than 30 and less than 50. 1. ____________

2. The set of composites greater than 35 and less than 45. 2. ____________

3. The set of even multiples of 11. 3. ____________

4. The set of odd multiples of 11. 4. ____________

5. The set of prime factors of 154. 5. ____________

6. The set of composite factors of 154. 6. ____________

7. The odd factors of 70. 7. ____________

8. The even divisors of 70. 8. ____________

9. The composite divisors of 61. 9. ____________

10. The set of even prime numbers. 10. ____________

11. The factors of 54. 11. ____________

12. The prime factors of 54. 12. ____________

13. The composite factors of 54. 13. ____________

14. The even factors of 54. 14. ____________

15. The odd factors of 54. 15. ____________

UNIT 2 POSTTEST

(1–4) List the members of the set described. (2.1)

1. The set of natural numbers greater than 13. 1. ____________

2. The set of digits less than 8. 2. ____________

3. The set of natural numbers n for which $\frac{n}{6}$ is a natural number. 3. ____________

4. The set of digits d for which $\frac{d}{3}$ is a digit. 4. ____________

(5–8) List the elements in each described subset of the set of natural numbers. (2.2)

5. The factors of 42. 5. ____________

6. The divisors of 15. 6. ____________

7. The multiples of 9. 7. ____________

8. The factors of 6 greater than 7. 8. ____________

(9–12) List the elements in each described subset of the set of natural numbers. (2.3)

9. The set of odd numbers between 100 and 110. 9. ____________

10. The set of even divisors of 48. 10. ____________

11. The primes greater than 40 but less than 50. 11. ____________

12. The even composite factors of 20. 12. ____________

UNIT 3

BASIC AXIOMS

UNIT 3 PRETEST

(1–5) Identify each statement as an example of the reflexive axiom, the symmetric axiom, the transitive axiom, or as a false statement by writing the word "reflexive," "symmetric," "transitive," or "false." (3.1)

1. If $20 = 3x - 7$, then $3x - 7 = 20$. 1. ______________
2. $3x - 7 = 7 - 3x$ for all values of x. 2. ______________
3. $3x - 7 = 3x - 7$ for all values of x. 3. ______________
4. If $x = y$ and $y = 5$, then $x = 5$. 4. ______________
5. If $\frac{5x}{5} = 6$ and $x = \frac{5x}{5}$, then $x = 6$. 5. ______________

(6–11) Use the closure and commutative axioms to rewrite each of the following in compliance with algebraic conventions. (3.2)

6. ts	7. $4 + x$	8. $y5$
______________	______________	______________
9. $t + n$	10. $(y + x)\ 6$	11. $8 + yx$
______________	______________	______________

(12–15) Apply an associative axiom to the following, and compute the value of both the original and the final expressions. (3.3)

12. $96 + (4 + 38) =$

13. $(86 \cdot 25)\ 4 =$

14. $(87 + 75) + 25 =$

15. $125\ (8 \cdot 79) =$

(16–19) Use the commutative and associative axioms to rearrange the terms of a sum or the factors of a product in compliance with algebraic conventions. (3.4)

16. $ba5$ __________ 17. $r + 7 + s$ __________

18. $6 + y2x$ __________ 19. $yx + 8 + ba$ __________

(20–21) Use the distributive axiom to express the indicated product as a sum. (3.5)

20. $7(x + 2y) =$ __________ 21. $3y(x + 1) =$ __________

(22–23) Use the distributive axiom to express the indicated sum as a product. (3.5)

22. $2a + 8b =$ __________ 23. $3x + 5x =$ __________

(24–25) Use the distributive axiom to perform the given calculation in a different way than the one indicated. (3.5)

24. $25(40 + 8) =$ 25. $49(45) + 49(55) =$

__________ __________

3.1 EQUALITY AXIOMS

OBJECTIVE: To identify a given statement as an example of the reflexive axiom, the symmetric axiom, the transitive axiom, or as a false statement.

In order to develop the subject of algebra, some assumptions must be made. **Axiom** is a mathematical word that means an assumption. Stated another way, an axiom is a statement accepted without proof as a true statement. Usually, the axioms are thought of as obvious facts about the subject matter.

There are three important axioms that deal with the equality relation: the reflexive axiom, the symmetric axiom, and the transitive axiom.

The Reflexive Axiom

$$A = A$$

The reflexive axiom states that any equation having the form $A = A$ is accepted as a true statement. As examples, $5 = 5$ is true, $2x = 2x$ is true, and $3(x - 2) = 3(x - 2)$ is true.

The Symmetric Axiom

$$\text{If } A = B, \text{ then } B = A.$$

The symmetric axiom states that if an equation having the form $A = B$ is true, then the equation having the form $B = A$ is also true. In other words, the sides of an equation may be exchanged. As an example, if $3 = 2x - 5$ is true, then $2x - 5 = 3$ is true.

The Transitive Axiom

$$\text{If } A = B \text{ and } B = C, \text{ then } A = C.$$

The transitive axiom means that if two quantities are each equal to the same quantity, then these two quantities are equal to each other. As examples,

$$\text{if } 2(3 + 4) = 2(7) \text{ and } 2(7) = 14, \text{ then } 2(3 + 4) = 14;$$

$$\text{if } y = 10 \text{ and } 3x - 1 = y, \text{ then } 3x - 1 = 10.$$

SAMPLE PROBLEM 1

Write an equation of the form $A = A$ if A has the value $4x - 5$.

SOLUTION

$4x - 5 = 4x - 5$

This problem illustrates the reflexive axiom.

SAMPLE PROBLEM 2

Write an equation of the form $B = A$, given the following equation having the form $A = B$: $7 = 6x + 1$.

SOLUTION

$6x + 1 = 7$

This problem illustrates the symmetric axiom.

SAMPLE PROBLEM 3

Write an equation of the form $A = C$ for this pair of equations having the form $A = B$ and $B = C$: $(x - 7)^2 = x^2 - 14x + 49$ and $x^2 - 14x + 49 = 36$.

SOLUTION

$(x - 7)^2 = 36$

This problem illustrates the transitive axiom.

SAMPLE PROBLEM 4

Write an equation of the form $A = C$ for this pair of equations having the form $A = B$ and $B = C$: $3(x + 7) = 42$ and $3x + 21 = 3(x + 7)$.

SOLUTION

The given pair of equations may also be written:

$3x + 21 = 3(x + 7)$ and $3(x + 7) = 42$.

Therefore, $3x + 21 = 42$.

This problem also illustrates the transitive axiom.

EXERCISES 3.1

(1–5) Write the reflexive axiom, an equation of the form $A = A$, for each of the following values for A.

A.

1. 5 __________
2. $3 + 7$ __________
3. $x + 2$ __________
4. $x + y$ __________
5. $2(x - 5)$ __________

B.

1. $2(x + 4)$ __________
2. $7xy$ __________
3. $(2x - y)^2$ __________
4. $10 - (x - 3)$ __________
5. $10 - 3(x + 2)$ __________

(6–10) Rewrite the following equations by applying the symmetric axiom; that is, for each given equation having the form $A = B$, write an equation having the form $B = A$.

A.

6. $5 + 3 = 8$ 6. ____________

7. $4 = 7 - 3$ 7. ____________

8. $6 = 5x - 1$ 8. ____________

9. $10 = 2x$ 9. ____________

10. $0 = x - 5$ 10. ____________

B.

6. $0 = x + 7$ 6. ____________

7. $1 = \frac{7}{7}$ 7. ____________

8. $\frac{3}{3} = 1$ 8. ____________

9. $2\,(x + y) = 2x + 2y$ 9. ____________

10. $x^2 + 2xy + y^2 = (x + y)^2$ 10. ____________

(11–15) Rewrite the following equations by applying the transitive axiom; that is, for each pair of equations having the form $A = B$ and $B = C$, write an equation having the form $A = C$.

A.

11. $3\,(10 - 6) = 3\,(4)$ and $3\,(4) = 12$ 11. ____________

12. $2\,(5^2) = 2\,(25)$ and $2\,(25) = 50$ 12. ____________

13. $x^2 + 6x + 9 = 25$ and $(x + 3)^2 = x^2 + 6x + 9$ 13. ____________

14. $x\,(2 + x) = 15$ and $x\,(x + 2) = x\,(2 + x)$ 14. ____________

15. $3x + x = 20$ and $4x = 3x + x$ 15. ____________

B.

11. $8 - x - 3 = 7$ and $5 - x = 8 - x - 3$ 11. ____________

12. $2\,(x - 1) = 8$ and $2x - 2 = 2\,(x - 1)$ 12. ____________

13. $4\left(\frac{x}{4}\right) = 3$ and $x = 4\left(\frac{x}{4}\right)$ 13. ____________

14. $\frac{3x}{3} = 2$ and $x = \frac{3x}{3}$ 14. ____________

15. $(x + 5) - 5 = 9$ and $x = (x + 5) - 5$ 15. ____________

(16–22) For each given statement, if the statement is true, write "reflexive," "symmetric," or "transitive" to indicate which axiom (or axioms) the statement represents; if the statement is false, write "false."

A.

16. $7 = 7$ 16. ____________

17. $2 = 5$ 17. ____________

18. If $2 + 4 = 6$, then $6 = 2 + 4$. 18. ____________

19. $8 - 3 = 3 - 8$ 19. ____________

20. If $(x + 2) - 2 = 6$ and $x = (x + 2) - 2$, then $x = 6$. 20. ____________

21. If $7 - 2 = 5$, then $5 = 2 - 7$. 21. ____________

22. If $x = 3\left(\frac{x}{3}\right)$ and $3\left(\frac{x}{3}\right) = 4$, then $x = 4$. 22. ____________

B.

16. If $7 - x = x + 4$, then $x + 4 = 7 - x$. 16. ____________

17. $3(7 - 2) = 3(2 - 7)$ 17. ____________

18. $2(x - 1) = 2(x - 1)$ 18. ____________

19. If $xy = x + y$ and $x + y = 6$, then $xy = 6$. 19. ____________

20. If $10 - (5 - 2) = 7$, then $7 = (10 - 5) - 2$. 20. ____________

21. If $(2)(3)(4) = 2(3 \cdot 4)$, and $2(3 \cdot 4) = 24$, then $(2)(3)(4) = 24$. 21. ____________

22. If $5 = x$, then $x = 5$. 22. ____________

3.2 CLOSURE AND COMMUTATIVE AXIOMS

OBJECTIVE: To use the closure and commutative axioms for addition and multiplication and to rewrite a given expression in compliance with algebraic conventions.

The equality axioms provide information about the equality relation. There are other axioms that deal with the operations of addition and multiplication. For now, the numbers involved in these operations will be restricted to the set of natural numbers, N. Later, as the subject matter of algebra is developed, more numbers will be introduced, and the universal set of numbers will be extended.

The Closure Axioms

The **closure axiom for addition** states that the sum of any two natural numbers is a natural number. If a and b are any two natural numbers, then $a + b$ is a natural number. For example, $5 + 7$ is a natural number, $38 + 792$ is a natural number, and $x + 5$ is a natural number if x is a natural number.

The **closure axiom for multiplication** states that the product of two natural numbers is a natural number. If a and b are any two natural numbers, then ab is a natural number. For example, $5\ (7)$ and $38\ (792)$ are natural numbers. If x is a natural number, then $5x$ is also a natural number.

Subtraction and division are *not* closed with respect to the set of natural numbers. For example, $2 - 7$ and $\frac{2}{7}$ are not natural numbers.

Whether or not a set is closed with respect to a certain operation depends on both the set and the operation. As our study of algebra progresses, we will work with sets of numbers that are closed with respect to addition, subtraction, multiplication, and division.

The Commutative Axioms

The **commutative axiom for addition** states that the order of the terms of a sum may be changed without changing the value of the sum. The axiom in symbols, for all natural numbers a and b, is

$$a + b = b + a.$$

For example, $3 + 5 = 5 + 3$ and $4 + x = x + 4$.

The **commutative axiom for multiplication** states that the order of the factors of a product may be changed without changing the value of the product. The axiom in symbols, for all natural numbers a and b, is

$$ab = ba.$$

For example, $3 \cdot 5 = 5 \cdot 3$ and $x5 = 5x$.
Also, $(4 + n)\,7 = 7\,(4 + n) = 7\,(n + 4)$.

Subtraction and division are *not* commutative operations. For example, $7 - 5 \neq 5 - 7$ and $\frac{6}{2} \neq \frac{2}{6}$. (The symbol $\neq$ is read "does not equal.")

Since it is possible to write sums and products in different ways, certain conventions are used to make it easier to compare one person's work with another's.

Algebraic Conventions

1. Letters are written in alphabetical order, when possible.
2. In a product, a single letter or a single numeral is written to the left of an expression enclosed by grouping symbols.
3. In a sum, a literal term is written to the left of a constant or numerical term.
4. In a product, a numerical factor is written to the left of another factor.

For Sample Problems 1–4, use the commutative axioms to rewrite each of the following in compliance with the stated algebraic conventions.

SAMPLE PROBLEM 1

yx

SOLUTION

xy

SAMPLE PROBLEM 2

$y + x$

SOLUTION

$x + y$

SAMPLE PROBLEM 3

$(3 + x)\,2$

SOLUTION

$2\,(x + 3)$

Note that, since $3 + x$ is a natural number by the closure axiom for addition, the commutative axiom for multiplication applies and $(3 + x)\,2 = 2\,(3 + x)$. Then, $2\,(3 + x) = 2\,(x + 3)$, since $3 + x = x + 3$ by the commutative axiom for addition.

SAMPLE PROBLEM 4

$4 + yx$

SOLUTION

$4 + yx = yx + 4$ **commutative axiom, addition**

$= xy + 4$ **commutative axiom, multiplication**

EXERCISES 3.2

Use the commutative axiom(s) to rewrite each of the following in compliance with the stated algebraic conventions.

A.

1. dc CD
2. $d + c$ C+D
3. $x7$ 7x
4. $(y + 5)\,2$ ______
5. $5 + y$ ______
6. $3 + (x + y)$ (x+y)+3
7. $5 + (a - b)$ a-b +5
8. $(y + 6) + x$ ______
9. $(z - 7) + y$ z+y-7
10. $2 + yx$ ______

B.

1. $4 + dc$ cd+4
2. $(yx)\,8$ 8xy
3. $4 + (d + c)$ ______
4. $(c - d)\,4$ ______
5. $5\,(b + a)$ ______
6. $(y + x)\,3$ 3(x+y)
7. $y\,(3x)$ ______
8. $ts + 7$ ______
9. $t + sr$ ______
10. $b\,(4 + a)$ ______

3.3 ASSOCIATIVE AXIOMS

OBJECTIVE: To apply the associative axioms for addition or multiplication and to compute the value of a given expression in two different ways.

The **associative axiom for addition** states that the terms of a sum may be regrouped without changing the value of the sum. The axiom in symbols is

$$a + (b + c) = (a + b) + c \text{ for all natural numbers } a, b, \text{ and } c.$$

The **associative axiom for multiplication** states that the factors of a product may be regrouped without changing the value of the product. The axiom in symbols is

$$a\,(bc) = (ab)\,c \text{ for all natural numbers } a, b, \text{ and } c.$$

For Sample Problems 1–4, apply the associative axiom to each of the following, and then compute the value of both the original and the final expressions.

SAMPLE PROBLEM 1

$3 + (7 + 9)$

SOLUTION

$$3 + (7 + 9) = (3 + 7) + 9$$

$$3 + 16 = 10 + 9$$

$$19 = 19$$

SAMPLE PROBLEM 2

$4\,(25 \cdot 9)$

SOLUTION

$$4\,(25 \cdot 9) = (4 \cdot 25)\,9$$

$$4\,(225) = (100)\,9$$

$$900 = 900$$

SAMPLE PROBLEM 3

(87 + 85) + 15

SOLUTION

$$(87 + 85) + 15 = 87 + (85 + 15)$$

$$172 + 15 = 87 + 100$$

$$187 = 187$$

SAMPLE PROBLEM 4

(17 · 8) 125

SOLUTION

$$(17 \cdot 8)\ 125 = 17\ (8 \cdot 125)$$

$$(136)\ 125 = 17\ (1000)$$

$$17{,}000 = 17{,}000$$

Subtraction and division are *not* associative operations. For example, $15 - (10 - 2) \neq (15 - 10) - 2$, since $7 \neq 3$.

Also, $24 \div (6 \div 2) \neq (24 \div 6) \div 2$, since $8 \neq 2$.

EXERCISES 3.3

Apply an associative axiom to each of the following, and then compute the value of both the original and the final expressions.

A.

1. 4 + (6 + 7) =
   (4+6)+7
   10+7=17
2. (3 + 2) + 6 =
   3+(2+6)
   3+8=11
3. (3 + 9) + 5 =
   3+(9+5)
   3+14=17

B.

1. 15 + (10 + 5) =
   (15+10)+5
   25+5 = 30
2. 8 + (7 + 4) =
   (8+7)+4
   15+4= 16
3. (68 + 35) + 65 =
   65 + 35 = 100
   68+(35+65)
   68+100 = 168

A.

4. $25 + (75 + 87) =$

5. $3\,(7 \cdot 4) =$

6. $2\,(50 \cdot 8) =$

7. $(21 \cdot 4)\,5 =$

8. $(57 \cdot 25)\,4 =$

9. $(45 + 55) + 87 =$

10. $(96 \cdot 8)\,125 =$

B.

4. $(59 + 49) + 51 =$

5. $3\,(4 \cdot 5) =$

6. $4\,(25 \cdot 9) =$

7. $(47 \cdot 2)\,50 =$

8. $(92 \cdot 4)\,25 =$

9. $8\,(125 \cdot 25) =$

10. $43 + (57 + 39) =$

3.4 COMMUTATIVE AND ASSOCIATIVE AXIOMS

OBJECTIVE: To rearrange the terms of a sum or the factors of a product by using the commutative and associative axioms and by complying with algebraic conventions.

Combining the Commutative and Associative Axioms

When the commutative and associative axioms for addition are combined, they imply that the *terms of a sum can be rearranged* in any order without changing the value of the sum.

When the commutative and associative axioms for multiplication are combined, they imply that the *factors of a product can be rearranged* in any order without changing the value of the product.

Conventions. The terms of a sum are written so that letters are in alphabetical order (reading from left to right) and so that a term involving one or more letters is to the left of a numerical term.

The factors of a product are written so that letters are in alphabetical order (reading from left to right) and so that a numerical factor is written to the left of a literal factor.

For Sample Problems 1 and 2, rearrange the terms and factors to comply with the stated conventions.

SAMPLE PROBLEM 1

$b + 5 + a$

SOLUTION

$a + b + 5$

ANALYSIS

$(b + 5) + a = a + (b + 5)$ **commutative, addition**

$= (a + b) + 5$ **associative, addition**

$= a + b + 5$ **convention (adding from left to right)**

SAMPLE PROBLEM 2

$x3y$

SOLUTION

$3xy$

ANALYSIS

$(x3)\,y = (3x)\,y$ **commutative axiom, multiplication**

$= 3xy$ **convention (multiplying from left to right)**

In working a problem, the analysis, shown in Sample Problems 1 and 2, is not usually written but is done mentally. How the written work usually looks is shown in Sample Problems 3–5.

For Sample Problems 3–5, rearrange the terms or factors to comply with the stated conventions.

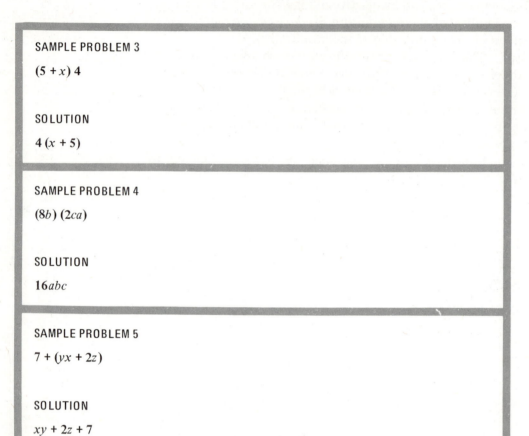

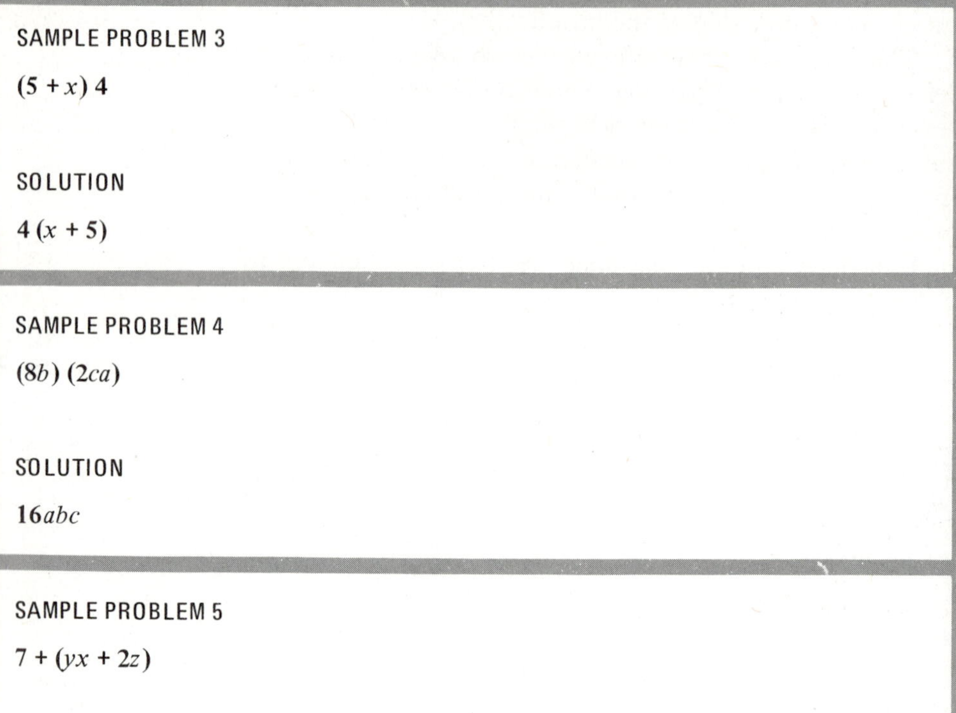

SAMPLE PROBLEM 3

$(5 + x)\,4$

SOLUTION

$4\,(x + 5)$

SAMPLE PROBLEM 4

$(8b)\,(2ca)$

SOLUTION

$16abc$

SAMPLE PROBLEM 5

$7 + (yx + 2z)$

SOLUTION

$xy + 2z + 7$

EXERCISES 3.4

Rearrange the terms of a sum and the factors of a product to comply with the conventions stated in this section.

A.		B.	
1. $2 + y + x$	x+y+2	1. $yx2$	2xy
2. $cd\,(3)$	3cd	2. $d + (b + c)$	b+c+d
3. $sr5$	5rs	3. $y + 5 + x$	x+y+5

A.

4. $w\,(2l)$ 2Lw

5. $n + l + m$ l+M+N

6. $2 + t + r$ ______

7. $n\,(7 + k)$ N(K+7)

8. $y + 3 + x$ 3+x+y

9. $(4y)\,x$ ______

10. $(b + 1) + (a + c)$ ______

11. $(2x)\,(3x)$ $6x^2$

12. $(3q)\,(5p)$ 15pq

13. $(y + x)\,2$ ______

14. $5 + yx + z$ ______

15. $(y + 2 + x)\,5$ ______

16. $8 + 5nm$ ______

17. $(6 + dc) +$ $(4 + nt)$ ______

B.

4. zxy ______

5. $n\,(1 + m)$ ______

6. $n7k$ ______

7. $m5n$ ______

8. $10 + t + 4 + s$ ______

9. $2x\,(4y)$ ______

10. $(y + 2) + (x + 1)$ ______

11. $(5y)\,(3x)$ ______

12. $(2z)\,(xy)$ ______

13. $(b + a)\,6$ ______

14. $8 + ba + dc$ ______

15. $9 + tsr$ ______

16. $(4 + x + y)\,3$ ______

17. $(yx + 5) +$ $(ba + 8)$ 2b+xy+13

3.5 THE DISTRIBUTIVE AXIOM

OBJECTIVES: To use the distributive axiom to express an indicated product as a sum or to express a sum of products as a product.

To use the distributive axiom to perform an indicated calculation in a different way than the one indicated.

In arithmetic, the product 23 (15) is worked as follows:

$$\begin{array}{r} 23 \\ \underline{15} \\ 115 \\ \underline{230} \\ 345 \end{array}$$

Written horizontally, the problem is worked as follows:

$$23\,(15) = 23\,(10 + 5) = 23\,(10) + 23\,(5) = 230 + 115 = 345$$

In other words, the factor 23 is distributed to the terms of the sum, 10 and 5, and then the two resulting products are added. When this number property is generalized, it is called the distributive axiom.

The distributive axiom in symbols, for all natural numbers a, b, and c, is

$$a\,(b + c) = ab + ac.$$

The distributive axiom relates the operations of addition and multiplication. It states that a product of the form $a\,(b + c)$ can be replaced by a sum of products, $ab + ac$.

Also, by using the symmetric axiom of equality, it follows that

$$ab + ac = a\,(b + c)$$

and that a sum of product, $ab + ac$, can be replaced by a product of the form $a\,(b + c)$.

As an example,

$$3\,(4 + 5) = 3\,(4) + 3\,(5)$$

$$3\,(9) = 12 + 15$$

$$27 = 27$$

Similarly,

$$67\,(25) + 67\,(75) = 67\,(25 + 75)$$

$$= 67\,(100) = 6700$$

SAMPLE PROBLEM 1

Apply the distributive axiom and express the indicated product as a sum: 8 (x + 2).

SOLUTION

$8\,(x + 2) = 8\,(x) + 8\,(2) = 8x + 16$

For Sample Problems 2–4, apply the distributive axiom and express the indicated sum as a product.

SAMPLE PROBLEM 2

$4x + 4y$

SOLUTION

$4x + 4y = 4(x + y)$

SAMPLE PROBLEM 3

$5x + 30$

SOLUTION

$5x + 30 = 5x + 5\,(6) = 5\,(x + 6)$

SAMPLE PROBLEM 4

$7y + 7$

SOLUTION

$7y + 7 = 7y + 7\,(1) = 7\,(y + 1)$

For Sample Problems 5–6, use the distributive axiom,

$$a\,(b + c) = ab + ac,$$

to perform the indicated calculation.

SAMPLE PROBLEM 5

25 (40 + 6)

SOLUTION

25 (40 + 6) = 25 (40) + 25 (6) = 1000 + 150 = 1150

SAMPLE PROBLEM 6

89 (47) + 89 (53)

SOLUTION

89 (47) + 89 (53) = 89 (47 + 53)
= 89 (100) = 8900

SAMPLE PROBLEM 7

Write $6\,(x + y + 4)$ as a sum.

SOLUTION

$6\,(x + y + 4) = 6x + 6y + 24$

ANALYSIS (done mentally)

$6\,(x + y + 4)$	$= 6\,(x + y + 4)$	considering $x + y$ as a single number
	$= 6\,(x + y) + 6\,(4)$	using the distributive axiom
	$= 6x + 6y + 24$	using the distributive axiom again

SAMPLE PROBLEM 8

Write $4x + 2x$ as a product.

SOLUTION

$4x + 2x = 6x$

ANALYSIS (done mentally)

$4x + 2x$	$= x4 + x2$	commutative axiom, multiplication
	$= x(4 + 2)$	distributive axiom
	$= x(6)$	simplifying
	$= 6x$	commutative axiom, multiplication

SAMPLE PROBLEM 9

Write $5x + 3x + x$ as a product.

SOLUTION

$$5x + 3x + x = 5x + 3x + 1x$$
$$= (5 + 3 + 1)x = 9x$$

EXERCISES 3.5

(1–10) Apply the distributive axiom and express each indicated product as a sum.

A.

1. $4(x + 7)$ ____________
2. $x(y + 2z)$ ____________
3. $ab(r + s)$ ____________
4. $5x(x + 1)$ ____________
5. $8(x + y + z)$ ____________
6. $2x(x + 6)$ ____________
7. $3y(2x + y)$ ____________
8. $x(x + y + 1)$ ____________

B.

1. $6(x + 4y)$ ____________
2. $5x(x + 9)$ ____________
3. $3x(x + 2y)$ ____________
4. $2a(b + 1)$ ____________
5. $9k(r + 2s + 1)$ ____________
6. $4y(y + 10)$ ____________
7. $7x(4x + 5y)$ ____________
8. $8y(x + 4y)$ ____________

9. $6x(x + 2y + 5)$ ________	9. $2xy(3x + 2y + 4)$ ________
10. $xy(2x + y + 5)$ ________	10. $4x(x + y + z)$ ________

(11–20) Apply the distributive axiom and express each indicated sum as a product.

A.	B.
11. $9x + 9y$ ________	11. $xy + 2xz$ ________
12. $xy + y$ ________	12. $4x + 8$ ________
13. $6rs + 56$ ________	13. $x^2 + x$ ________
14. $3x + 4x$ ________	14. $5y + y$ ________
15. $7r + 14s + 7t$ ________	15. $8x + 8y + 8$ ________
16. $5x + 5$ ________	16. $18x + 3$ ________
17. $5x + 5x$ ________	17. $3x + 3x$ ________
18. $24y + 12$ ________	18. $30y + 50$ ________
19. $2x^2 + 6x$ ________	19. $10y^2 + 5y$ ________
20. $y + 8y$ ________	20. $x + 7x$ ________

(21–25) Use the distributive axiom to perform the indicated calculation.

A.	B.
21. 5 (60 + 7) ________	21. 4 (25 + 9) ________
22. 72 (75) + 72 (25) ________	22. 96 (28) + 96 (72) ________
23. 25 (20 + 8) ________	23. 36 (85) + 64 (85) ________
24. 97 (64) + 3 (64) ________	24. 70 (90 + 2) ________
25. 125 (80 + 4) ________	25. 59 (35) + 59 (25) + 59 (40) ________

UNIT 3 POSTTEST

(1–5) Identify each statement as an example of the reflexive axiom, the symmetric axiom, the transitive axiom, or as a false statement by writing the word "reflexive," "symmetric," "transitive," or "false." (3.1)

1. If $(x - 3) + 3 = 4$ and $x = (x - 3) + 3$, then $x = 4$. 1. ____________

2. If $35 - 4x = 3x$, then $3x = 35 - 4x$. 2. ____________

3. If $\frac{x}{6} = 2$, then $\frac{6}{x} = 2$. 3. ____________

4. $8 - 2 = 2 - 8$ 4. ____________

5. $8 - 2x = 8 - 2x$ 5. ____________

(6–11) Use the closure and commutative axioms to rewrite each of the following in compliance with algebraic conventions. (3.2)

6. $5 + n$ ____________ 7. $x8$ ____________ 8. $z + y$ ____________

9. yx ____________ 10. $6 + sr$ ____________ 11. $(3 + x)\,9$ ____________

(12–15) Apply an associative axiom to the following, and compute the value of both the original and the final expressions. (3.3)

12. $25\,(4 \cdot 56) =$

13. $69 + (31 + 97) =$

14. $(76 + 89) + 11 =$

15. $(95 \cdot 125)\,8 =$

(16–19) Use the commutative and associative axioms to rearrange the terms of a sum or the factors of a product in compliance with algebraic conventions. (3.4)

16. $4 + y + x$ ____________ 17. $c9x$ ____________

18. $nk + 3 + dc$ ____________ 19. $12 + xa7$ ____________

(20–21) Use the distributive axiom to express the indicated product as a sum. (3.5)

20. $4b\ (ac + 1) =$ ______________ 21. $5x\ (x + 4) =$ ______________

(22–23) Use the distributive axiom to express the indicated sum as a product . (3.5)

22. $7y + 2y =$ ______________ 23. $6x + 15y =$ ______________

(24–25) Use the distributive axiom to perform the given calculation in a different way than the one indicated. (3.5)

24. $25\ (79) + 75\ (79) =$

25. $80\ (60 + 5) =$

UNIT 4

INTEGERS

Unit 4
PRETEST

1. If +750 means an altitude of 750 feet above sea level, state the meaning of −750 and of 0. (4.1)

   −750 means ______________________

   0 means ______________________

2. State the coordinates of points A, B, and C on the number line below. (4.2)

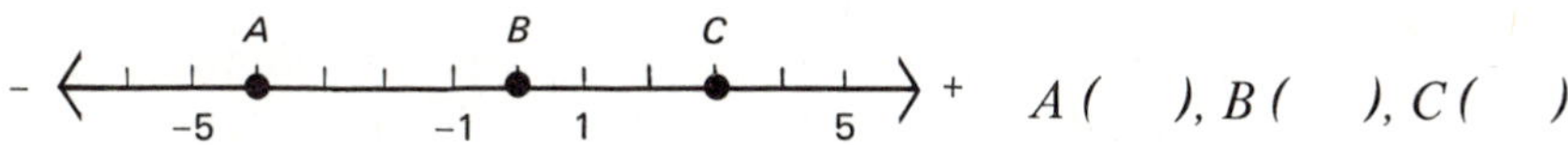

A(   ), B(   ), C(   )

3. State the coordinates of points D, E, and F on the number line shown at the right. (4.2)

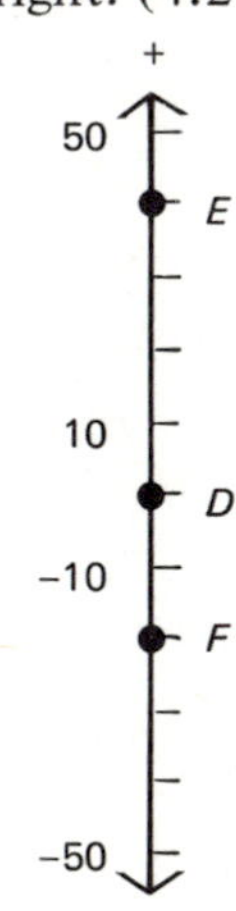

D(   ), E(   ), F(   )

(4–7) Rewrite each of the following as a single integer. (4.3)

4. |−20| = ______________

5. |−6| + |+ 13| = ______________

6. |−30| − |−5| = ______________

7. |−4| · |+9| = ______________

(8–9) Translate into an algebraic statement. (4.4)

8. Three more than x is less than 9.

9. The product of 3 and x is positive.

(10–13) Insert the symbol $<$ or the symbol $>$ between each given pair of integers, so that the resulting statement is true. State the algebraic meaning of the resulting statement. (4.4)

10. +7 ____ +5, ________________________________

11. -7 ____ +5, ________________________________

12. -7 ____ -5, ________________________________

13. -5 ____ -7, ________________________________

(14–17) Express each of the following in symbols and do the addition, using a positive number for a bank deposit and a negative number for a withdrawal from the bank. (4.5)

14. A deposit of \$120 followed by a withdrawal of \$30.

15. A withdrawal of \$30 followed by a deposit of \$120.

16. A withdrawal of \$45 followed by a withdrawal of \$40.

17. A deposit of \$200 followed by a withdrawal of \$250.

(18–23) Add. (4.6)

18. $(+8) + (-20) =$ ______ 19. $(-15) + (-14) =$ ______

20. $(-50) + (+70) =$ ______ 21. $(+30) + (-30) =$ ______

22. $(-45) + 0 =$ ______ 23. $(-13) + (+5) =$ ______

(24–29) Subtract. (4.7)

24. $(+40) - (+15) =$ ______ 25. $(+15) - (+40) =$ ______

26. $(+35) - (-5) =$ ______ 27. $(-8) - (-58) =$ ______

28. $(+70) - (+70) =$ ______ 29. $(-25) - (-25) =$ ______

(30–35) Multiply. (4.8)

30. $(-30)(-4) =$ ______ 31. $(-50)(+6) =$ ______

32. $(-9)(+8) =$ ______ 33. $(-9)(0) =$ ______

34. $(-6)(1) =$ ______ 35. $(-1)(+20) =$ ______

(36–39) Divide. (4.9)

36. $\frac{-54}{-6} =$ ______ 37. $\frac{-42}{+7} =$ ______

38. $\frac{+25}{-25} =$ ______ 39. $\frac{0}{-36} =$ ______

(40–43) Evaluate. (4.10)

40. $\frac{25 - 20}{3 - 8} =$ ______ 41. $-4(-7 + 3) =$ ______

42. $(9 - 20)(4 - 10) =$ ______ 43. $2(-5)^2 =$ ______

(44–45) Evaluate. (4.11)

44. xyz for $x = -5$, $y = -6$, and $z = -7$

45. $\dfrac{b^2 - 4ac}{2a}$ for $a = -2$, $b = -6$, and $c = -5$

4.1 SIGNED NUMBERS, INTERPRETATIONS

OBJECTIVE: To state the meaning of a negative number or zero for a given interpretation.

Signed numbers are numbers that are tagged with a + or − sign. These numbers came into existence in order to measure quantities opposite in nature. The numbers x and $-x$ are called **opposites**, or **additive inverses**, of each other. For example, +5 and −5 are opposites of each other. The number +5 is called a positive number, and the number −5 is called a negative number. It is agreed that +5 and 5 are different names for the same number. In general, if p is a positive number, then $+p = p$. Similarly, $+n = n$ and $+5 = 5$.

The number zero, 0, is neither positive nor negative. Zero is located midway between the positive numbers and the negative numbers on a scale of measurement.

For Sample Problems 1-3, state the meaning of −25 if +25 has the value given.

SAMPLE PROBLEM 1

A gain of $25.

SOLUTION

−25 means a loss of $25.

SAMPLE PROBLEM 2

A temperature of 25° above 0° centigrade.

SOLUTION

−25 means a temperature of 25° below 0° centigrade.

SAMPLE PROBLEM 3

A velocity of 25 miles per hour of a car moving due east.

SOLUTION

−25 means a velocity of 25 miles per hour of a car moving due west.

SAMPLE PROBLEM 4

State the meaning of 0 if +30 means the distance of a car 30 miles due north of a city, and if −30 means the distance of a car 30 miles due south of the city.

SOLUTION

0 means the car is neither north nor south of the city but is in the city.

EXERCISES 4.1

State the meaning of each of the following numbers for the given interpretation.

A.

1. −50 if +50 means a bank deposit of $50.

    ____

2. −8 if +8 means a gain of 8 yards in a football game.

    ____

3. −200 if +200 means an altitude 200 feet above sea level.

    ____

4. −45 if +45 means a velocity of 45 m.p.h. in the north direction.

    ____

5. 0 if +4 means a gain in weight of 4 pounds and −4 means a loss in weight of 4 pounds.

    ____

6. $-\frac{3}{4}$ if $+\frac{3}{4}$ means an increase of 75¢ in the price of a stock.

    ____

7. −1250 if +1250 means A.D. 1250.

    ____

8. −20 if +20 means a force of 20 pounds downward.

    ____

9. 0 if +60 means a rotation of 60° counterclockwise and −60 means a rotation of 60° clockwise.

    ____

10. −8 if +8 means a current of 8 amps. for a charging battery.

    ____

B.

1. −30 if +30 means 30° longitude west of Greenwich.

    ____

2. −20 if +20 means a temperature of 20° above zero.

    ____

3. 0 if +10 means 10° north latitude and −10 means 10° south latitude.

    ____

4. −5 if +5 means 5 yards gained.

    ____

5. 0 if +3 and −3 mean an increase and a decrease respectively.

    ____

6. −100 if +100 means a profit of $100.

    ____

7. −15 if +15 means an acceleration of 15 m.p.h.

    ____

8. −24 if +24 means the distance of an image 24 cm. in front of a mirror.

____

9. −32 if +32 means 32 points won in a card game where points may be won or lost.

____

10. −17 if +17 means a score of 17 points above the average score on a certain psychology test.

____

4.2 THE SET OF INTEGERS, THE NUMBER LINE

OBJECTIVES: To state the coordinate of a point on a horizontal or vertical number line.

To graph a point, given its coordinate, on a horizontal or vertical number line.

The set of natural numbers is not closed with respect to subtraction. For example, although 7 − 2 is the natural number 5, 2 − 7 is not a natural number, and 7 − 7 is not a natural number. Actually, 2 − 7 = −5 and 7 − 7 = 0. The numbers −5 and 0 are called **integers.**

By including 0 and the negatives of the natural numbers in the universal set of numbers, the subtraction operation will be closed. In other words, the answer to every subtraction problem will be a number in the universal set.

The **set of integers,** I, is the set consisting of the natural numbers, the number 0 (zero), and the negatives of the natural numbers. In symbols,

$$I = \{x | x = n, x = 0, \text{ or } x = -n \text{ where } n \text{ is a natural number}\}$$

Also,

$$I = \{\ldots, -5, -4, -3, -2, -1, 0, 1, 2, 3, 4, 5, \ldots\}$$

The natural numbers, being integers, are also called the **positive integers.** The negatives of the natural numbers are called the **negative integers.** Zero is neither positive nor negative.

Properties of Zero

There is exactly one integer 0 (zero), called the **additive identity**, such that for all integers x, $x + 0 = x$ and $0 + x = x$.

As examples, $5 + 0 = 5$, $-5 + 0 = -5$, $0 + 6 = 6$, and $0 + (-7) = -7$.

For each integer x, there is exactly one integer $-x$, called the **additive inverse** of x, such that $x + (-x) = 0$ and $-x + x = 0$.

As examples, $3 + (-3) = 0$, $-4 + 4 = 0$, and $0 + 0 = 0$. Note that zero is the additive inverse of itself.

It is useful to picture the integers on a number line as shown in Figure 4.1.

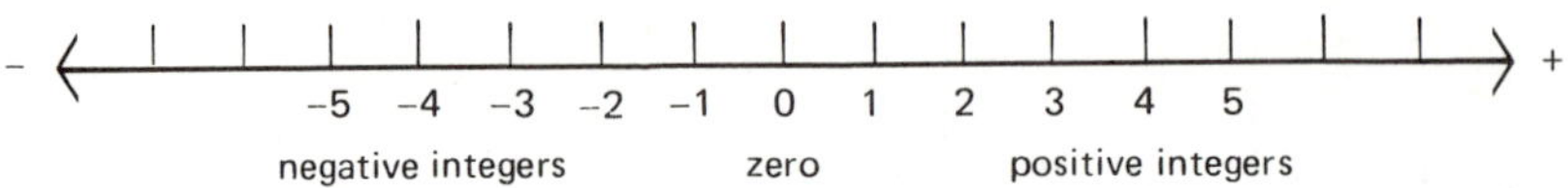

FIGURE 4.1 A number line.

A **number line** is a line whose points are named by using numbers.

The **coordinate** of a point on a number line is the number that names the point.

The **graph** of a number is the point on the number line that is named by the number.

The **origin** is the point whose coordinate is 0 (zero).

The **positive numbers** are assigned to the points on one side of the origin, and the **negative numbers** are assigned to the points on the other side of the origin.

SAMPLE PROBLEM 1

State the coordinates of points *A*, *B*, *C*, and *D* on the number line below.

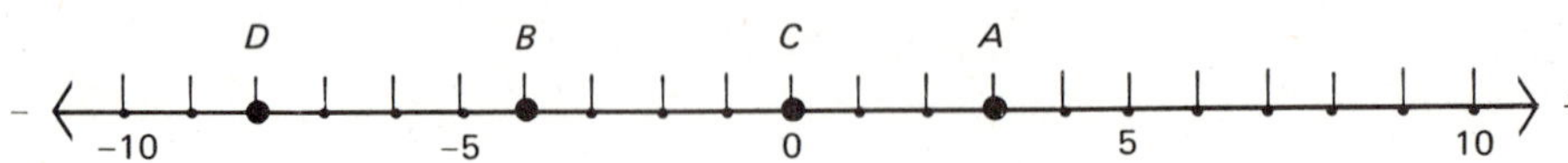

SOLUTION

A (3), *B* (-4), *C* (0), *D* (-8)

The notation A (3) is used to designate the point whose geometric name is A and whose algebraic name is 3. Similarly, B (−4) means the point whose geometric name is B and whose algebraic name is −4.

SAMPLE PROBLEM 2

State the coordinates of points A, B, C, and D on the number line below.

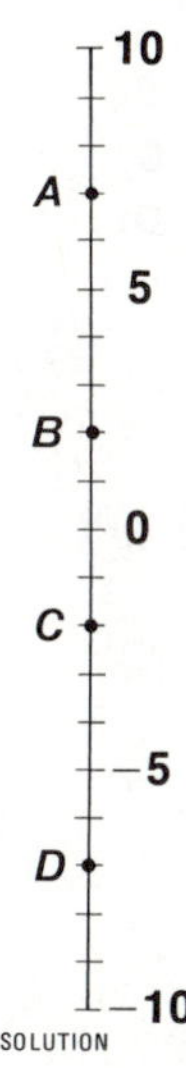

SOLUTION

A (+7), B (+2), C (−7), D (−2)

A.

EXERCISES 4.2

(1–5) State the coordinates of the points A, B, C, and D on each number line.

1.

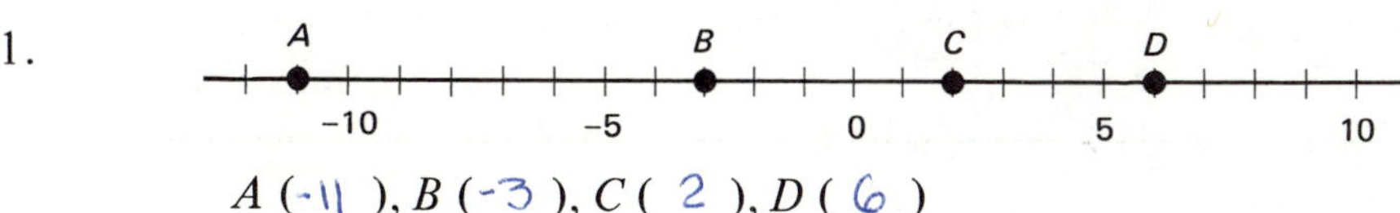

A (-11), B (-3), C (2), D (6)

2.

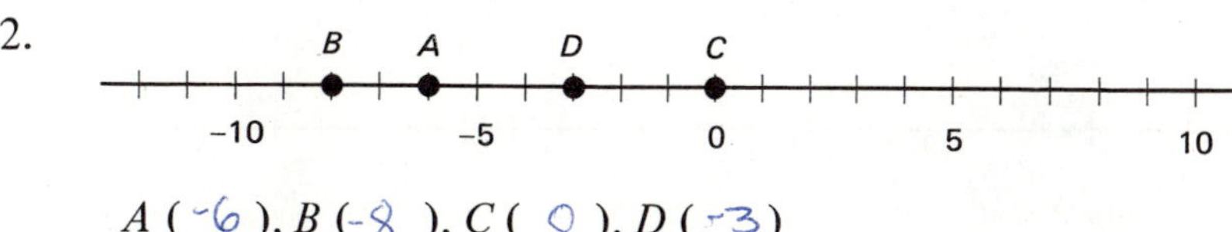

A (-6), B (-8), C (0), D (-3)

3.

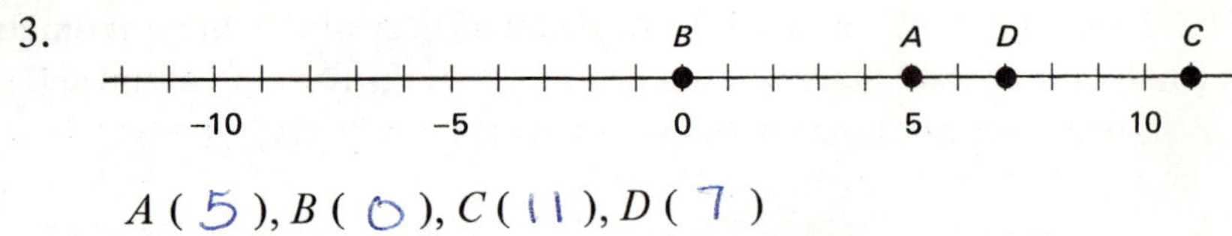

A (5), B (0), C (11), D (7)

4.   5.

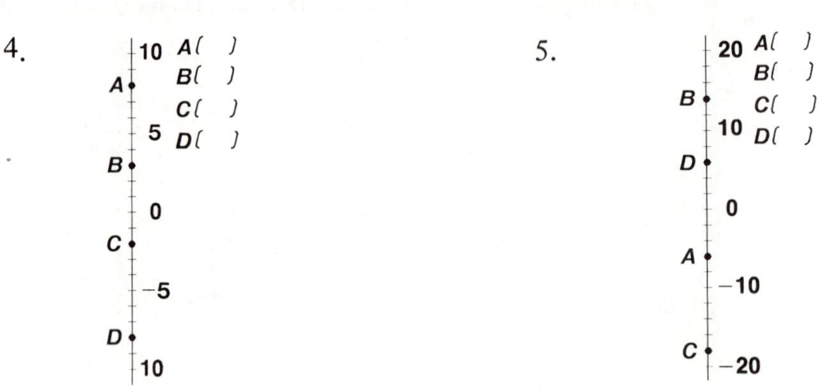

(6–15) Graph each of the following points on a horizontal number line whose positive direction is to the right. Graph five points on one number line. Place a solid dot at the correct position on the number line, and write the letter name of the point above the dot.

6. A (6)   7. B (−6)   8. C (13)   9. D (8)   10. E (−8)

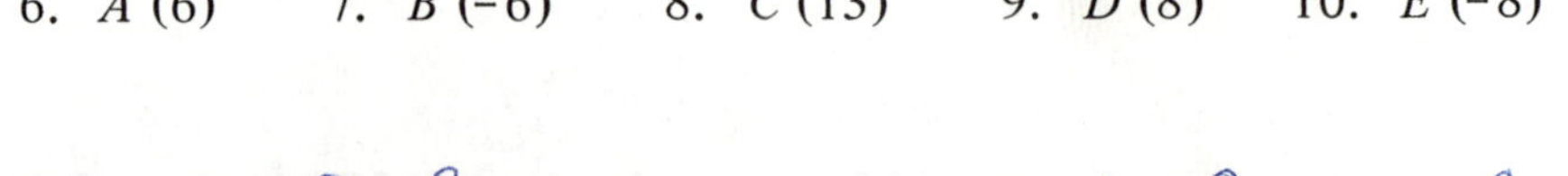

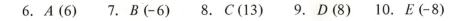

11. A (4)   12. B (−4)   13. C (−6)   14. D (0)   15. E (6)

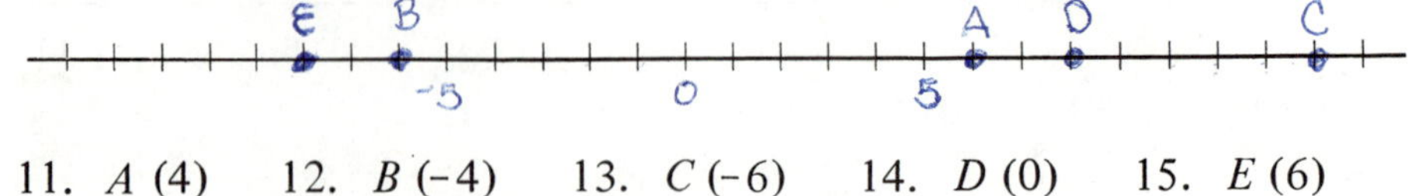

(16–25) Graph each of the following points on a vertical number line whose positive direction is upward. Graph five points on one number line. Place a solid dot at the correct position on the number line, and write the letter name of the point to the right of the dot.

16. A (−3)
17. B (0)
18. C (14)
19. D (3)
20. E (−7)

21. A (150)
22. B (−25)
23. C (50)
24. D (−75)
25. E (0)

B.

(1–5) State the coordinates of the points A, B, C, and D on each number line.

1.

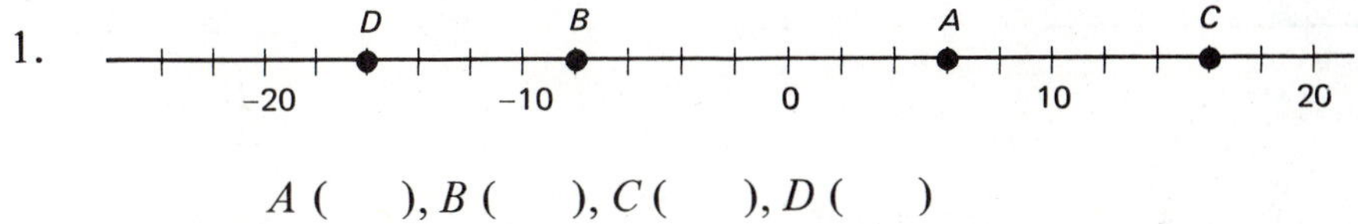

A (   ), B (   ), C (   ), D (   )

2.

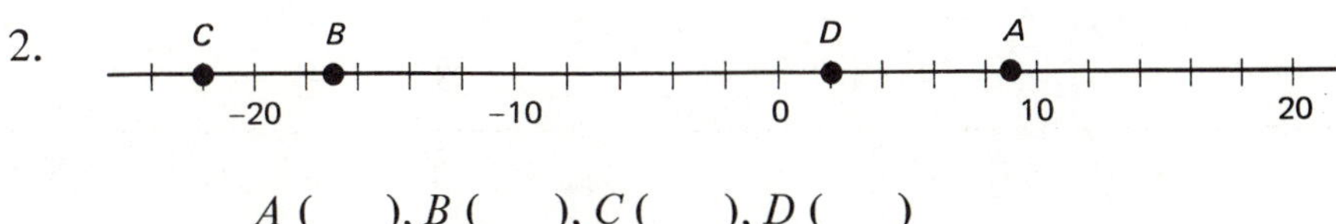

A (   ), B (   ), C (   ), D (   )

3.

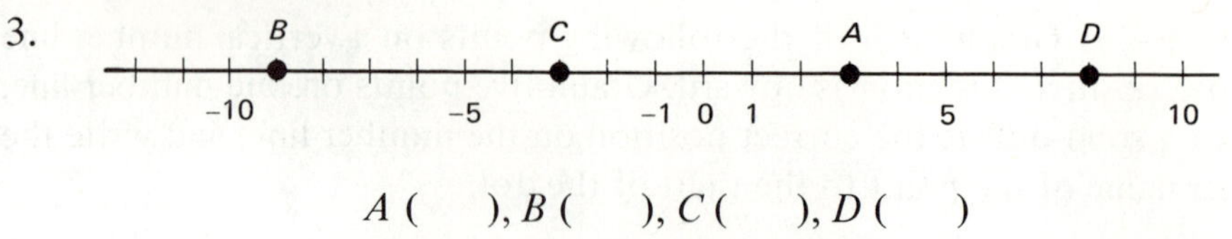

A ( ), *B* ( ), *C* ( ), *D* ( )

4.

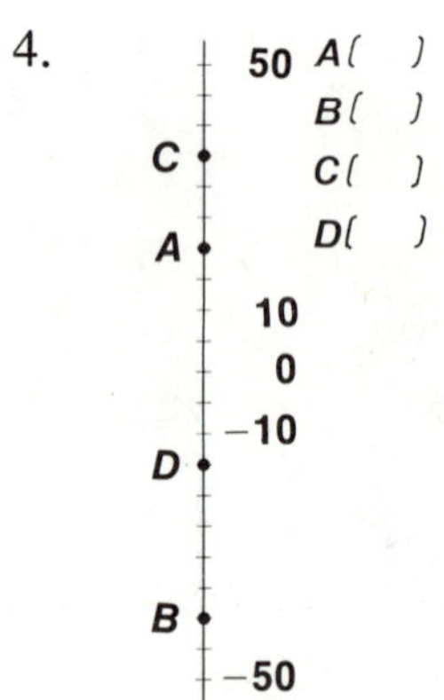

5.

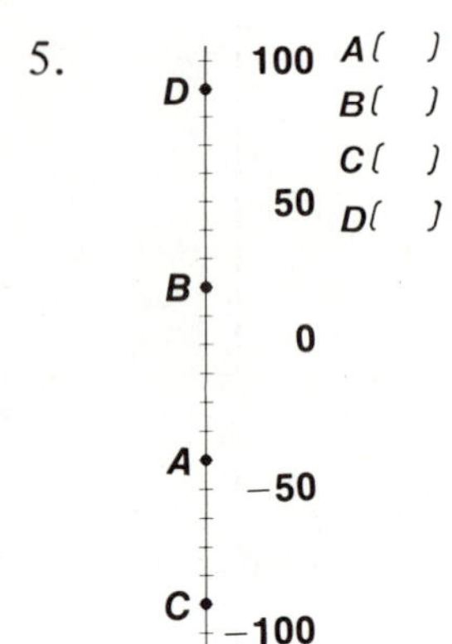

(6–25) Graph each point on the number line indicated.

6. *A* (−25) 7. *B* (25) 8. *C* (−30) 9. *D* (15) 10. *E* (−10)

11. *A* (250) 12. *B* (−150) 13. *C* (−25) 14. *D* (100) 15. *E* (50)

16. A (-10)
17. B (-50)
18. C (20)
19 D (30)
20. E (-40)

21. A (0)
22. B (8)
23. C (-4)
24. D (-12)
25. E (4)

4.3 ABSOLUTE VALUE

OBJECTIVE: To rewrite an expression involving absolute value symbols as a single integer.

The points 6 and -6 on a number line are the same distance from the origin, but they are on opposite sides of the origin. The algebraic sign of a number shows on which side of the origin the corresponding point is located. Distance is either positive or zero. To designate the distance of a point from the origin, regardless of the side of the origin on which the point lies, the absolute value symbol is used. In symbols, it is written

$|x|$ and is read "the absolute value of x."

As examples, $|6|$ represents the distance (6 units) from the origin to the point 6; $|-6|$ represents the distance (6 units) from the origin to the point -6.

The absolute value of a number can be thought of as the "size" of a number without regard to its position on the number line. The distance of a point from the origin is one application of absolute value. Absolute value is also useful for explaining how to add, subtract, multiply, and divide integers.

The absolute value of a number is written in symbols as $|x|$. As examples, $|5| = 5$ and $|-5| = 5$.

Definition of Absolute Value. If p is a positive number, then $|p| = p$, $|-p| = p$, and $|0| = 0$.

SAMPLE PROBLEM 1

Simplify $|-7|$.

SOLUTION

$|-7| = 7$

SAMPLE PROBLEM 2

Simplify $|-3| + |-6|$.

SOLUTION

$|-3| + |-6| = 3 + 6 = 9$

SAMPLE PROBLEM 3

Simplify $|-8| - |+3|$.

SOLUTION

$|-8| - |+3| = 8 - 3 = 5$

SAMPLE PROBLEM 4

Simplify $3\,|-5|$.

SOLUTION

$3\,|-5| = 3\,(5) = 15$

SAMPLE PROBLEM 5

Simplify $\frac{|-12|}{3}$.

SOLUTION

$\frac{|-12|}{3} = \frac{12}{3} = 4$

EXERCISES 4.3

Simplify each of the following. (Rewrite each as a single integer.)

A.

1. $|6|$
2. $|-7|$
3. $|0|$
4. $|-4|$
5. $|-8|$
6. $|+9|$
7. $|-13| + |5|$
8. $|-1| + |-2|$
9. $|-6| + |+1|$
10. $|10| - |-7|$
11. $|-10| - |-3|$
12. $|-9| - |+5|$
13. $2\,|-6|$
14. $7\,|-2|$
15. $|-5| \cdot |-4|$
16. $3\,|7 - 7|$
17. $-\,|-5|$
18. $3\,|-4| - 2\,|-6|$
19. $\dfrac{|-9|}{3}$
20. $\dfrac{|48|}{|-6|}$
21. $\dfrac{10}{|-10|}$

B.

1. $|-2| + |-7|$
2. $|+81| - |+61|$
3. $|10| + |-7|$
4. $|-10| - |3|$
5. $|-10| + |-3|$
6. $|12| - |-5|$
7. $4\,|-4|$
8. $|-5|\,4$
9. $|-2| \cdot |+30|$
10. $|5| \cdot |-8|$
11. $4\,|-5| - 5\,|-3|$
12. $20 - 2\,|-3| \cdot |-2|$
13. $\dfrac{|-35|}{|5|}$
14. $\dfrac{|-8|}{8}$
15. $\dfrac{|-4| + |-8|}{|-2|}$

4.4 ORDER RELATIONS, TRICHOTOMY AXIOM

OBJECTIVE: To insert an order or an equal relation symbol between two integers and to state the algebraic and geometric meanings of the resulting statement.

Given two sets of objects, there is a natural way to compare the number of objects in one set with those in the other. Either the two sets have the same number of objects, or the first set has fewer objects than the second set, or the first set has more objects than the second set.

If a is the number of objects in the first set and if b is the number of objects in the second set, then there are three possibilities:

1. a is equal to b. In symbols, $a = b$.
2. a is less than b. In symbols, $a < b$.
3. a is greater than b. In symbols, $a > b$.

These three types of comparisons are called relations. They include the equal relation (=) and the order relations (< and >). As examples:

$2 < 7$ (read "two is less than seven")

$8 > 5$ (read "eight is greater than five")

A number line shows a natural way to order the set of integers. A number to the left of another number on a horizontal number line is less than the number to its right. Similarly, the number on the right is greater than the number on its left. These ideas are summarized in the following table.

Relation in Symbols	*Algebraic Meaning*	*Geometric Meaning (for a horizontal number line with positive direction to the right)*
$a < b$	a is less than b	point a is left of point b
$a > b$	a is greater than b	point a is right of point b
$a = b$	a equals b	point a is the same as point b

The Trichotomy Axiom. If x and y are any integers, then exactly one of the following is valid:

$$x < y \quad \text{or} \quad x > y \quad \text{or} \quad x = y$$

For Sample Problems 1–5, insert an order symbol or an equal sign between each pair of numbers, so that the resulting statement is true. State the algebraic meaning of the relation, and state the geometric meaning.

SAMPLE PROBLEM 1

+6 _____ +2

SOLUTION

+6 > +2
+6 is greater than +2.
Point +6 is to the right of point +2.

SAMPLE PROBLEM 2

−6 ______ −2

SOLUTION

−6 < −2
−6 is less than −2.
Point −6 is to the left of point −2.

SAMPLE PROBLEM 3

−5 _____ 9

SOLUTION

−5 < 9
−5 is less than 9.
Point −5 is to the left of point 9.

SAMPLE PROBLEM 4

0 ______ −4

SOLUTION

0 > −4
0 is greater than −4.
Point 0 is to the right of point −4.

SAMPLE PROBLEM 5

-2 ______ -2

SOLUTION

$-2 = -2$
-2 equals -2.
Point -2 is the same as point -2.

For Sample Problems 6–9, translate each verbal statement into an algebraic (symbolic) statement.

SAMPLE PROBLEM 6

The product of 3 and x is greater than 12.

SOLUTION

$3x > 12$

SAMPLE PROBLEM 7

Twice the sum of n and 4 is less than 15.

SOLUTION

$2\,(n + 4) < 15$

SAMPLE PROBLEM 8

The product of x and y is positive.

SOLUTION

$xy > 0$

SAMPLE PROBLEM 9

Eight more than the sum of r and s is negative.

SOLUTION

$r + s + 8 < 0$

1 For two positive numbers use the natural order
2. Any positive number is greater than zero
3. Any negative number is less " "
4. Any negative number is less than any positive number
5. The smaller of two negative numbers is the number with the larger absolute value

EXERCISES 4.4

A.

(1–10) Insert an order symbol or an equal sign between each given pair of numbers, so that the resulting statement is true. State the algebraic meaning of the relation, and state the geometric meaning.

1. +4 __<__ +7, ______________________________

2. +8 __>__ +3, ______________________________

3. −4 __>__ −7, ______________________________

4. −8 __<__ −3, ______________________________

5. −2 __<__ +5, ______________________________

6. +2 __>__ −5, ______________________________

7. 0 __>__ −4, ______________________________

8. −6 __=__ −6, ______________________________

9. 6 __>__ −3, ______________________________

10. −6 __<__ 0, ______________________________

(11–16) Translate each verbal statement into an algebraic (symbolic) statement.

11. The sum of x and 7 is greater than 12. 11. $x+7>12$

12. The product of y and 5 is less than 30. 12. $5y<30$

13. The sum of n and t is negative. 13. ______________

14. Four more than x is greater than 10. 14. ______________

15. Five less than y is less than 8. 15. ______________

16. The absolute value of x is positive. 16. ______________

B.

(1–10) Insert an order symbol or an equal sign between each given pair of numbers, so that the resulting statement is true. State the algebraic meaning of the relation, and state the geometric meaning.

1. 2 ____ -3, ______________________________

2. 4 ____ 0, ______________________________

3. 0 ____ 5, ______________________________

4. 7 ____ 5, ______________________________

5. -12 ____ -14, ______________________________

6. -3 ____ -3, ______________________________

7. -12 ____ 0, ______________________________

8. 7 ____ 7, ______________________________

9. -8 ____ 4, ______________________________

10. -4 ____ -9, ______________________________

(11–16) Translate each verbal statement into an algebraic statement.

11. The product of t and 6 is less than 24. 11. $6t < 24$

12. The sum of x and y is greater than the product of x and y. 12. ____________

13. The product of 4 and y is negative. 13. ____________

14. The absolute value of the product of a and b is positive. 14. ____________

15. Seven less than x is greater than y. 15. ____________

16. Eight more than x is less than y. 16. ____________

4.5 ADDITION OF INTEGERS, APPLIED

OBJECTIVE: To find the sum of two integers whose applied meaning is stated.

The rules for adding integers are determined by a desire to retain the arithmetical meaning of the sum of two natural numbers and to have the closure, commutative, and associative axioms for addition remain valid for the set of integers.

The identity axiom for addition states the meaning of the addition of 0; that is, $x + 0 = x$ and $0 + x = x$.

Since a gain of 2 yards followed by a gain of 5 yards results in a gain of 7 yards,

$$(+2) + (+5) = +7.$$

Similarly, a gain of 5 yards followed by a gain of 2 yards results in a gain of 7 yards and

$$(+5) + (+2) = +7.$$

It follows that $(+2) + (+5) = (+5) + (+2)$.

Since a loss of 2 yards followed by a loss of 5 yards results in a loss of 7 yards,

$$(-5) + (-2) = -7.$$

And as before, $(-2) + (-5) = -7$ and $(-5) + (-2) = (-2) + (-5)$.

In general, if a and b are natural numbers, then

$$(+a) + (+b) = +(a+b) \quad \text{and} \quad (-a) + (-b) = -(a+b).$$

Since a gain of 5 yards followed by a loss of 2 yards results in a net gain of 3 yards,

$$(+5) + (-2) = +3.$$

Similarly, a loss of 2 yards followed by a gain of 5 yards results in a net gain of 3 yards:

$$(-2) + (+5) = +3.$$

Again, the commutative axiom for addition is valid and

$$(+5) + (-2) = (-2) + (+5).$$

Finally, a loss of 5 yards followed by a gain of 2 yards results in a net loss of 3 yards, as does a gain of 2 yards followed by a loss of 5 yards:

$$(-5) + (+2) = -3 \quad \text{and} \quad (+2) + (-5) = -3$$

$$\text{thus, } (-5) + (+2) = (+2) + (-5).$$

In general, if a and b are natural numbers, then

$$(+a) + (-b) = +(a - b) \quad \text{if } a > b$$

and

$$(+a) + (-b) = -(b - a) \quad \text{if } b > a.$$

For Sample Problems 1–4, express the statement in symbols, and perform the addition, using a positive number for a gain and a negative number for a loss.

SAMPLE PROBLEM 1

A gain of $8 followed by a loss of $3.

SOLUTION

$(+8) + (-3) = +5$

SAMPLE PROBLEM 2

A gain of $4 followed by a loss of $10.

SOLUTION

$(+4) + (-10) = -6$

SAMPLE PROBLEM 3

A loss of 5 inches followed by a loss of 7 inches.

SOLUTION

$(-5) + (-7) = -12$

SAMPLE PROBLEM 4

A loss of 9 yards on a football play followed by neither a gain nor a loss on the next play.

SOLUTION

(−9) + 0 = −9

EXERCISES 4.5

Express in symbols and do the addition.

A.

(1–14) Use a positive number for a gain and a negative number for a loss.

1. A gain of $6 followed by a gain of $3. — 1. (+6)+(+3) =+9
2. A gain of $8 followed by a loss of $6. — 2. (+8)+(−6)=+2
3. A loss of $4 followed by a loss of $5. — 3. (−4)+(−5)= −9
4. A loss of $7 followed by a gain of $10. — 4. (−7)+(+10)=+3
5. A loss of $7 followed by a gain of $3. — 5. (−7)+(+3)= −4
6. A gain of $5 followed by a loss of $9. — 6. (+5)+(−9)= −4
7. A gain of $8 followed by neither a gain nor a loss. — 7. (+8)+(0)=+8
8. A loss of $6 followed by neither a gain nor a loss. — 8. (−6)+(0)= −6
9. A gain of 3 yards on a football play followed by a gain of 7 yards on the next play. — 9. (+3)+(+7)=+10
10. A loss of 8 yards on a football play followed by a loss of 7 yards on the next play. — 10. (−8)+(−7)= −15
11. A gain of 6 yards on a football play followed by a loss of 15 yards on the next play. — 11. (+6)+(−15)= −9

12. A gain of 14 yards on a football play followed by a loss of 10 yards on the next play. — 12. (+14)+(−10)=+4

13. A loss of 4 yards on a football play followed by neither a gain nor a loss on the next play. — 13. (−4)+(0)=−4

14. No gain nor loss in yardage on one football play followed by no gain nor loss in yardage on the next play. — 14. 0

B.

(1–6) Use a positive number for a deposit in a bank account and a negative number for a withdrawal.

1. A deposit of \$100, then a withdrawal of \$20. — 1. (+100)+(−20)=80

2. A deposit of \$35, then a deposit of \$40. — 2. (+35)+(+40)=75

3. A withdrawal of \$25, then a deposit of \$15. — 3. (−25)+(+15)=−10

4. A withdrawal of \$12, then a withdrawal of \$30. — 4. (−12)+(−30)=−42

5. A deposit of \$250, then a withdrawal of \$400. — 5. (+250)+(−400)=−150

6. A withdrawal of \$46, then a deposit of \$70. — 6. (−46)+(+70)=+24

(7–12) Use a positive number for an increase and a negative number for a decrease.

7. An increase in weight of 3 pounds followed by an increase of 6 pounds. — 7. ______

8. A decrease in weight of 4 pounds followed by a decrease of 1 pound. — 8. ______

9. An increase in weight of 2 pounds followed by a decrease of 5 pounds. — 9. ______

10. An increase in temperature of 10° followed by a decrease of 3°. — 10. ______

11. An increase in temperature of 8° followed by a decrease of 12°. — 11. ______

12. A decrease in temperature of 23° followed by an increase of 6°. 12. ______________

4.6 ADDITION OF INTEGERS

OBJECTIVE: To find the sum of two integers.

The rules for adding integers are now summarized.

1. The sum of 0 and any integer is the integer; $x + 0 = x$ and $0 + x = x$.
2. The sum of an integer and its negative is 0; $x + (-x) = 0$ and $-x + x = 0$.
3. To add numbers having the same sign, add their absolute values, and put their common sign before this sum.
4. To add numbers having opposite signs, subtract their absolute values, the smaller from the larger, and put the sign of the number having the larger absolute value before this difference.

SAMPLE PROBLEMS	SOLUTIONS
Add:	
1. $(+7) + (+2)$	$= +(7 + 2) = +9 = 9$
2. $(-5) + (-3)$	$= -(5 + 3) = -8$
3. $(+7) + (-3)$	$= +(7 - 3) = +4 = 4$
4. $(+3) + (-7)$	$= -(7 - 3) = -4$
5. $(-7) + (+2)$	$= -(7 - 2) = -5$
*6. $(-12) + 15$	$= (-12) + (+15) = +(15 - 12) = 3$
*7. $20 + (-60)$	$= (+20) + (-60) = -(60 - 20) = -40$
*8. $7 + (-7)$	$= 7 + (-7) = 0$
*9. $-50 + 50$	$= -50 + 50 = 0$
*10. $6 + 0$	$= 6 + 0 = 6$
11. $0 + (-9)$	$= 0 + (-9) = -9$
12. $0 + 0$	$= 0 + 0 = 0$

*Note that n and $+n$ name the same number.

EXERCISES 4.6 Add.

A.

1. (+3) + (+5)
2. (+7) + (+4)
3. (−2) + (−8)
4. (−43) + (−51)
5. (+11) + (+13)
6. (−3) + (−4)
7. (−6) + (−8)
8. (−11) + (−15)
9. (+20) + (+50)
10. (−25) + (−75)
11. (+4) + (−3)
12. (+8) + (−11)
13. (−12) + (+5)
14. (−17) + (+12)
15. (+15) + (−8)
16. (+10) + (−15)
17. (−11) + (+6)
18. (−7) + (+10)
19. (+1) + (−1)
20. (−6) + (+6)
21. 0 + (−5)
22. (−4) + 0

B.

1. (−11) + 10
2. (−11) + (−10)
3. (−4) + 4
4. (+13) + 0
5. 8 + (−7)
6. 8 + (−9)
7. (−7) + (−6)
8. (+7) + (−6)
9. (−7) + (+6)
10. (−13) + (+13)
11. (−13) + 14
12. (−5) + 4
13. (+8) + (−8)
14. (+7) + (+5)
15. (+25) + (+75)
16. (−30) + (−70)
17. (−3) + (+2)
18. (−2) + (−3)
19. 10 + (−10)
20. −8 + 8
21. −50 + 35
22. 75 + (−25)
23. (+32) + 0
24. 0 + (−32)

4.7 SUBTRACTION OF INTEGERS

OBJECTIVE: To find the difference between two integers.

In arithmetic, $7 - 2$ is defined as the number that must be added to 2 to obtain 7.

$$7 - 2 = 5 \quad \text{because} \quad 5 + 2 = 7.$$

In general, $a - b = d$ **if and only if** $d + b = a$.

Addition and subtraction are called inverse operations of each other, since they are related in this way.

Since $7 + (-2) = 5$, it follows that $7 - 2 = 7 + (-2)$. This result holds in general; that is, $a - b = a + (-b)$. Adding $a + (-b)$ to b, $a + (-b) + b = a + (-b + b) = a + 0 = a$. Since this result agrees with the meaning of subtraction in arithmetic, it can be used as the definition of subtraction.

Definition of Subtraction. If x and y are any integers, then

$$x - y = x + (-y).$$

This definition is often stated more informally as follows.

To subtract signed numbers:

Change the sign of the subtrahend and add the resulting numbers according to the rules for the addition of integers.

If the subtrahend is zero, then $n - 0 = n$.

It is helpful to think of the subtraction of a negative number as the removal of a debt. For example, suppose a man owes \$20,000 on a house. By a stroke of good luck, he is able to pay off \$5000 of this debt. In other words, from $-20{,}000$ he removes -5000.

$$-20{,}000 - (-5000) = -20{,}000 + 5000 = -15{,}000$$

As a result, the man now owes only \$15,000 on the house.

In this example, the statement that $-(-5000) = +5000$ has the practical meaning that decreasing the debts by \$5000 is equivalent to increasing the assets by \$5000.

As another example, suppose a certain business has a net profit of \$50,000 after all bills are paid. After an audit, it is discovered that a bill for \$400 has been paid twice. To find the correct net profit, it is necessary to remove one debt of \$400 from the \$50,000. In other words, from 50,000 it is necessary to subtract −400.

$$50{,}000 - (-400) = 50{,}000 + 400 = 50{,}400$$

For Sample Problems 1–8, subtract and check by the inverse operation, addition.

SAMPLE PROBLEMS	SOLUTIONS	CHECK
1. 7 − 2	= 7 + (−2) = 5	5 + 2 = 7
2. 7 − (−2)	= 7 + (+2) = 9	9 + (−2) = 7
3. 2 − 7	= 2 + (−7) = −5	−5 + 7 = 2
4. −2 − 7	= −2 + (−7) = −9	−9 + 7 = −2
5. −2 − (−7)	= −2 + (+7) = 5	5 + (−7) = −2
6. 5 − 0	= 5	5 + 0 = 5
7. −3 − 0	= −3	−3 + 0 = −3
8. 0 − 5	= 0 + (−5) = −5	−5 + 5 = 0

EXERCISES 4.7

(1–20) Subtract.

A.

1. 5 − 3
2. 11 − (−10)
3. 4 − 6
4. 12 − 10
5. 6 − (−4)
6. 5 − 9
7. −4 − 3
8. −2 − 8
9. 3 − (−13)
10. −3 − (−13)
11. −5 − (−7)
12. −8 − (−2)
13. −1 − 1
14. 7 − (−7)
15. −6 − 6
16. 4 − 4
17. −5 − (−5)
18. 0 − 7
19. 0 − (−6)
20. −3 − (−3)

(21–24) Find the difference between the high temperature and the low temperature recorded for each of the cities listed below.

	City	*High*	*Low*
21.	Los Angeles	63	55
22.	Anchorage	-9	-16
23.	Minneapolis	15	-5
24.	Bismarck	15	0

(25–28) Find the difference in longitudes between each given pair of cities. A positive number means degrees east longitude (east of Greenwich, England), and a negative number means degrees west longitude.

25. Manila, +121; and Hong Kong, +114
26. Tokyo, +140; and San Francisco, -122
27. Boston, -71; and Denver -105
28. Paris, +2; and Mexico City - 99

B.

(1–20) Subtract.

1. -4 - 6
2. 7 - 0
3. -5 - (-5)
4. -5 - 5
5. 5 - (-5)
6. 0 - 4
7. -3 - 0
8. 0 - (-2)
9. 9 - 9
10. -9 - 0
11. 5 - 0
12. -4 - (-9)
13. -8 - 8
14. -5 - (-6)
15. 6 - 4
16. 6 - (-4)
17. 6 - 0
18. 0 - 6
19. 4 - (-6)
20. 4 - 6

(21–24) On a certain day, a newspaper listed the closing prices in dollars for the day and the net changes in dollars for the day of different stocks, as shown below. For each stock, subtract the "net" from the "close" to find the closing price for the previous day. Note, $\frac{1}{4}$ means 25¢ and $\frac{1}{8}$ means $12\frac{1}{2}$¢.

	Stock	*Close*	*Net*
21.	Texaco	$29\frac{3}{4}$	$+2\frac{3}{4}$
22.	General Motors	48	+1
23.	Coca Cola	$9\frac{1}{8}$	$-4\frac{5}{8}$
24.	Pan Am	4	$-\frac{1}{4}$

(25–28) Find the difference in elevation between each pair of geographic locations whose elevations are given below. A positive number means an elevation above sea level, and a negative number means an elevation below sea level.

23. Mount Whitney, California, + 14,495 feet
    Death Valley, California, - 282 feet

26. Qattara Depression, Egypt, -436 feet
    Dead Sea (surface level), - 1292 feet

27. Mount Everest, + 29,028 feet
    San Francisco, 0 feet

28. Brawley, California, - 119 feet
    El Centro, California, - 45 feet

4.8 MULTIPLICATION OF INTEGERS

OBJECTIVE: To find the product of two integers.

In arithmetic 3 X 4 means the sum of three fours; that is,

$$3 \times 4 = 4 + 4 + 4.$$

The product of two integers is defined so that this meaning is retained and also so that the closure, commutative, associative, and distributive axioms remain valid.

Thus, $(+3)(+4) = (+4) + (+4) + (+4) = +12$
and $(+3)(-4) = (-4) + (-4) + (-4) = -12$.

Since the commutative axiom for multiplication is to remain valid, $(-3)(+4)$ must be the same as $(+4)(-3)$:

$$(-3)(+4) = (+4)(-3) = (-3) + (-3) + (-3) + (-3) = -12.$$

Before considering the product $(-3)(-4)$, it is necessary to consider products of the form $x \cdot 0$ where x is any integer. Since $3 \cdot 0 = 0 + 0 + 0 = 0$ and, similarly, $n \cdot 0 = 0$ for all natural numbers n, it is agreed that $x \cdot 0 = 0 \cdot x = 0$ for all integers x.

Returning to the product $(-3)(-4)$, let $x = (-3)(-4)$, then x has an additive inverse $-x$, so that $x + (-x) = 0$. For the distributive axiom to be valid,

$$\begin{aligned}(-3)(-4) + (+3)(-4) &= (-3 + 3)(-4)\\ &= 0(-4) = 0.\end{aligned}$$

Therefore, if $x = (-3)(-4)$, then $-x = (+3)(-4) = -12$ and $x = +12$. Thus, $(-3)(-4) = +12$.

These results are generalized in the following equations. Let x and y be any integers, Then,

$$x \cdot 0 = 0 \cdot x = 0$$

$$x(-y) = -xy$$

$$-x(y) = -xy$$

$$(-x)(-y) = +xy = xy$$

Stated less formally, the multiplication of integers can be expressed as follows:

1. The product of any number and 0 is 0.
2. The product of two numbers having different signs is negative (–).
3. The product of two numbers having the same sign is positive (+).

For Sample Problems 1–10, multiply.

SAMPLE PROBLEMS	SOLUTIONS
1. (+5) (+3)	= +15 = 15
2. (−5) (−3)	= +15 = 15
3. (+5) (−3)	= −15
4. (−5) (+3)	= −15
5. (+3) (0)	= 0
6. (0) (+3)	= 0
7. (−3) (0)	= 0
8. (0) (−3)	= 0
9. $(+5)^2$	= (+5) (+5) = +25 = 25
10. $(-5)^2$	= (−5) (−5) = +25 = 25

EXERCISES 4.8

A.

(1–30) Multiply.

1. (+5) (−2)
2. (−5) (−2)
3. (+2) (−5)
4. (+5) (+2)
5. (−3) (−9)
6. (+8) (−5)
7. (+4) (+1)
8. (−4) (+1)
9. (−1) (−1)
10. (−4) (−9)
11. (−7) (+8)
12. (−4) (25)
13. (−3) (−3)
14. (−25) (−4)
15. (+8) (+125)
16. (+125) (−8)
17. (16) (−4)
18. (+3) (−3)
19. (−10) (100)
20. (100) (−100)
21. 5 (−6)
22. 4 (−25)
23. −6 (7)
24. −2 (9)
25. (−9) (−9)
26. (60) (60)
27. $(-10)^2$
28. $(+20)^2$
29. (−12) (0)
30. (0) (−8)

(31–35) In physics, the torque T due to a force F at a directed distance d from a fixed position is given by the formula:

$$T = Fd$$

Torque measures the tendency of a mass to rotate.
If T is positive, the torque is counterclockwise.
If T is negative, the torque is clockwise.
If F is positive, the force is downward.
If F is negative, the force is upward.
If d is positive, the distance is to the right of a fixed position.
If d is negative, the distance is to the left of a fixed position.

Find T for each of the following, and state whether the torque is clockwise or counterclockwise.

31. $F = +50$ pounds, $d = +3$ feet

32. $F = -50$ pounds, $d = +2$ feet

33. $F = +30$ grams, $d = -12$ centimeters

34. $F = -16$ grams, $d = -30$ centimeters

35. $F = -25$ grams, $d = 0$

(36–40) The directed distance d of an airplane from a certain airport is given by the formula: $d = vt$, where v is the velocity and t is the directed time.

If d is positive, the plane is east of the airport.
If d is negative, the plane is west of the airport.
If v is positive, the plane is moving toward the east.
If v is negative, the plane is moving toward the west.
If t is positive, then t gives the number of hours after midnight.
If t is negative, then t gives the number of hours before midnight.

Assuming that each plane is at the airport at midnight, find the position of each plane having the given velocity at the given time. State whether the plane is east or west of the airport.

36. $v = +500$ m.p.h. at 2 A.M. ($t = +2$)

37. $v = +350$ m.p.h. at 9 P.M. ($t = -3$)

38. $v = -400$ m.p.h. at 10 P.M. ($t = -2$)

39. $v = -150$ m.p.h. at 1 A.M. ($t = +1$)

40. $v = -275$ m.p.h. at midnight.

B.

(1–30) Multiply.

1. 6 (0)
2. (−5) (0)
3. −2 (−54)
4. 125 (−8)
5. (−25) (−4)
6. (0) (5)
7. (0) (−6)
8. (−7) (0)
9. (0) (0)
10. (+1) (−15)
11. (−25) (1)
12. (−1) (−10)
13. (−46) (100)
14. (−58) (−1000)
15. $(+6)^2$
16. $(+8)^2$
17. $(-7)^2$
18. $(-9)^2$
19. (−30) (+20)
20. (+50) (−60)
21. 6 (−25)
22. −7 (40)
23. −9 (−800)
24. −16 (−9)
25. $-2\,(-3)^2$
26. $(-2 \cdot 3)^2$
27. $(0)\,(-5)^2$
28. $(3 \cdot 0)\,(-7)$
29. $(-4)\,(-4)^2$
30. $(-6 \cdot 6)\,(-6)$

(31–35) For a certain television set, the change C in the capacitance of a condenser is given by $C = KT$, where K is a constant and T is the change in temperature. Find C for each given set of values.

31. $K = -4$, $T = +20°$
32. $K = 3$, $T = -4°$
33. $K = -5$, $T = -3°$
34. $K = 7$, $T = 12°$
35. $K = -6$, $T = 0°$

(Note: In order to simplify these problems (31–35), the data and units have been modified from those commonly used.)

(36–40) In chemistry, a subscript at the lower right of a symbol for an atom indicates the number of atoms. For example, sulfuric acid is expressed in symbols as H_2SO_4, meaning 2 atoms of hydrogen (H), 1 atom of sulfur (S), and 4 atoms of oxygen (O). The valence of an ion, a positive or negative integer indicating how the ion combines chemically, is the sum of the valences of the atoms composing the ion. For example, the valence of the sulfate ion, SO_4, equals (valence of S) + 4 (valence of O) = (+6) + 4 (−2) = 6 − 8 = −2. Find the valence of the ions in Exercises 36–40, given the valence of the component atoms.

36. Nitrate ion, NO_3; valence of N = +5, valence of O = −2

37. Phosphate ion, PO_4; valence of P = +5, valence of O = −2

38. Bichromate ion, Cr_2O_7; valence of Cr = +5, valence of O = −2

39. Ammonium ion, NH_4; valence of N = −3, valence of H = +1

40. Permanganate ion, MnO_4; valence of Mn = +7, valence of O = −2

4.9 DIVISION OF INTEGERS

OBJECTIVE: To find the quotient of two integers.

In arithmetic, $\frac{12}{3} = 4$, since $3 \cdot 4 = 12$. This idea is retained in defining the quotient of two integers.

Definition of Division. Let x and y be integers. Then $\frac{x}{y} = q$ if and only if $yq = x$ for exactly one integer q.

As examples,

$$\frac{+20}{+4} = +5, \quad \text{since} \quad (+4)(+5) = +20$$

$$\frac{-20}{-4} = +5, \quad \text{since} \quad (-4)(+5) = -20$$

$$\frac{+20}{-4} = -5, \quad \text{since} \quad (-4)(-5) = +20$$

$$\frac{-20}{+4} = -5, \quad \text{since} \quad (+4)(-5) = -20$$

$$\frac{0}{-4} = 0, \quad \text{since} \quad (-4)\ (0) = 0$$

However, $\frac{4}{0}$ is undefined, since there is no integer q for which $0 \cdot q = 4$ because $0 \cdot q = 0$ for all integers q.

$\frac{0}{0}$ is undefined for another reason. Since $0 \cdot q = 0$ for all integers q, one and only one integer cannot be selected as the quotient.

The indicated quotient of two nonzero integers is not always an integer. For example, $\frac{2}{7}$ and $\frac{-3}{6}$ are not integers. This shows that the set of integers is not closed with respect to the division operation.

Division and multiplication are said to be inverse operations of each other, since they are related by the definition of division.

The following informal statements are helpful in determining the sign of an indicated quotient:

The quotient of two numbers having the same sign is positive (+).

The quotient of two numbers having different signs is negative (–).

For Sample Problems 1–7, divide and check by the inverse operation, multiplication.

PROBLEMS	SOLUTIONS	CHECK
1. $\frac{+12}{+3}$	$= +4$	$(+4)(+3) = +12$
2. $\frac{-12}{-3}$	$= +4$	$(+4)(-3) = -12$
3. $\frac{+12}{-3}$	$= -4$	$(-4)(-3) = +12$
4. $\frac{-12}{+3}$	$= -4$	$(-4)(+3) = -12$
5. $\frac{0}{+3}$	$= 0$	$(0)(+3) = 0$
6. $\frac{0}{-3}$	$= 0$	$(0)(-3) = 0$
7. $\frac{+3}{0}$ and $\frac{-3}{0}$ are undefined.		

Divide and check by the inverse operation, multiplication.

EXERCISES 4.9

A.

1. $\frac{+15}{+5}$
2. $\frac{+15}{+3}$
3. $\frac{+15}{-3}$
4. $\frac{-15}{-5}$
5. $\frac{-42}{+21}$
6. $\frac{+30}{-3}$
7. $\frac{-45}{-9}$
8. $\frac{-21}{+7}$
9. $\frac{+8}{-2}$
10. $\frac{-8}{+4}$
11. $\frac{42}{-7}$
12. $\frac{-1000}{10}$
13. $\frac{-100}{-10}$
14. $\frac{-7}{-7}$
15. $\frac{7}{-7}$
16. $\frac{-21}{21}$
17. $\frac{-5}{-5}$
18. $\frac{0}{+3}$
19. $\frac{0}{-5}$
20. $\frac{-200}{8}$

B.

1. $\frac{352}{-11}$
2. $\frac{800}{-10}$
3. $\frac{0}{9}$
4. $\frac{-63}{-9}$
5. $\frac{-10{,}000}{-10}$
6. $\frac{-72}{+8}$
7. $\frac{0}{-4}$
8. $\frac{65}{-5}$
9. $\frac{-68}{17}$
10. $\frac{-84}{-6}$
11. $\frac{100{,}000}{-100}$
12. $\frac{-1{,}000{,}000}{-10{,}000}$
13. $\frac{-25}{25}$
14. $\frac{-6}{-6}$

15. $\frac{19}{-1}$

16. $\frac{-27}{1}$

17. $\frac{-16}{-1}$

18. $\frac{100}{-25}$

19. $\frac{-1000}{-8}$

20. $\frac{-10,000}{16}$

4.10 COMBINED OPERATIONS

OBJECTIVE: To evaluate a numerical expression involving operations on integers.

Properties of the set of integers are summarized below as an aid in doing calculations.

Let x, y, and z be any integers:

Addition Axioms

Closure:	$x + y$ is an integer
Commutative:	$x + y = y + x$
Associative:	$(x + y) + z = x + (y + z)$
Identity:	$x + 0 = 0 + x = x$
Inverse:	$x + (-x) = -x + x = 0$

Multiplication Axioms

Closure:	xy is an integer
Commutative:	$xy = yx$
Associative:	$(xy)\,z = x\,(yz)$
Identity:	$x\,(1) = 1\,(x) = x$

Definition of Subtraction

$$x - y = x + (-y)$$

Definition of Division

$\frac{x}{y} = z$ if and only if $yz = x$ for one and only one integer z.

Distributive Axiom

$x\ (y + z) = xy + xz$

The following convention is restated for convenience.

Convention

(1) The operations within the innermost set of grouping symbols are done first, following the procedure in (2). (Recall that the bar used in division is a grouping symbol.)

(2) Unless grouping symbols indicate otherwise, the operations are done in the following order:

First, squaring and/or cubing, as read from left to right.

Second, multiplication and/or division, as read from left to right.

Third, addition and/or subtraction, as read from left to right.

SAMPLE PROBLEM 1

Evaluate $20 - 3\ (-7 + 4)$.

SOLUTION

$$20 - 3\ (-7 + 4) = 20 - 3\ (-3)$$
$$= 20 - (-9) = 20 + 9 = 29$$

SAMPLE PROBLEM 2

Evaluate $\dfrac{20 - 50}{-7 + 1}$.

SOLUTION

$$\frac{20 - 50}{-7 + 1} = \frac{-30}{-6} = 5$$

SAMPLE PROBLEM 3

Evaluate $(-2)^2 - (-2)^3$.

SOLUTION

$$(-2)^2 - (-2)^3 = 4 - (-8) = 4 + 8 = 12$$

SAMPLE PROBLEM 4

Evaluate $4 - 2(3 - 6)(4 - 9)$.

SOLUTION

$$4 - 2(3 - 6)(4 - 9) = 4 - 2(-3)(-5)$$
$$= 4 - 2(15) = 4 - 30 = -26$$

EXERCISES 4.10

Evaluate.

A.

1. $5 - (-6 + 6)$
2. $2(-6) + (-4)(-3)$
3. $5(4 - 4)$
4. $-6(-8 + 8)$
5. $\frac{7 - 7}{3}$
6. $-2(3 - 2) - (-3)(2 - 3)$
7. $\frac{3 - 4}{4 - 3}$
8. $\frac{6 - 2}{2 - 6}$
9. $(0 - 2)(5 - 3)$
10. $0 - 3(2 - 4)$
11. $(-3)(-3)(-3)$
12. $(-5)(-5)(-5)(-5)$
13. $-5(7 - 11)$
14. $7 - [3 + 3(-3)]$
15. $8 - [2(9) - 4(4)]$
16. $\left(\frac{7 - 3}{3 - 5}\right)\left(\frac{-7 + 7}{4 - 6}\right)$
17. $\left(\frac{-4 - 6}{5 - 3}\right)\left(\frac{11 - 4}{5 + 2}\right)$
18. $3\left(\frac{21 - 12}{-8 + 5}\right)$

19. $-5\left(\frac{5-3}{3-5}\right)$

20. $9-(5-7)$

B.

1. $12-2(-6-4)$
2. $(-6+4)^2$
3. $(-6+4)^3$
4. $\frac{3(6-10)}{-6}$
5. $\frac{-2(5-8)}{0-3}$
6. $(9-8)(8-7)(7-6)$
7. $15-6(5-2)$
8. $-8-2(-4+7)$
9. $\frac{9-16}{3-4}$
10. $\frac{25-4}{5-8}$
11. $(-2)(-3)(-4)$
12. $(-3)(-2)+(3)(-2)-(-3)(2)$
13. $2(-10)^2$
14. $-2(10)^2$
15. $(-4)^3$
16. $-(-4)^3$
17. $(-5)^2-(-3)^2$
18. $(-10)^3+(-5)^3$
19. $(-1)^2-4(-2)(-1)$
20. $(5-9)^2-(3-5)^2$

4.11 EVALUATION

OBJECTIVE: To evaluate a formula by substituting integral values.

It is sometimes necessary to substitute one or more negative values in an expression or a formula. The following examples illustrate such evaluations.

SAMPLE PROBLEM 1

Evaluate $b^2 - 4ac$ for $b = -2, a = 3, c = -5$.

SOLUTION

$$b^2 - 4ac = (-2)^2 - 4(3)(-5)$$
$$= 4 - 4(-15)$$
$$= 4 - (-60) = 4 + 60 = 64$$

SAMPLE PROBLEM 2

Evaluate $C = \frac{5(F - 32)}{9}$ where C is degrees centigrade and $F = -22$ degrees Fahrenheit.

SOLUTION

$$C = \frac{5(F - 32)}{9}$$
$$= \frac{5(-22 - 32)}{9} = \frac{5(-54)}{9} = 5(-6) = -30° \text{ centigrade}$$

SAMPLE PROBLEM 3

Evaluate $P = \frac{100(p + q)}{pq}$ for $p = 10, q = -20$.

SOLUTION

$$P = \frac{100(p + q)}{pq}$$
$$= \frac{100(10 + (-20))}{(10)(-20)} = \frac{100(-10)}{-200} = \frac{-1000}{-200} = 5$$

The formula in Sample Problem 3 is used for corrective lenses where:

P = power in diopters (positive for farsighted persons and negative for nearsighted persons).

p = object distance from the lens in centimeters.

q = image distance from the lens in centimeters (positive when on the opposite side of the lens from the object and negative when on the same side).

EXERCISES 4.11

Evaluate.

A.

1. (Average of three numbers)
$\dfrac{x+y+z}{3}$ for $x = 40$, $y = -50$, $z = -20$

2. (Part of quadratic formula)
$b^2 - 4ac$ for $a = 5$, $b = 4$, $c = -2$
$(4)^2 - 4(5)(-2)$
$16 - 4(5)(-2)$
$16 - (20)(-2)$
$16 - (-40)$
$16 + (+40) = 56$

3. (Analytic geometry)
$\dfrac{m-n}{1+mn}$ for $m = -2$, $n = 1$

4. (Conversion from degrees Fahrenheit to degrees centigrade)
$C = \dfrac{5\,(F-32)}{9}$ for $F = -4°$

5. (Corrective lenses – see Sample Problem 3)
$P = \dfrac{100\,(p+q)}{pq}$
Find P for $p = 25$, $q = -4$. Is the person farsighted or nearsighted?

6. (Center of gravity)
$x = \dfrac{md + MD}{m + M}$
Find x for $m = 15$, $d = 4$, $M = 5$, $D = -20$.

7. (Flow of liquids)
$P = \dfrac{d\,(V^2 - v^2)}{2}$
Find P for $d = 50$, $V = 2$, and $v = 4$.

8. (Mirrors and photography)
$f = \dfrac{pq}{p+q}$
Find f for $p = 45$ cm. and $q = -180$ cm.
$f = \dfrac{(45)(-180)}{(45)+(-180)}$

B.

1. (Average of four numbers)
$\frac{x + y + z + w}{4}$ for $x = 9, y = -14, z = -18, w = 15$

2. (Trigonometry)
$a^2 + b^2 - ab$ for $a = 6, b = -3$

3. (Analytic geometry)
$\frac{3x + 4y - 24}{5}$ for $x = -6, y = 3$

4. (Conversion from degrees centigrade to degrees Fahrenheit)
$F = \frac{9C}{5} + 32$
Find F for $C = -5°$.

5. (Corrective lenses — see Sample Problem 3)
$P = \frac{100(p + q)}{pq}$
Find P for $p = 100$ and $q = -25$. Is the person farsighted or near-sighted?

6. (Acoustics)
$F = \frac{fV}{v + V}$
Find F for $f = 250$, $v = -1320$, and $V = 1100$.

7. (Radio and television)
$r = \frac{R(E - G)}{E}$
Find r for $R = 2$, $E = -12$, and $G = -9$.

8. (Mirrors and photography)
$f = \frac{pq}{p + q}$
Find f for $p = 6$ and $q = -3$.

UNIT 4 POSTTEST

1. If +45 means a latitude of 45° north of the equator, state the meaning of −45 and of 0. (4.1)

   −45 means 45° South

   0 means the equator

2. State the coordinates of points *A*, *B*, and *C* on the number line which follows. (4.2)

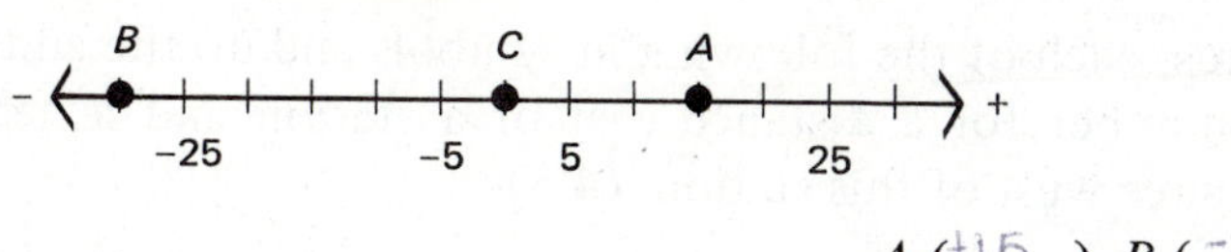

   A (+15), *B* (−30), *C* (0)

3. State the coordinates of points *P*, *Q*, and *R* on the number line at right. (4.2)

   P (+80), *Q* (0), *R* (−40)

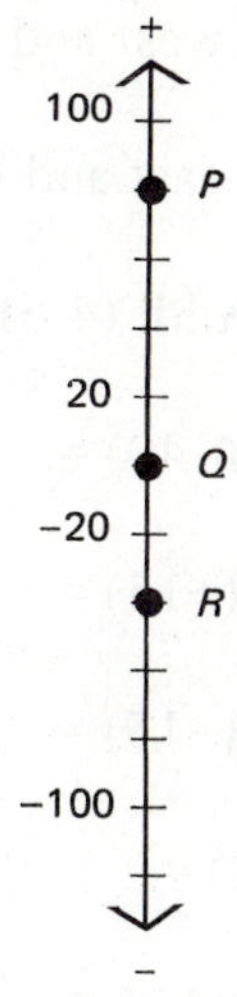

(4–7) Rewrite each of the following as a single integer. (4.4)

4. |−7| = 7

5. |−8| + |−6| = 14

6. |−5| · |−3| = 15

7. |−20| − |+9| = 11

(8–9) Translate into algebraic symbols. (4.3)

8. Six less than *y* is greater than 5. 

9. Nine more than *y* is negative.

(10–13) Insert the symbol < or the symbol > between each given pair of integers, so that the resulting statement is true. State the algebraic meaning of the resulting statement. (4.4)

10. +4 ______ +8, ______________________________

11. +4 ______ -8, ______________________________

12. -4 ______ -8, ______________________________

13. -8 ______ -4, ______________________________

(14–17) Express each of the following in symbols and do the addition, using a positive number for a distance east of a station and a negative number for a distance west of this station. (4.5)

14. 80 miles east and then 50 miles west. ______________

15. 70 miles west and then 25 miles west. ______________

16. 95 miles west and then 110 miles east. ______________

17. 46 miles east and then 46 miles west. ______________

(18–23) Add. (4.6)

18. (-17) + (-40) = ______________

19. (-28) + (+15) = ______________

20. (+28) + (-15) = ______________

21. 0 + (-12) = ______________

22. (-150) + 125 = ______________

23. (-78) + (+78) = ______________

(24–29) Subtract. (4.7)

24. (+80) - (+100) = ______________

25. (-17) - (+8) = ______________

26. (-18) - (-6) = ______________

27. (-17) - (-17) = ______________

28. 40 - 85 = ______________

29. 56 - 56 = ______________

(30–35) Multiply. (4.8)

30. $(-8)(+125) =$ ⁻1000

31. $(-25)(-6) =$ + 150

32. $(+23)(-27) =$ − 621

33. $(-35)(-1) =$ ⁺35

34. $(0)(-34) =$ 0

35. $(+1)(-17) =$ ⁻17

(36–39) Divide. (4.9)

36. $\frac{-72}{-12} =$ 6

37. $\frac{-36}{+4} =$ -9

38. $\frac{+39}{-3} =$ -13

39. $\frac{0}{-39} =$ 0

(40–43) Evaluate. (4.10)

40. $(-4)(-38)(-25) =$

41. $\frac{5^2 - (-3)^2}{2 - 6} =$ $\frac{25-9}{-4}$ $\frac{16}{-4}$ -4

42. $(-6)^2 - 4(2)(-1) =$ 36-4 2(-1) 36 + 8 - 44

43. $(10 - 2)(10 - 16) =$ (8)(-6) = -48

(44–45) Evaluate. (4.11)

44. $(x - y)^3 - (y - x)^3$ for $x = 7, y = -3$

([7]-[-3])³ - ([-3]-[7])³

(10)³ (-10)³ 1000 + ⁺1000 = 2000

45. $\frac{ab}{a + b}$ for $a = 6, b = -10$

UNIT 5

LINEAR EQUATIONS IN ONE VARIABLE

Unit 5
PRETEST

(1–4) Simplify. (5.1)

1. $4x + 5x =$ __________ 2. $8y - 2y =$ __________

3. $6y + y =$ __________ 4. $7x - 6x =$ __________

(5–8) Collect like terms. (5.2)

5. $7x - 14 - 3x + 9 =$ __________
6. $4x - 7y - 7x =$ __________
7. $x^2 - 5x + 3x - 15 =$ __________
8. $2x^2 + yx - 6xy - 3y^2 =$ __________

(9–14) Remove parentheses. (5.3)

9. $4(x + 6) =$ __________
10. $-2(x + 7) =$ __________
11. $-6(y - 2) =$ __________
12. $-(y - 8) =$ __________
13. $-(5x + 7) =$ __________
14. $-(x + 2y - 4) =$ __________

(15–18) Remove parentheses and collect like terms. (5.4)

15. $8x - 6(x + 3) =$ __________
16. $14 - (x - 2) =$ __________
17. $4(3x - 5) - 6(2x + 3) =$ __________
18. $2x(x + 4y) - 5y(x + 4y) =$ __________

(19–22) Solve and check. (5.5)

19. $x + 5 = 12$

20. $3x = 21$

21. $x - 7 = 9$

22. $\frac{x}{4} = 8$

(23–26) Solve and check. (5.6)

23. $4x - 9 = 19$

24. $\frac{x + 20}{2} = 6$

25. $\frac{3(x - 4)}{2} = 9$

26. $8 = 5(x - 6) + 78$

(27–30) Solve and check. (5.7)

27. $5(x - 2) = 3(x + 4)$

28. $4x - (x + 10) = 4(x - 2)$

29. $8 - (x - 6) = 5 - x$

30. $3(5x - 4) - 5(3x - 4) = 8$

(31–32) Solve for the indicated letter. (5.8)

31. $2x + 5y = 20$ for y

32. $A = \frac{r(r + h)}{2}$ for h

$a(b+c) = ab + ac$
$(b+c)a = ba + ca$

5.1 DISTRIBUTIVE PROPERTIES

OBJECTIVE: To write the indicated sum of two or three like terms as a single term.

Equations such as $x = 4$, $x + 3 = 10$, $3x - 7 = 8$, $2x + 6 = 0$, $3x - 2(x - 3) = 0$, and $2(x - 4) = 9 - (x + 2)$ are examples of a linear equation in the variable x.

A **solution of a linear equation in one variable** is a numerical value that makes the equation true when the variable is replaced by this value.

For example, if $x = 7$, then $x + 3 = 10$ becomes $7 + 3 = 10$, a true statement, and 7 is a solution of $x + 3 = 10$.

The distributive axiom plays an important role in the solution of many linear equations. This axiom states that multiplication can be distributed over an addition. In symbols,

$$a(b + c) = ab + ac.$$

This property also holds for a sum of more than two terms; that is,

$$a(b + c + d) = ab + ac + ad.$$

Multiplication can also be distributed over a subtraction. In symbols,

$$a(b - c) = ab - ac.$$

The distributive properties which follow can be derived from the distributive axiom. These properties are useful for rewriting special sums or differences as single terms.

Distributive Properties. For all integers a, b, c, and d:

$$a(b + c) = ab + ac \quad \text{and} \quad (b + c)a = ba + ca$$

$$a(b - c) = ab - ac \quad \text{and} \quad (b - c)a = ba - ca$$

$$ad + bd + cd = (a + b + c)d$$

For Sample Problems 1–4, rewrite each as a single term.

SAMPLE PROBLEM 1

$3x + 4x$

SOLUTION

$(3 + 4)x = 7x$

SAMPLE PROBLEM 2

$5x - x$

SOLUTION

$5x - 1x = (5 - 1)\,x = 4x$

SAMPLE PROBLEM 3

$7x + 5x + x$

SOLUTION

$(7 + 5 + 1)\,x = 13x$

SAMPLE PROBLEM 4

$9x - 10x + 2x$

SOLUTION

$(9 - 10 + 2)\,x = (1)\,x = x$

EXERCISES 5.1

Rewrite each of the following as a single term.

A.

1. $5x + 2x$	1. ________
2. $6y - 3y$	2. ________
3. $-3x + 5x$	3. ________
4. $6y - 5y$	4. ________
5. $2x - 2x$	5. ________
6. $-7z + z$	6. ________
7. $5r - 5s$	7. ________
8. $3x + 2x + 5x$	8. ________
9. $7y - 4y - 2y$	9. ________
10. $7z - 5z + 6z$	10. ________
11. $2n + 3n - 6n$	11. ________
12. $6x + 2x - 6x$	12. ________

B.

1. $5x + x$ 1. ________
2. $7y - y$ 2. ________
3. $4y - 7y$ 3. ________
4. $-4x + 3x$ 4. ________
5. $-5x + 5x$ 5. ________
6. $4n + 3n$ 6. ________
7. $2x + 2y$ 7. ________
8. $15x - 2x - 12x$ 8. ________
9. $2t - 4t + 2t$ 9. ________
10. $12k - 5k + 3k$ 10. ________
11. $-4d - 5d - d$ 11. ________
12. $6x - 3x - 10x$ 12. ________

5.2 COLLECTING LIKE TERMS

OBJECTIVE: To simplify an expression by collecting the like terms.

Like terms are terms whose literal factors are identical, and like terms can be collected by using the distributive axiom.

$3x$ and $-7x$ are like terms and $3x + (-7x) = -4x$.

$3x^2$ and $5x^2$ are like terms and $3x^2 + 5x^2 = 8x^2$.

$2xy$ and $7xy$ are like terms and $2xy - 7xy = -5xy$.

$2xy$ and $7yx$ are like terms and $2xy + 7yx = 2xy + 7xy = 9xy$.

The terms $3x^2$ and $5x$ are unlike terms, and $3x^2 + 5x$ cannot be combined by adding 3 and 5. Similarly, $3x$ and 5 are unlike terms, and the sum $3x + 5$ must be left as a sum of two terms. As another example, $5x - 6y$ must be left as a difference of two unlike terms.

SAMPLE PROBLEM 1

Simplify $5x + 2 - 3x - 5$.

SOLUTION

$(5x - 3x) + (2 - 5) = 2x - 3$

SAMPLE PROBLEM 2

Simplify $2x^2 - 15x + 6x - 15$.

SOLUTION

$$2x^2 + (-15x + 6x) - 15 = 2x^2 + (-9x) - 15$$
$$= 2x^2 - 9x - 15$$

SAMPLE PROBLEM 3

Simplify $x^2 + 3xy - 2yx + y^2$.

SOLUTION

$$x^2 + (3xy - 2yx) + y^2 = x^2 + (3xy - 2xy) + y^2$$
$$= x^2 + (3 - 2)xy + y^2$$
$$= x^2 + 1xy + y^2$$
$$= x^2 + xy + y^2$$

EXERCISES 5.2

Simplify each of the following by collecting like terms.

A.

1. $4x + 6 - 2x - 1$ 1. ________
2. $6x - 4 - 5x + 2$ 2. ________
3. $4y - 3y - 6$ 3. ________
4. $7x + 5 - 8x$ 4. ________
5. $10 - x + 5x$ 5. ________
6. $x^2 - 3x + 2x - 6$ 6. ________
7. $3y^2 - y - 6y + 2$ 7. ________
8. $2x + 3y + 3x$ 8. ________
9. $2s + 2t - s - t$ 9. ________
10. $3x - y + 2y - 3x$ 10. ________

11. $5xy + 4yx$ 11. ______

12. $-yx - 4xy$ 12. ______

13. $x^2 + 4xy - 5yx + y^2$ 13. ______

14. $3xy - 4xz + 5zx + 6yx$ 14. ______

15. $8a^2b + 7a^2b^2 - 3ba^2$ 15. $5a^2b + 7a^2b^2$

B.

1. $5x + 7 + 3x - 9$ 1. ______

2. $3y - y + 5$ 2. ______

3. $6x - 2 - 6x$ 3. ______

4. $10 + 2x - 12$ 4. ______

5. $6 - 2t - 3t$ 5. ______

6. $y^2 + 3y - 2y - 6$ 6. ______

7. $5k^2 + 5k - 8k - 8$ 7. ______

8. $3x + 5y - 5x$ 8. ______

9. $5s + 3t - 3s - 3t$ 9. ______

10. $6n + 3m + 3n$ 10. ______

11. $7xy - 6yx$ 11. ______

12. $-yx - 4xy$ 12. ______

13. $r^2 - 7rs - sr - s^2$ 13. ______

14. $2ab + ac - 2ca - ba$ 14. ______

15. $6xyz - 4yzx - 5zxy$ 15. ______

5.3 REMOVING PARENTHESES

OBJECTIVE: To remove parentheses from a given expression by using the distributive properties.

The distributive properties $a\,(b + c) = ab + ac$ and $a\,(b - c) = ab - ac$ state that the products $a\,(b + c)$ and $a\,(b - c)$ can be rewritten as expressions that do not contain parentheses. This process, called removing parentheses, is useful in solving certain linear equations.

Parentheses can also be removed from expressions such as $-(b + c)$ and $-(b - c)$ by using the fact that $(-1)\,x = -x$ for any integer x. For example, $-5 = (-1)\,(5)$ and $-8 = (-1)\,(8)$. Since $b + c$ is an integer, by the closure axiom for addition,

$$-(b + c) = (-1)\,(b + c) = (-1)\,b + (-1)\,c = -b - c.$$

Similarly, since $b - c$ is an integer, and $b - c = b + (-c)$,

$$-(b - c) = (-1)\,(b + (-c)) = (-1)\,b + (-1)\,(-c) = -b + c.$$

In summary, the following distributive properties are valid:

$$a\,(b + c) = ab + ac$$

$$a\,(b - c) = ab - ac$$

$$-(b + c) = -b - c$$

$$-(b - c) = -b + c$$

$$a\,(b + c + d) = ab + ac + ad$$

In Sample Problems 1–7, remove parentheses.

PROBLEMS	SOLUTIONS
1. $5\,(x + 6)$	$= 5x + 5(6) = 5x + 30$
2. $3\,(x - 2)$	$= 3x - 3(2) = 3x - 6$
3. $-2\,(x + 4)$	$= -2x + (-2)\,(4) = -2x - 8$
4. $-4\,(2y - 7)$	$= -4(2y) + (-4)\,(-7) = -8y + 28$
5. $-(x + 5)$	$= (-1)\,(x + 5) = -x - 5$
6. $-(2y - 3)$	$= (-1)\,(2y - 3) = -2y + 3$
7. $2\,(x + y - 5)$	$= 2x + 2y - 10$

EXERCISES 5.3

Remove parentheses.

A.

1. $2(x+4)$	1. ________
2. $5(x-3)$	2. ________
3. $-3(x+6)$	3. ________
4. $-4(x-1)$	4. ________
5. $6(2x+3)$	5. ________
6. $7(3x+4)$	6. ________
7. $4(3y-5)$	7. ________
8. $y(y-4)$	8. ________
9. $-(6x+7)$	9. ________
10. $-9(2r+1)$	10. ________
11. $-8(t-2)$	11. ________
12. $-(4y-5)$	12. ________
13. $-(6y-9)$	13. ________
14. $-(5t+2)$	14. ________
15. $5(x+y-3)$	15. ________
16. $4(2x-3y+2)$	16. ________
17. $3(x+5y+7)$	17. ________
18. $10(5x-6y-8)$	18. ________
19. $-3(x+4y-1)$	19. ________
20. $-7(2x-5y+6)$	20. ________
21. $-(x+y+1)$	21. ________
22. $-(x-y+1)$	22. ________
23. $-(2x-y-2)$	23. ________
24. $-(4x+9y-8)$	24. ________
25. $2x(3x^2-2x-1)$	25. ________

B.

1. $3(y+6)$	1. ________
2. $-4(x+7)$	2. ________
3. $7(x-9)$	3. ________
4. $-6(x-8)$	4. ________
5. $2x(x-5)$	5. ________

6. $-3y(4 - y)$	6.	________
7. $-(5x + 8)$	7.	________
8. $-(6y - 9)$	8.	________
9. $5(2x - 3y + 6)$	9.	________
10. $-8(4x + 5y - 2)$	10.	________
11. $-(r - s - t)$	11.	________
12. $-(r - 2s + 1)$	12.	________
13. $2x(x^2 - x - 1)$	13.	________
14. $-4y(2 + y - y^2)$	14.	________
15. $-a(a + b + c)$	15.	$-a^2 - ab - ac$
16. $ab(a^2 - ab - b^2)$	16.	$a^3b - a^2b^2 - ab^3$
17. $xy(x^2 - 5xy - 6y^2)$	17.	$x^3y - 5x^2y^2 - 6xy^3$
18. $-2xy(x + 7y - 1)$	18.	________

5.4 REMOVING PARENTHESES AND COLLECTING LIKE TERMS

OBJECTIVE: To simplify an expression by removing parentheses and collecting like terms.

In solving some linear equations, it is necessary both to remove parentheses and to collect like terms. This can be done by using one or more of the distributive properties which are listed in the summary found in Section 5.3.

In Sample Problems 1–3, remove parentheses and collect like terms.

SAMPLE PROBLEM 1

$4x + 2(x - 5)$

SOLUTION

$4x + 2x - 10 = 6x - 10$

SAMPLE PROBLEM 2

$3(x-4)-2(x-7)$

SOLUTION

$$3x-12-2x+14=(3x-2x)+(-12+14)$$
$$=x+2$$

SAMPLE PROBLEM 3

$8-(y+6-2x)$

SOLUTION

$$8-y-6+2x=2x-y+2$$

EXERCISES 5.4

Remove parentheses and collect like terms.

A.

1. $10+3(x-2)$		1. $3x+4$	$10+3x-6$
2. $3x-(5x-4)$		2. $-2x+4$	$3x-5x+4$
3. $7+3(y-4)$		3. ________	
4. $2-(6t-5)$		4. $-6t+7$	$2-6t+5$
5. $8x-9(x-4)$		5. ________	
6. $3(2y-4)-4y$		6. ________	
7. $6z-(9z-5)$		7. ________	
8. $5-(7-8n)$		8. ________	
9. $9+(x-4)$		9. ________	
10. $2(3-x)-4(x-3)$		10. $-6x+18$	$6-2x-4x+12$
11. $3y-2(x-2y)$	$3y-2x+4y$	11. $-2x+7y$	
12. $2y^2-3(y-2)-y^2$	$2y^2-3y+6-y^2$	12. y^2-3y+6	

13. $5(a-b)+2(2a+3b)$ 13. ________

14. $4(3x-2y)-3(4x-3y)$ 14. ________

15. $7(r+s)-(r-s)$ 15. ________

16. $-(y-x)$ 16. ________

17. $x^2+x-(x^2-x)$ 17. ________

18. $-3(r+s-2t)$ 18. ________

19. $5x-(7y+5x-4)$ 19. ________

20. $\dfrac{5-4x-5}{-4}$ 20. ________

B.

1. $6x-2(x+5)$ 1. ________

2. $12+4(y-5)$ 2. ________

3. $8-5(x+2)$ 3. ________

4. $9t-7(t+1)$ 4. ________

5. $7x-6(x-1)$ 5. ________

6. $5y-(3y+4)$ 6. ________

7. $6-(4n+9)$ 7. ________

8. $4x-(3+3x)$ 8. ________

9. $4(x-3)-3(x-2)$ 9. ________

10. $2x+3(x+2y)$ 10. ________

11. $x^2-2(x+3)+5x$ 11. ________

12. $3(x+y)-2(x-y)$ 12. ________

13. $8(r-2s)-3(2r+s)$ 13. ________

14. $5(x+2y)-2(2x+5y)$ 14. ________

15. $c+d-(d-c)$ 15. ________

16. $-(9-x)$ 16. ______________

17. $-2(x-y+z)$ 17. ______________

18. $7-(y-4x-5)$ 18. ______________

19. $-(2x+3-3x)$ 19. ______________

20. $-(9-8x+7y)$ 20. ______________

5.5 SOLUTIONS, EQUIVALENCE THEOREMS, EQUATIONS

OBJECTIVE: To solve a linear equation using one equivalence theorem, to state which theorem was used, and to check the solution.

A **solution** of an equation in one variable is a constant such that the equation becomes a true statement when the variable is replaced by the constant.

A solution of an equation is also called a **root** of the equation. An equation may have no roots, exactly one root, or more than one root.

The **solution set** of an equation in one variable is the set of all solutions of the equation.

Equivalent equations are equations that have the same solution set.

The solution set of an equation having the form $x = a$ is $\{a\}$.

For each pair of equations in Sample Problems 1–2, determine whether the equations are equivalent or not. The set stated with each problem contains all the solutions to both equations.

EXAMPLE 1

$2x - 6 = 2$ and $x = 4$; $\{2, 4, 6, 8, 10\}$

SOLUTION

The only solution of $x = 4$ is 4.

For $x = 4$ and $2x - 6 = 2$,

$$2(4) - 6 = 2$$
$$8 - 6 = 2$$
$$2 = 2, \text{ true; 4 is a solution.}$$

For $x = 2, 6, 8,$ or 10, it may be seen that $2x - 6 = 2$ is false.

Thus, the equations are equivalent and $\{4\}$ is the solution set of each.

EXAMPLE 2

$x^2 = 9$ and $x = 3$; $\{3, -3\}$

SOLUTION

The solution set of $x = 3$ is $\{3\}$.

For $x = 3$ and $x^2 = 9$, $3^2 = 9$, $9 = 9$, true; 3 is a solution.

For $x = -3$ and $x^2 = 9$, $(-3)^2 = 9$, $9 = 9$, true; -3 is a solution.

The solution set of $x^2 = 9$ is $\{3, -3\}$.

Since the set $\{3\}$ is not the same as $\{3, -3\}$, the equations are not equivalent.

The **basic idea** in solving a linear equation is to change the given equation into an equation having the form $x = a$.

Method. Use an equivalence theorem that involves the operation which is inverse to the one stated in the problem.

Inverse Operations
(1) Addition and subtraction are inverse operations of each other.
(2) Multiplication and division are inverse operations of each other.

Equivalence Theorems. (These theorems state which operations may be performed on an equation without changing its solution set.) Let $A = B$ be an equation in one (or more) variable(s). Let C be any real number.

(1) **Addition Theorem.** $A + C = B + C$ is equivalent to $A = B$. (The same number may be added to each side.)

(2) **Subtraction Theorem.** $A - C = B - C$ is equivalent to $A = B$. (The same number may be subtracted from each side.)

(3) **Multiplication Theorem.** $AC = BC$ where $C \neq 0$ is equivalent to $A = B$. (Each side may be multiplied by a number different from zero.)

(4) **Division Theorem.** $\frac{A}{C} = \frac{B}{C}$ where $C \neq 0$ is equivalent to $A = B$. (Each side may be divided by a number different from zero.)

In Sample Problems 1–4, solve each equation, state which equivalence theorem was used, and check the answer by substituting into the original equation.

SAMPLE PROBLEMS	SOLUTIONS	CHECK
1. $x - 4 = 5$	$(x - 4) + 4 = 5 + 4$ $x = 9$	$9 - 4 = 5$ $5 = 5$, true

Addition theorem

Note that 4 must be *added* to $x - 4$ to *undo* the operation of *subtracting* 4 from x. Adding 4 to each side produces the equivalent equation $x = 9$ where 9 can be recognized as the solution.

2. $x + 3 = 8$	$(x + 3) - 3 = 8 - 3$ $x = 5$	$5 + 3 = 8$ $8 = 8$, true

Subtraction theorem

Note that 3 must be *subtracted* from $x + 3$ to *undo* the operation of *adding* 3 to x. Subtracting 3 from each side produces the equivalent equation $x = 5$ where 5 can be recognized as the solution.

3. $\frac{x}{4} = 3$	$\frac{x}{4}(4) = 3\,(4)$ $x = 12$	$\frac{12}{4} = 3$ $3 = 3$, true

Multiplication theorem

Note that $\frac{x}{4}$ must be *multiplied* by 4 to *undo* the operation of *dividing* x by 4. Multiplying each side by 4 produces the equivalent equation $x = 12$ where 12 can be recognized as the solution.

4. $3x = 21$	$\frac{3x}{3} = \frac{21}{3}$ $x = 7$	$3\,(7) = 21$ $21 = 21$, true

Division theorem

Note that $3x$ must be *divided* by 3 to *undo* the operation of *multiplying* x by 3. Dividing each side by 3 produces the equivalent equation $x = 7$ where 7 can be recognized as the solution.

EXERCISES 5.5

Solve each equation, state which equivalence theorem was used, and check the answer by substituting into the original equation.

A.

1. $x - 3 = 9$ 1. $x = 12$

$x - 3 + 3 = 9 + 3$

2. $x + 4 = 8$ — 2. ______

3. $8x = 24$ — 3. ______

4. $\frac{x}{3} = 12$ — 4. ______

5. $x + 7 = 2$ — 5. ______

6. $x - 1 = -3$ — 6. ______

7. $-2x = 16$ — 7. ______

8. $\frac{x}{5} = 10$ — 8. ______

9. $x + 1 = 1$ — 9. ______

10. $-x = -4$ — 10. ______

$-x = 5$
$\frac{-x}{-1} = \frac{5}{-1} = -5$

terms add together
Reciprocal multiply both side by res

11. $-3x = 18$ — 11. ____________

12. $-\frac{x}{2} = 6$ — 12. ____________

13. $y + 5 = -8$ — 13. ____________

14. $y - 6 = -8$ — 14. ____________

$y - 6 + 6 =$

15. $y - 3 = -3$ — 15. $y = 0$

$y - 3 + 3 = -3 + 3$

16. $-5t = -20$ — 16. ____________

17. $\frac{-t}{4} = -8$ — 17. ____________

18. $5u = 0$ — 18. ____________

19. $-x = 50$ — 19. $x = -50$

$\frac{-x}{-1} = \frac{50}{-1}$

20. $-10x = 4000$

20. ____________

B.

1. $x + 5 = 12$

2. $y - 7 = 8$

3. $\frac{x}{5} = 15$

4. $5x = 30$

5. $x - 4 = -9$

6. $-6y = 42$

7. $t + 8 = 4$

8. $\frac{x}{-4} = 20$

9. $\frac{u}{-2} = -12$

10. $x + 1 = 0$

11. $y + 6 = 6$

12. $-3x = -24$

13. $-5t = 0$

14. $x - 10 = 20$

15. $-10x = 20$

16. $\frac{u}{-10} = 20$

17. $-10v = -20$

18. $r + 5 = -5$

19. $s - 5 = -5$

20. $-x = 6$

5.6 MULTIPLE STEP EQUATIONS

OBJECTIVE: To solve a linear equation requiring the use of more than one equivalence theorem and to check the solution.

Basic Solution Method

(1) Analyze the equation to determine which operations were performed on the number named by the letter and in what order these operations were performed.

(2) Apply the inverse operations in the reverse order in which they were performed.

(3) Check the result by substituting into the original equation the value which has been found and by doing the operations in the exact order in which they are indicated.

SAMPLE PROBLEM 1

Solve and check $3x - 7 = 11$.

ANALYSIS

First, x was multiplied by 3.

Second, 7 was subtracted from the product.

SOLUTION PLAN

We want to undo the operations that were done on x. Our goal is to obtain an equivalent equation having the form $x = a$ where a is the solution.

First, we add 7 to each side to undo the operation of subtracting 7.

Second, we divide each side by 3 to undo the operation of multiplying x by 3.

SOLUTION

$3x - 7 = 11$	writing the original equation
$(3x - 7) + 7 = 11 + 7$	adding 7 to each side
$3x = 18$	simplifying
$\frac{3x}{3} = \frac{18}{3}$	dividing each side by 3
$x = 6$	simplifying

CHECK

$3x - 7 = 11$	writing the original equation
$3(6) - 7 = 11$	replacing x by the value found
$18 - 7 = 11$	doing the operations in specified order
$11 = 11$	true, the result checks

The solution is 6.

SAMPLE PROBLEM 2

Solve and check $2\left(\frac{x}{5} + 4\right) = 12$.

ANALYSIS

First, x was divided by 5.

Second, 4 was added to the quotient.

Third, the sum was multiplied by 2.

SOLUTION PLAN

Our goal is to get an equivalent equation having the form $x = a$.

First, divide each side by 2. This undoes the last operation, a multiplication by 2.

Second, subtract 4 from each side. This undoes the next to the last operation, an addition of 4.

Finally, multiply each side by 5. This undoes the first operation that was done on x, a division by 5.

SOLUTION

$2\left(\frac{x}{5} + 4\right) = 12$	writing the original equation
$\frac{2\left(\frac{x}{5} + 4\right)}{2} = \frac{12}{2}$	dividing each side by 2
$\frac{x}{5} + 4 = 6$	simplifying
$\left(\frac{x}{5} + 4\right) - 4 = 6 - 4$	subtracting 4 from each side
$\frac{x}{5} = 2$	simplifying
$5\left(\frac{x}{5}\right) = 5\ (2)$	multiplying each side by 5
$x = 10$	simplifying

CHECK

$2\left(\frac{x}{5} + 4\right) = 12$	writing the original equation
$2\left(\frac{10}{5} + 4\right) = 12$	replacing x by 10
$2\ (2 + 4) = 12$	doing the operations as specified
$2\ (6) = 12$	simplifying
$12 = 12$	true, result checks

The solution is 10.

SAMPLE PROBLEM 3

Solve and check $18 = 5\,(x + 3) + 23$.

ANALYSIS

By the symmetric axiom, the sides may be exchanged and the equation may be written: $5\,(x + 3) + 23 = 18$.

First, 3 was added to x.

Second, the sum was multiplied by 5.

Third, 23 was added to the product.

SOLUTION PLAN

First, subtract 23 from each side.

Second, divide each side by 5.

Third, subtract 3 from each side.

SOLUTION

$5\,(x + 3) + 23 = 18$	exchanging sides
$5\,(x + 3) + 23 - 23 = 18 - 23$	subtracting 23 from each side
$5\,(x + 3) = -5$	simplifying each side
$\dfrac{5\,(x + 3)}{5} = \dfrac{-5}{5}$	dividing each side by 5
$x + 3 = -1$	simplifying each side
$x + 3 - 3 = -1 - 3$	subtracting 3 from each side
$x = -4$	simplifying each side

CHECK

$18 = 5\,(x + 3) + 23$
$18 = 5\,(-4 + 3) + 23$
$18 = 5\,(-1) + 23$
$18 = -5 + 23$
$18 = 18$ true; the solution is -4.

EXERCISES 5.6

A.

(1–5) Write the analysis and solution plan for each of the following, then solve and check.

1. $3x - 5 = 7$

Analysis

Solution Plan

Solution

Check:

2. $\frac{3x + 7}{2} = 8$

Analysis

Solution Plan

Solution

Check:

3. $6(x - 5) = 42$

Analysis

Solution Plan

Solution

Check:

4. $4(x+3)=32$

Analysis

Solution Plan

Solution

Check:

5. $2(6-x)=8$

Analysis

Solution Plan

Solution

Check:

A.

(6–15) Solve and check. (Do the analysis and solution plan mentally or on scratch paper.)

6. $5x-2=13$

6.

7. $\frac{x}{5}-6=3$

7.

8. $2(x - 7) = 6$ 8.

9. $4\left(\frac{x}{3} - 1\right) = -12$ 9.

10. $6 + 2(x - 2) = 10$ 10.

11. $6 - 4(x + 3) = 2$ 11.

12. $10 = \left(\frac{x}{2} - 7\right) + 13$ 12.

13. $12 + 2(3x - 4) = 16$ 13.

14. $4(5x + 7) - 16 = 12$ 14.

15. $5(3x + 2) - 3 = 7$ 15.

B.

(1–5) Write the analysis and solution plan for each of the following, then solve and check.

1. $3\left(\frac{x}{5} + 4\right) = 12$

Analysis

Solution Plan

Solution

Check:

2. $\frac{2(x - 6)}{5} = 4$

Analysis

Solution Plan

Solution

Check:

3. $5\left(\frac{x}{4} + 3\right) = 10$

Analysis

Solution Plan

Solution

Check:

4. $8 = 2(x - 2) + 2$

Analysis

Solution Plan

Solution

Check:

5. $14 = 3(x + 2) + 2$

Analysis

Solution Plan

Solution

Check:

B.

(6–15) Solve and check. (Do the analysis and solution plan mentally or on scratch paper.)

6. $\frac{x+4}{3} = 3$ 6.

7. $\frac{3x-7}{5} = 4$ 7.

8. $3(2x+9) = 9$ 8.

9. $\frac{\frac{3x}{5} + 23}{4} = 5$ 9.

10. $7 = 3(x+2) + 4$ 10.

11. $5 + 6(x-3) = 23$ 11.

12. $8 = \left(\frac{x}{3} + 5\right) + 6$ 12.

13. $25 + 3(2x - 7) = 70$ 13.

14. $2(6x + 1) + 13 = 3$ 14.

15. $25 - 6(x + 1) = 19$ 15.

5.7 GENERAL LINEAR EQUATIONS

OBJECTIVE: To solve the general linear equation and check the solution.

Solving a general linear equation involves four basic steps:

(1) Remove parentheses.

(2) Collect like terms.

(3) Obtain the form $x = a$, if possible, by using the equivalence theorems.

(4) Check by substituting the value a into the original equation.

Not all linear equations can be reduced to the form $x = a$. Examples of this are shown in Sample Problems 3 and 4.

SAMPLE PROBLEM 1

Solve and check $4(x - 2) = 19 - (2x - 3)$.

SOLUTION

$4x - 8 = 19 - 2x + 3$	Remove parentheses.
$4x - 8 = 22 - 2x$	Collect like terms.
$6x - 8 = 22$	Add $2x$ to each side.
$6x = 30$	Add 8 to each side.
$\frac{6x}{6} = \frac{30}{6}$	Divide each side by 6.
$x = 5$	

CHECK

$$4(x - 2) = 19 - (2x - 3)$$
$$4(5 - 2) = 19 - (2 \cdot 5 - 3)$$
$$4(3) = 19 - (10 - 3)$$
$$12 = 19 - 7$$
$$12 = 12$$

The solution is 5.

SAMPLE PROBLEM 2

Solve and check $x + 7 - 2(x + 2) = 4(x - 8)$.

SOLUTION

$x + 7 - 2(x + 2) = 4(x - 8)$	
$x + 7 - 2x - 4 = 4x - 32$	Remove parentheses.
$-x + 3 = 4x - 32$	Collect like terms.
$4x - 32 = -x + 3$	Exchange sides.*
$4x = -x + 35$	Add 32 to each side.
$5x = 35$	Add x to each side.
$x = 7$	Divide each side by 5.

CHECK

$$x + 7 - 2(x + 2) = 4(x - 8)$$
$$7 + 7 - 2(7 + 2) = 4(7 - 8)$$
$$14 - 2(9) = 4(-1)$$
$$14 - 18 = -4$$
$$-4 = -4\text{, true; value checks.}$$

The solution is 7.

---

*This step is optional. It is useful to exchange sides when the coefficient of the variable on the right side is positive and greater than the coefficient of the variable on the left side. This avoids working with a minus sign, a common source of error.

SAMPLE PROBLEM 3

Solve and check $3(x + 5) = 25 - (10 - 3x)$.

SOLUTION

$3(x + 5) = 25 - (10 - 3x)$
$3x + 15 = 25 - 10 + 3x$ Remove parentheses.
$3x + 15 = 15 + 3x$ Collect like terms.

Because $3x + 15 = 15 + 3x$ for all integers x, since it is an example of the commutative axiom for addition, the original equation is also true for all integers x. The solution set is I, the set of integers.

If the commutative axiom were not apparent, and if the equivalence theorems were applied, then

$3x + 15 = 15 + 3x$

$3x = 3x$ Subtract 15 from each side.

$0 = 0$ Subtract $3x$ from each side.

Certainly at this point, it can be noted that $0 = 0$ is true and that it is true for all integers x.

CHECK

Select any value for x, say $x = 2$.

$3(x + 5) = 25 - (10 - 3x)$
$3(2 + 5) = 25 - (10 - 3 \cdot 2)$
$3(7) = 25 - 4$
$21 = 21$, and 2 checks.

SAMPLE PROBLEM 4

Solve and check $6x - 5(x - 2) = x + 26$.

SOLUTION

$6x - 5(x - 2) = x + 26$
$6x - 5x + 10 = x + 26$ Remove parentheses
$x + 10 = x + 26$ Collect like terms.
$x = x + 16$ Subtract 10 from each side.
$0 = 16$ Subtract x from each side.

Since $0 = 16$ is false, this means that the original equation is also false for all values of x. The solution set is $\emptyset$, the empty set.

CHECK

Since there is no number that can be substituted to obtain a true statement, another method must be used. One way is to rework the problem, very slowly and carefully, to see if the same result is obtained a second time.

SAMPLE PROBLEM 5

Solve and check $\frac{3x - 5}{2} = 5x + 1$.

SOLUTION

$2\left(\frac{3x - 5}{2}\right) = 2(5x + 1)$	Multiply each side by 2.
$3x - 5 = 10x + 2$	Simplify.
$3x - 5 + 5 = 10x + 2 + 5$	Add 5 to each side.
$3x = 10x + 7$	Simplify.
$3x - 10x = 10x + 7 - 10x$	Subtract $10x$ from each side.
$-7x = 7$	Simplify.
$\frac{-7x}{-7} = \frac{7}{-7}$	Divide each side by -7.
$x = -1$	Simplify.

CHECK

$$\frac{3x - 5}{2} = 5x + 1$$
$$\frac{3(-1) - 5}{2} = 5(-1) + 1$$
$$\frac{-8}{2} = -5 + 1$$
$$-4 = -4, \text{ true}$$

The solution is -1.

EXERCISES 5.7

(1–15) Solve and check by completing the format.

A.

1. Given: $4(2x - 5) = 5x - (3x + 2)$

   Remove parentheses.

   Collect like terms.

   Use equivalence theorems and simplify.

   Check.

2. Given: $5 + 4x = 5(x - 2) + 8$

   Remove parentheses.

   Collect like terms.

   Use equivalence theorems and simplify.

   Check.

3. Given: $20 - (5x - 6) = 6 - 10(x + 1)$

   Remove parentheses.

   Collect like terms.

   Use equivalence theorems and simplify.

   Check.

4. $3x - 5 = 3 - x$ 4.

5. $4x - 7 = 7x + 2$ 5.

6. $2(y - 3) = 3(13 - y)$ 6.

7. $10 - (y + 4) = 2y - 15$ 7.

8. $7x - 5(x + 2) = 6$ 8.

9. $8x - 4(2x - 3) = 12$ 9.

10. $5x - 3 = 2(2x + 5)$ 10.

11. $5 - 2z = 24 - 3(z + 2)$ 11.

12. $2(x + 4) = 5x - 2(x - 3)$ 12.

13. $10 - 3t = 19 - (3t + 4)$ 13.

14. $5(8x + 7) - 8(5x - 2) = 45$ 14.

15. $2y - 7(y + 6) = 6(3y - 7)$ 15.

B.

1. Given: $6(x + 9) = 5x - 2(x - 6)$

   Remove parentheses.

   Collect like terms.

   Use equivalence theorems and simplify.

   Check.

2. Given: $2(x - 5) = 10 - (8 - 5x)$

   Remove parentheses.

   Collect like terms.

   Use equivalence theorems and simplify.

   Check.

3. $2x + 19 = 5x + 4$ 3. ____

4. $5x + 4 = 3x - 8$ 4.

5. $5(2y + 2) = 2(5y - 1)$ 5.

6. $7 - (y - 5) = 3 - 2y$ 6.

7. $4x - 3(x - 5) = 25$ 7.

8. $5(x - 4) - 6(x + 1) = 4$ 8.

9. $8x + 7 = 5(2x - 3)$ 9.

10. $7 - 3z = 12z - 8(z - 7)$ 10.

11. $8x - (2x + 2) = 2(3x - 1)$ 11.

12. $3(5t + 6) - 2(10t - 3) = 24$ 12.

13. $32 - (7x + 3) = 8(3 - x)$ 13.

14. $3(n + 2) - 4(n - 4) = 5(n - 10)$ 14.

15. $7(n - 20) - 5(n - 22) = 2n$ 15.

5.8 LITERAL EQUATIONS

OBJECTIVE: To solve a linear literal equation.

A **literal equation** is an equation that contains two or more letters. For some applications, it is useful to solve such an equation for one of the letters. This is accomplished by treating the letter which is to be solved for as the variable and by treating all other letters as constants. The methods used for solving linear literal equations are the same as those used for solving linear equations in one variable.

SAMPLE PROBLEM 1

Solve $7x + 4y = 28$ for y.

SOLUTION

$$\begin{array}{rl} 7x + 4y = 28 & \\ -7x \qquad\quad -7x & \\ \hline 4y = 28 - 7x & \text{subtracting } 7x \text{ from each side} \\ \dfrac{4y}{4} = \dfrac{28 - 7x}{4} & \text{dividing each side by 4} \\ y = \dfrac{28 - 7x}{4} & \end{array}$$

SAMPLE PROBLEM 2

Solve $C = \dfrac{k - 5n}{k}$ for n.

SOLUTION

$$\begin{array}{rl} Ck = k - 5n & \text{multiplying each side by } k \\ Ck + 5n = k & \text{adding } 5n \text{ to each side} \\ 5n = k - Ck & \text{subtracting } Ck \text{ from each side} \\ n = \dfrac{k - Ck}{5} & \text{dividing each side by 5} \end{array}$$

SAMPLE PROBLEM 3

Solve $C = \dfrac{5(F-32)}{9}$ for F.

SOLUTION

A. Mental Analysis

(1) Determine what was done to the number F.

First, 32 was subtracted.

Second, the difference was multiplied by 5 and divided by 9.

(2) Perform the inverse operations in the reverse order.

B. Written Solution

$$C = \frac{5}{9}(F-32)$$

$$9C = 5(F-32) \qquad \text{multiplying each side by 9}$$

$$\frac{9C}{5} = F - 32 \qquad \text{dividing each side by 5}$$

$$\frac{9C}{5} + 32 = F \qquad \text{adding 32 to each side}$$

$$F = \frac{9C}{5} + 32 \qquad \text{exchanging sides}$$

EXERCISES 5.8

Solve each equation for the variable indicated.

A.

1. $2x + y = 10$ for y

2. $4x - y = 5$ for y

3. $3x + 2y = 12$ for y

B.

1. $3x + y = 7$ for y

2. $2x - y = 6$ for y

3. $2x + 3y + 12 = 0$ for x

4. $5x - 3y - 15 = 0$ for y

5. $ax + by + c = 0$ for y

6. (Uniform motion)
$d = rt$ for t

7. (Perimeter of triangle)
$p = a + b + c$ for c

8. (Area of triangle)
$A = \frac{bh}{2}$ for h

9. (Simple interest)
$A = P + Prt$ for t

10. (Arithmetic progression)
$S = \frac{A + L}{2}$ for L

4. $5x - 2y - 10 = 0$ for x

5. $mx - y + b = 0$ for y

6. (Ohm's law)
$E = IR$ for I

7. (Perimeter of rectangle)
$P = 2W + 2L$ for W

8. (Area of trapezoid)
$A = \frac{1}{2}h\,(a + b)$ for b

9. (Arithmetic progression)
$L = a + (n - 1)\,d$ for d

10. (Ideal gas law)
$PV = RT$ for R

11. (Arithmetic progression)
$L = a + (n - 1)\,d$ for n

11. (Business profit)
$P = S - C$ for C

12. (Tension in cable)
$T = m\,(g - a)$ for a

12. (Expansion of gases)
$L = a\,(1 + ct)$ for c

13. (Specific gravity)
$S = \frac{A - W}{W}$ for A

13. (Rockets–missiles)
$L = \frac{cdAv^2}{2}$ for A

14. (Economics–depreciation)
$R = \frac{C - S}{n}$ for S

14. (Psychology–intelligence quotient)
$Q = \frac{100\,M}{C}$ for M

15. (Photography)
$(a + b)\,f = ab$ for f

15. (Chemistry–ionization)
$K = \frac{a^2 c}{1 - a}$ for c

UNIT 5 POSTTEST

(1–4) Simplify. (5.1)

1. $x + 7x =$ ______________
2. $5y - 4y =$ ______________
3. $6x + 9x - 8x =$ ______________
4. $2y - 3y - y =$ ______________

(5–8) Collect like terms. (5.2)

5. $10 - 2x - 3x - 12 =$ ______________
6. $2x^2 + 2x - 5x - 5 =$ ______________
7. $x^2 - 4yx + 2xy - 8y^2 =$ ______________
8. $6x^2y + 4xy^2 - 5yx^2 =$ ______________

(9–14) Remove parentheses. (5.3)

9. $6(x - 3) =$ ______________
10. $-5(x + 7) =$ ______________
11. $-(4x + 9) =$ ______________
12. $-(8 - 6x) =$ ______________
13. $4(x - 2y - 6) =$ ______________
14. $-2x(x + y - 1) =$ ______________

(15–18) Remove parentheses and collect like terms. (5.4)

15. $30 - 4(x + 5) =$ ______________
16. $5(x - 2) - (x - 6) =$ ______________
17. $3x(2x - 1) - 4(2x - 1) =$ ______________
18. $6x(x - 2y) + y(x - 2y) =$ ______________

(19–22) Solve and check. (5.5)

19. $x - 35 = 65$
20. $x + 17 = 13$
21. $-6x = 48$
22. $\frac{-x}{5} = 45$

(23–26) Solve and check. (5.6)

23. $\frac{x - 25}{3} = 5$

24. $7x + 15 = 1$

25. $4\left(\frac{x}{3} - 2\right) = 4$

26. $4 = \frac{6(x + 3)}{5} - 8$

(27–30) Solve and check. (5.7)

27. $7x - 3(x + 5) = x + 3$

28. $4(x - 2) - 6(x - 4) = 26$

29. $2(6 - x) = 16 - 2(x + 2)$

30. $2x - (x - 1) = 5 - (5 - x)$

(31–32) Solve for the indicated letter. (5.8)

31. $s = \frac{a + b + c}{2}$ for c

32. $9x - 2y = 18$ for y

UNIT 6

STATED PROBLEMS, LINEAR EQUATIONS IN ONE VARIABLE

Unit 6
PRETEST

1. If 7 is subtracted from 4 times a certain number, then the result is 8 more than the number. Find the number. (6.1)

2. A certain test shows that one object is twice as old as another. When the same test was performed 5 years ago, the object was 3 times as old as the other. Find the present ages of the objects. (6.2–6.3)

3. A farmer has 180 ft. of fencing. He wants to enclose a rectangular plot of land, so that the length will be 6 ft. longer than the width. Find the length and width of fencing that he should use. (6.4)

4. The treasurer of a club collected $84.00 to pay the exact cost of 20 dinners. Some had ordered the beef dinner at $4.50 each, and some had ordered the fish dinner at $3.75 each. How many dinners of each kind should be ordered? (6.5–6.7)

5. Two airplanes left an airport, one traveling east and the other west. At the end of 2 hours they were 2500 mi. apart. If the westbound plane traveled 50 m.p.h. faster than the eastbound plane, find the rate of each. (6.9)

6. Two hours after a ship left a port, a helicopter left the same port to overtake the ship. The ship traveled 40 m.p.h. and the helicopter traveled 200 m.p.h. How long did it take the helicopter to overtake the ship? (6.10)

7. A jet airplane averaged 640 m.p.h. on a certain trip because of a tail wind. On the return trip a head wind reduced the speed to 576 m.p.h., and the time was increased by 1 hr. Find the time for each trip and find the distance of the trip one way. (6.11)

6.1 NUMBER PROBLEMS

OBJECTIVE: To solve a number problem stated in words.

In many cases, an applied problem is stated in words that can be translated into symbols forming an equation. The solution of the equation then provides an answer to the problem. Number relations are basic to many problems in many areas, such as business, economics, science, and engineering.

Some number relations occur quite often in applications. Knowing how to solve a general number problem provides preparation for solving problems dealing with specific subject matter.

One way of solving stated number problems is described by the following method.

Method of Short Statements

(1) Choose a letter, such as x or n, as the name of the number whose value is to be found.

(2) Rewrite the problem as shortly and simply as possible.

(3) Translate the words into symbols. Use the list of basic translations provided, if necessary.

(4) Solve the resulting linear equation.

Basic Translations:

x added to y	$x+y$	x subtracted from y	$y-x$
the sum of x and y	$x+y$	the difference between x and y when x is subtracted from y	$y-x$
x plus y	$x+y$	x minus y	$x-y$
x more than y	$x+y$	x less than y	$y-x$
x increased by y	$x+y$	x decreased by y	$x-y$
the product of x and y	xy	the quotient of x divided by y	$\frac{x}{y}$
x times y	xy		
x multiplied by y	xy	x divided by y	$\frac{x}{y}$
twice x, x doubled	$2x$	one-half of x	$\frac{x}{2}$
thrice x, x tripled	$3x$	two-thirds of x	$\frac{2x}{3}$

is, are, is equal to, is the same as, becomes, was, will be: =

SAMPLE PROBLEM 1

One-third the sum of a certain number increased by 4 is 12 less than twice this number. Find the number.

SOLUTION

(1) Let x = the number.

(2) Rewrite problem as a short statement.

$\frac{1}{3}$(sum of x increased by 4) = 12 less than $2x$

(3) Translate into symbols.

$$\frac{1}{3}(x + 4) = 2x - 12$$

(4) Solve.

$$x + 4 = 3(2x - 12)$$
$$3(2x - 12) = x + 4$$
$$6x - 36 = x + 4$$
$$5x = 40$$
$$x = 8$$

SAMPLE PROBLEM 2

Seven more than the product of a number and 5 is twice the difference of the number decreased by 4. Find the number.

SOLUTION

(1) Let n = the number.

(2) Rewrite the problem as a short statement.

7 more than $5n$ = 2 (n decreased by 4)

(3) $7 + 5n = 2(n - 4)$

$$7 + 5n = 2n - 8$$
$$3n = -15$$
$$n = -5$$

SAMPLE PROBLEM 3

When the sum of a number and 6 is subtracted from 15, the result is 3. Find the number.

SOLUTION

(1) Let x = the number.

(2) $(x + 6)$ subtracted from $15 = 3$

(3) $15 - (x + 6) = 3$

(4) $15 - x - 6 = 3$

$$9 - x = 3$$

$$-x = -6$$

$$x = 6$$

EXERCISES 6.1

A.

1. Five more than a certain number is equal to 5 times the difference obtained when 3 is subtracted from the number. Find the number.

2. One more than 4 times a certain number is equal to 3 less than twice the number. Find the number.

3. Four times the difference of a number decreased by 4 is the same as 5 times the difference of the number decreased by 6. Find the number.

4. Ten times a number decreased by twice the difference when 1 is subtracted from the number is equal to 58. Find the number.

5. Five increased by the product of 4 and a certain number is the same as 17 decreased by twice the number. Find the number.

B.

1. When the product of 3 and the result of subtracting a certain number from 7 is increased by 12, the result is equal to the product of 8 and the number. Find the number.

2. One-half the sum of 7 and a certain number is equal to 1 more than the given number. Find the number.

3. Five more than 3 times a given number is the same as 3 less than 5 times the given number. Find the number.

4. Four is the sum of a certain number and twice the difference obtained when 5 is subtracted from 3 times the number. Find the number.

5. Six times the difference of a number decreased by 3 equals 5 times the difference of the number minus 5. Find the number.

6.2 AGE PROBLEMS, COMPONENT EXPRESSIONS

OBJECTIVE: To express in symbols common phrases that occur in age problems.

Age problems have little practical value, but they are useful in developing techniques that can be used in solving applied problems. As puzzles, these problems have fascinated people for hundreds of years. Number problems exist which are dated as early as 1650 B.C. in the Rhind papyrus, an Egyptian manuscript, and age problems occur in the *Greek Anthology*, a Greek work dated at about A.D. 500. Even today, age and number problems appear on the puzzle pages of various newspapers.

The following table summarizes common phrases used in age problems.

Age now	Age k years ago	Age k years from now	k years younger	k years older
a	$a - k$	$a + k$	$a - k$	$a + k$

A. **EXERCISES 6.2**

Express each of the following in symbols.

1. The age of a person 5 years ago if his present age is 30. 1. ____________
2. The age of a person 5 years ago if his present age is x. 2. ____________
3. The age of a person 5 years ago if he is now $x + 2$. 3. ____________
4. The age of a person 12 years from now if his present age is 25. 4. ____________
5. The age of a person 12 years from now if his present age is x. 5. ____________
6. The age of a person 12 years from now if his present age is $3x$. 6. ____________
7. The age of a person 7 years younger than a man 40 years old. 7. ____________

8. The age of a person x years younger than a man 40 years old. 8. ____________
9. The age of a person 7 years younger than a man x years old. 9. ____________
10. The age of a person 10 years older than a child 8 years old. 10. ____________
11. The age of a person 10 years older than a child y years old. 11. ____________
12. The age of a person x years older than a child 8 years old. 12. ____________
13. The age of a person 6 years younger than a man $x + 3$ years old. 13. ____________
14. The age of a person 6 years older than a child whose age is $2x$. 14. ____________

B.

Complete each table.

1. Person	Age now	Age 3 yr. ago	Age 7 yr. from now
Girl	x		
Sister	$x - 5$		
Mother	$4x$		

2. Building	Age now	Age 10 yr. ago	Age 50 yr. from now
Museum	x		
Church	$x + 20$		
Library	$x - 30$		

3. Person	Age now	Age 6 yr. ago	Age 9 yr. from now
Boy	14		
Brother, x yr. younger			
Sister, y yr. older			

4. Object	Age now	Age 8 yr. ago	Age 4 yr. from now
Table	x		
Chair, twice as old as table			
Desk, 5 yr. older than table			

6.3 AGE PROBLEMS

OBJECTIVE: To solve a stated age problem.

Method

1. Let x = present age. (The equation is usually simplest when x is the youngest age.)
2. Make a table of the ages for the times mentioned in the problem.

3.–4. Form an equation by translating a sentence from the problem, using expressions from the table.

5. Solve the equation.

SAMPLE PROBLEM 1

Ten years from now John will be twice as old as he was 7 years ago. How old is John now?

SOLUTION

(1) Let x = John's age now.

(2) Make a table as follows:

Age now	Age 10 years from now	Age 7 years ago
x	$x + 10$	$x - 7$

(3) Rewrite problem as a short statement. (John's age 10 years from now) will be twice (John's age 7 years ago).

(4) Translate the short statement into symbols, using the table in (2).

$$(x + 10) = 2\,(x - 7)$$

(5) Solve the equation.

$$2\,(x - 7) = x + 10$$

$$2x - 14 = x + 10$$

$$x = 24$$

SAMPLE PROBLEM 2

A boy's sister is 3 years younger than he is. The boy's father is 26 years older than the boy. Four years ago, the father's age was 3 times the sum of the ages of his two children then. Find the present age of the father and his two children.

SOLUTION

Person	Age now	Age 4 years ago
Boy	x	$x - 4$
Sister	$x - 3$	$(x - 3) - 4 = x - 7$
Father	$x + 26$	$(x + 26) - 4 = x + 22$

(Father's age 4 years ago) was 3 times (sum of ages of children 4 years ago)

$x + 22 = 3\,(x - 4 + x - 7)$

EQUATION

$$x + 22 = 3\,(2x - 11)$$

$$6x - 33 = x + 22$$

$$5x = 55$$

$$x = 11, x - 3 = 8, x + 26 = 37$$

EXERCISES 6.3

A.

1. Twenty years from now, a man will be 3 times as old as he was 8 years ago. Find his present age.

2. Nine years ago, Mary was one-half as old as she is now. How old is Mary now?

3. Fourteen years ago, Jack was one-third as old as he will be 8 years from now. Find Jack's present age.

4. In 20 years, a building will be twice as old as it is now. In how many years will the building be 3 times as old as it is now?

5. A man is now 3 times as old as his daughter. Fourteen years from now the man will be twice the age of his daughter then. How old is the man now?

B.

1. When Mr. and Mrs. Smith were married, Mr. Smith was 2 years older than Mrs. Smith. On their 25th wedding anniversary, the sum of their ages was 100. How old were Mr. and Mrs. Smith when they were married?

2. Bill is 2 years older than his brother and 5 years younger than his sister. Five years ago the sum of the ages of Bill and his brother was equal to their sister's age then. How old is Bill now?

3. A good wine should age 20 years before it is ready to drink. When David was born his father had a case of Amontillado laid down for him. This wine will be ready to drink when David's age equals his present age plus one-fourth of his present age. How old is David?

4. Anne has a sister one-half her age and a sister twice her age. If the combined ages of the three girls are 2 less than twice the age of the oldest sister, how old is Anne?

5. Vanessa has a book that is 54 years older than she is. In 5 years from now, the book will be 4 times as old as Vanessa. In how many years from now will the book be 100 years old?

6.4 GEOMETRIC PROBLEMS

OBJECTIVE: To solve a stated problem that involves geometric figures.

Basic Technique

(1) Make a well-labeled sketch of the figure involved in the problem.
(2) Use one letter only to mark lengths (or angles) on the figure.
(3) To find the equation, use the geometric formula that applies to the problem.
(4) Solve the equation.

Geometric Figures and Formulas

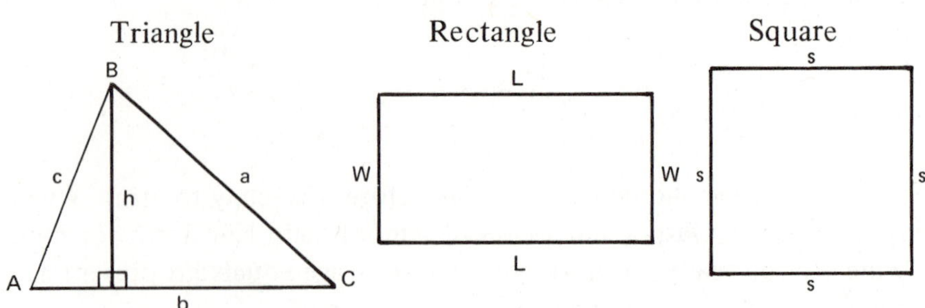

Angles. $A + B + C = 180°$

Perimeter. $p = a + b + c$	$p = 2W + 2L$	$p = 4s$
Area. $A = \dfrac{bh}{2}$	$A = WL$	$A = s^2$

SAMPLE PROBLEM 1

A piece of wire 48 in. long is to be bent into the shape of a rectangle whose length is 3 times its width. Find the dimensions of the rectangle.

SOLUTION

Let x = the width in inches, then $3x$ = the length in inches.

$$P = 2W + 2L = 48$$
$$2x + 2(3x) = 48$$
$$2x + 6x = 48$$
$$8x = 48$$
$$x = 6 \text{ in., width}$$
$$3x = 18 \text{ in., length}$$

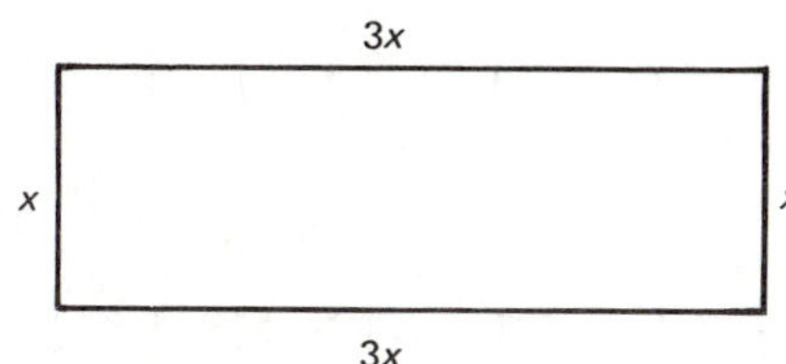

SAMPLE PROBLEM 2

The cross section of an irrigation ditch has the shape of an isosceles triangle (two equal sides) whose depth is 5 in. less than its width. By keeping the same width but digging the ditch 6 in. deeper, the area of the cross section can be increased by 36 sq. in. Find the width and the depth of the ditch before digging.

SOLUTION

Let x = the width, then $x - 5$ = the depth.

Old area $= \dfrac{x(x-5)}{2}$ and new area $= \dfrac{x(x-5+6)}{2}$.

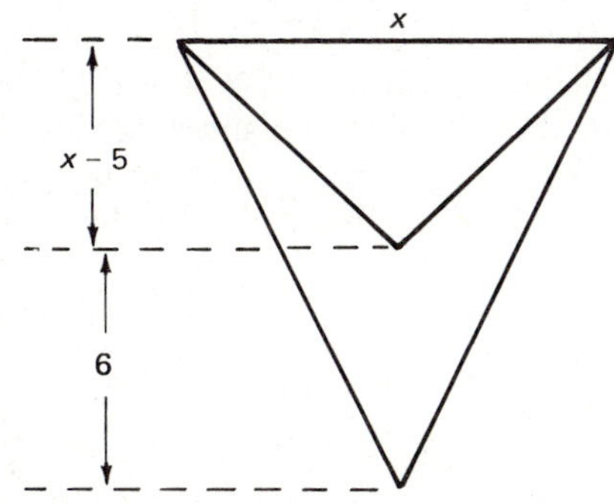

$$\frac{x(x+1)}{2} = \frac{x(x-5)}{2} + 36$$
$$x(x+1) = x(x-5) + 72$$
$$x^2 + x = x^2 - 5x + 72$$
$$x = -5x + 72$$
$$6x = 72$$
$$x = 12 \text{ in., width}$$
$$x - 5 = 7 \text{ in., depth}$$

A.

1. Find the area of a square whose perimeter is 32 in.

2. Find the dimensions of a rectangle whose length is twice its width and whose perimeter is 54 in.

3. A farmer uses 180 ft. of fencing to enclose a rectangular area that is 10 ft. longer than it is wide. Find the length and width of the rectangular area.

4. Find the number of degrees in each angle of a triangle if one angle is twice the smallest and the largest angle is three times the smallest.

5. A piece of wire 36 cm. long is to be bent into the shape of a triangle. The longest side is to be 6 cm. longer than the shortest side, and the third side is to be 3 cm. longer than the shortest side. Find the length of each side of the triangle.

6. A lot has the shape of a right triangle, with one leg (a side forming the right angle) 5 ft. longer than the other leg. A 5-ft. wide sidewalk along the longest side of the lot (the hypotenuse of the right triangle) decreases the length of each leg by 5 ft. and decreases the area of the lot by 900 sq. ft. Find the length of the legs of the triangular lot before the sidewalk was installed.

B.

1. A 36-in. by 48-in. rectangular picture is to have a frame of uniform width. The perimeter of the framed picture determines the length of the material needed for the frame. Find the width of the frame if the perimeter of the framed picture is 176 in.

2. A certain city code requires that the window space in a room be $\frac{1}{8}$ of the floor area. For a rectangular room 44 ft. long and 20 ft. wide, the owner specifies that 2 windows each measuring 5 ft. by 3 ft. be installed. How many additional windows each measuring 4 ft. by 4 ft. will be needed?

3. Sixty feet of wood molding are used along one length and the two widths of a rectangular room. If the room is twice as long as it is wide, find the dimensions of the room.

4. On a blueprint, the two equal sides of a triangle are each twice as long as the third side. If 160 cm. of metal stripping is specified to go around this triangle, find the length of each side.

5. A triangle has a 5-ft. base and a 3-ft. altitude (height). How much should the altitude be increased in order to double the area of the triangle?

6. One angle of a triangle is twice the size of a second angle. The third angle measures 60°. Find the number of degrees in the other two angles.

6.5 MIXTURE PROBLEMS, BASIC IDEA 1

OBJECTIVE: To express a value in symbols by using the product rule.

Many practical problems can be classified as mixture problems, the topic of the next three sections. The product rule plays an important role in the solution of these problems. Several examples of this rule are shown in the Table of Special Cases.

The Product Rule: (unit value) · (amount) = value

Many mixture problems involve percentages. Per cent means the number of parts per 100. For example, 20% means 20 parts per 100. Thus, $20\% = \frac{20}{100}$. The decimal 0.20 also means 20 parts per 100. Thus, $0.20 = \frac{20}{100}$ and 20% = 0.20. When a decimal is multiplied by 100, the decimal point is moved two places to the right. For example, 0.20 (100) = 20. Similarly,

$$3\% = 0.03 \quad \text{and } 0.03(100) = 3$$

$$15\% = 0.15 \quad \text{and } 0.15(100) = 15$$

$$4\tfrac{1}{2}\% = 0.045 \quad \text{and } 0.045(100) = 4.5$$

$$40\% = 0.40 \quad \text{and } 0.40(100) = 40$$

Table of Special Cases

Subject Matter	*Unit Value* •	*Amount* =	*Value*
Coins:	No. of cents in 1 coin	No. of coins	Total value of coins
Nickels	5	n	$5n$
Dimes	10	n	$10n$
Quarters	25	n	$25n$
Dollars	100	n	$100n$
Money invested:	Per cent as decimal	Investment in dollars	Income (interest) in dollars
$500, at 6%	0.06	500	0.06 (500)
Sales:	Price per unit weight (or volume)	Weight (or volume)	Cost
Walnuts	40¢ per lb.	3 lb.	40 (3) cents
Solutions:	Per cent of component as decimal	Volume of solution (or weight)	Volume (or weight) of component
3% salt solution	0.03	20 cc.	0.03 (20)
Digits:	Place value	Digit	Number represented
Tens digit is 7	10	7	10 (7)
Hundreds digit is 3	100	3	100 (3)
Salary:	Daily (hourly) wage	Number of days (hours)	Salary received
	$2.15 per hr.	50 hr.	2.15 (50) dollars
Work:	Output per unit time	Time working	Total output
Water pipe	2 cu. ft. per min.	30 min.	2 (30) cu. ft. of water
Time:	Days per wk.	No. of wk.	Total no. of days
	7	w	$7w$
	Min. per hr.	No. of hr.	Total no. of min.
	60	h	$60h$
	Sec. per min.	No. of min.	Total no. of sec.
	60	m	$60m$

EXERCISES 6.5

Basic Idea 1: (unit value) • (amount) = value
Express each of the following in symbols.

A.

1. The value in cents of 4 quarters. 1. ____________
2. The value in cents of x quarters. 2. ____________
3. The income in dollars from $2000 invested at 5%. 3. ____________
4. The income in dollars from $5000 invested at $6\frac{1}{2}$%. 4. ____________

5. The income in dollars from d dollars invested at 5%. 5. ______________
6. The income in dollars from $x + 500$ dollars invested at 5%. 6. ______________
7. The cost in cents of 5 lb. walnuts if the price is 15¢ per lb. 7. ______________
8. The cost in cents of y lb. walnuts if the price is 15¢ per lb. 8. ______________
9. The cost in cents of $20 - x$ lb. walnuts if the price is 15¢ per lb. 9. ______________
10. The cost in cents of 5 lb. walnuts if the price is c cents per lb. 10. ______________
11. The volume of acid in 40 cc. of a 5% acid solution. 11. ______________
12. The volume of acid in x cc. of a 5% acid solution. 12. ______________
13. The volume of acid in $x + 15$ cc. of a 5% acid solution. 13. ______________
14. The volume of acid in $50 - x$ cc. of a 5% acid solution. 14. ______________

B.

1. The weekly salary in dollars of a man who works 40 hr. at $3.10 per hr. 1. ______________
2. The weekly salary in dollars of a man who works h hr. at $3.10 per hr. 2. ______________
3. The weekly salary in dollars of a man who works 40 hr. at x dollars per hr. 3. ______________
4. The weight of copper in 60% copper ore weighing 20 lb. 4. ______________
5. The weight of copper in 60% copper ore weighing y lb. 5. ______________
6. The weight of copper in 60% copper ore weighing $x + 20$ lb. 6. ______________
7. The number of days in x wk. 7. ______________
8. The number of seconds in y min. 8. ______________
9. The number of seconds in h hr. 9. ______________
10. The volume of water delivered by a pipe working for s sec. if its output is 3 cu. ft. per sec. 10. ______________
11. The volume of water delivered by a pipe working for 50 sec. if its output is x cu. ft. per sec. 11. ______________

6.6 MIXTURE PROBLEMS, BASIC IDEA 2

OBJECTIVE: To express in symbols the amount of a mixture given the amounts of the components or the amount of a component given the amount of the mixture and the amount of the other component.

A second important idea involved in a mixture problem is that the sum of the amounts of the components of a mixture is equal to the amount of the mixture. If 30 lb. of almonds are combined with 50 lb. of walnuts, then there are 30 + 50, or 80, lb. of nuts in the mixture. In general, if a lb. of one ingredient are mixed with b lb. of another ingredient, then $a + b$ lb. are in the mixture.

Basic Idea: Sum of the amounts of the components = amount of mixture.

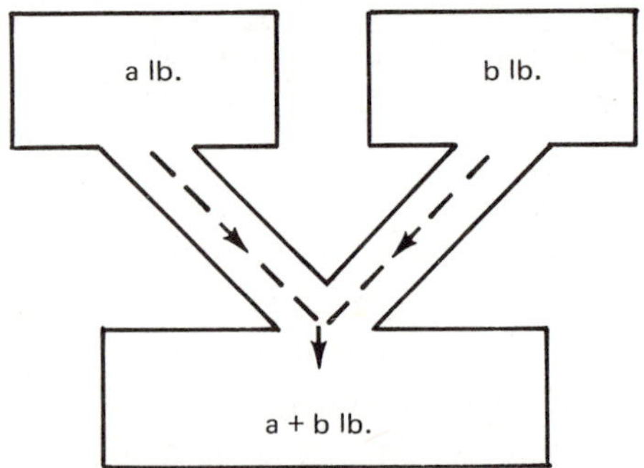

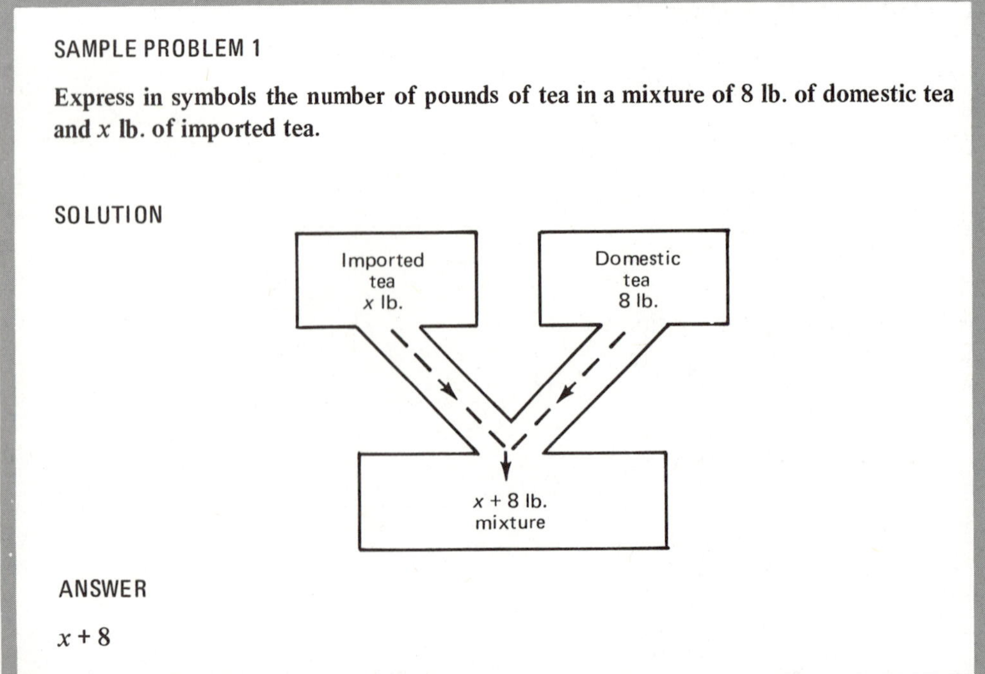

SAMPLE PROBLEM 1

Express in symbols the number of pounds of tea in a mixture of 8 lb. of domestic tea and x lb. of imported tea.

SOLUTION

ANSWER

$x + 8$

SAMPLE PROBLEM 2

Express in symbols the number of pounds of imported tea in a 20-lb. mixture consisting of imported tea and x lb. of domestic tea.

SOLUTION

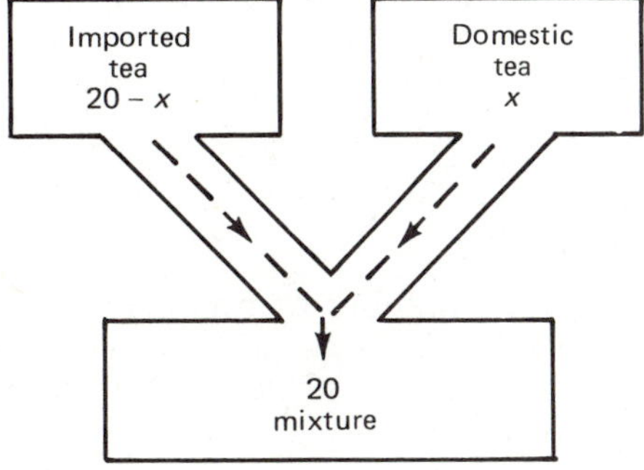

ANSWER

$20 - x$

NOTE

$(20 - x) + x = 20$

SAMPLE PROBLEM 3

Express in symbols the amount of money invested in bonds if \$5000 is invested in stocks and a total of x dollars is invested in bonds and stocks.

SOLUTION

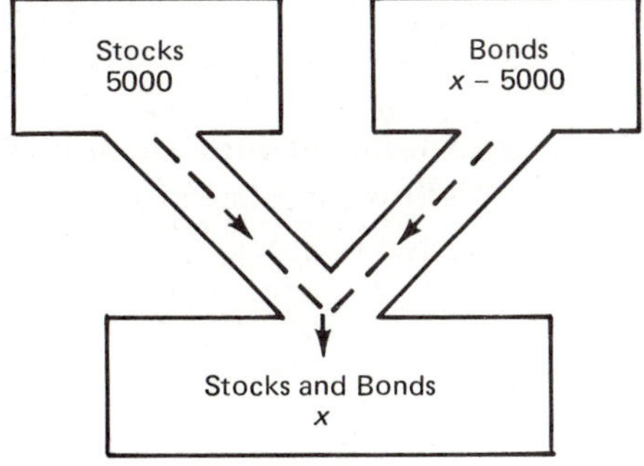

ANSWER

$x - 5000$

EXERCISES 6.6

Basic Idea 2: The sum of the amounts of the components is equal to the amount of the mixture.

Express each of the following in symbols. If necessary, draw a diagram like the ones shown in the first four problems.

A.

1. The number of pounds in a mixture of 5 lb. of walnuts and 3 lb. of peanuts.

1. 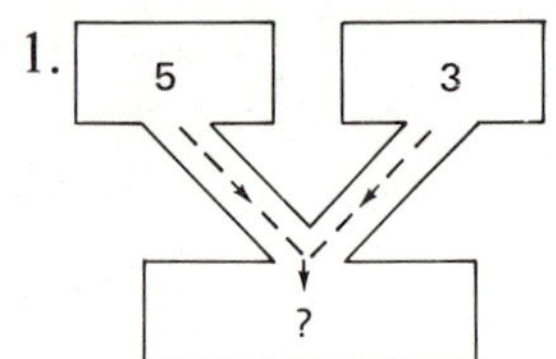

2. The number of pounds in a mixture of x lb. of walnuts and 3 lb. of peanuts.

2. 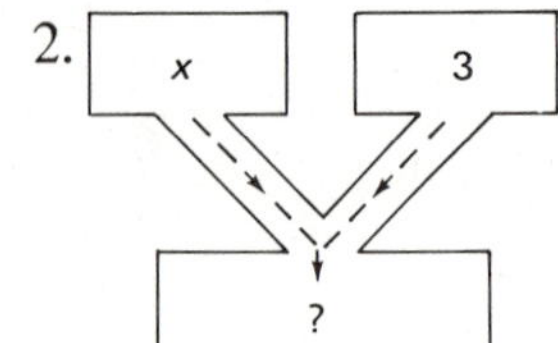

3. The number of pounds of walnuts in a 12-lb. mixture consisting of walnuts and 3 lb. of peanuts.

3. 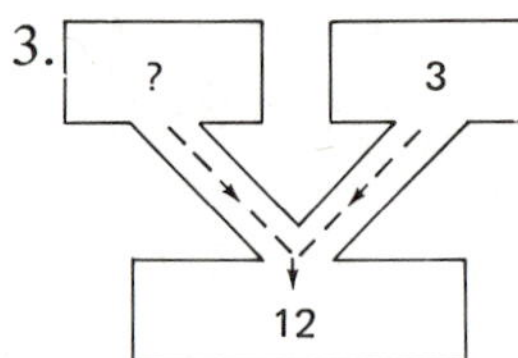

4. The number of pounds of walnuts in a 12-lb. mixture consisting of walnuts and x lb. of peanuts.

4. 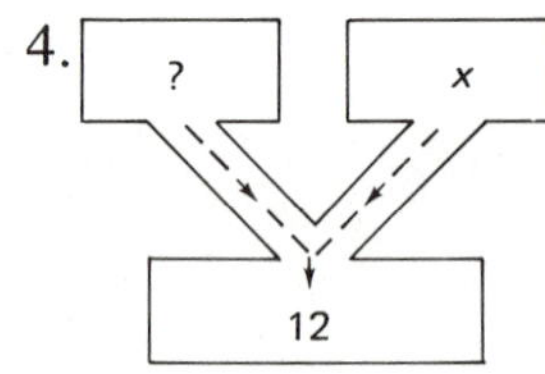

5. The total amount of money invested if $2000 is invested in bonds and $3000 is invested in stocks.

5.

6. The total amount of money invested if \$2000 is invested in bonds and d dollars is invested in stocks. 6.

7. The amount of money invested in bonds if a total of \$5000 is invested in both stocks and bonds and \$3000 of this is invested in stocks. 7.

8. The amount of money invested in bonds if a total of \$5000 is invested in both stocks and bonds and x dollars of this is invested in stocks. 8.

9. The volume of the mixture obtained when 20 cc. of water is added to 40 cc. of a certain acid solution. 9.

10. The volume of the mixture obtained when x cc. of water is added to 40 cc. of a certain acid solution. 10.

11. The volume of water added to 50 cc. of a certain acid solution to obtain a mixture whose volume is 75 cc. 11.

12. The volume of water added to 50 cc. of a certain acid solution to obtain a mixture whose volume is x cc. 12.

13. The total number of yards of material in a purchase consisting of y yd. of linen and 5 yd. of cotton. 13.

14. The number of yards of lace in a purchase of 15 yd. of trimming consisting of lace and x yd. of fringe. 14.

15. The number of quarts in a mixture consisting of x qt. of cream and twice as much milk as cream. 15.

B.

1. The total amount of money spent if $25 was spent on stamps and x dollars was spent on stationery. 1.

2. The amount of alcohol that must be added to 60 cc. of a certain solution to obtain a mixture whose volume is x cc. 2.

3. The amount of water that must be added to x gal. of a salt solution to obtain a mixture whose volume is 100 gal. 3.

4. The amount of copper in 40 gm. of an alloy consisting of copper and x gm. of tin. 4.

5. The amount of copper in x gm. of an alloy consisting of copper and 15 gm. of tin. 5.

6. The number of quarts in a mixture consisting of y qt. of cream and 6 qt. more milk than cream. 6.

7. The total number of coins in a mixture of d dimes, 3 times as many nickels as dimes, and 5 times as many cents as dimes. 7.

8. The number of stamps in a packet consisting of x air mail stamps and 8 less regular stamps than air mail stamps. 8.

9. The total amount of money invested if d dollars are invested in bonds, if the amount invested in stocks is twice the amount invested in bonds, and if the amount invested in mortgages is $2000 more than the amount invested in bonds. 9.

10. The total amount of fruit in a mixture of pears, apples, and oranges if the number of pears is x, if there are 3 times as many apples as pears, and if there are 5 less oranges than apples. 10.

6.7 MIXTURE PROBLEMS, COMPLETE SOLUTION

OBJECTIVE: To solve a stated mixture problem.

A table such as that shown in Sample Problem 1 is a convenient way to summarize the information stated in a mixture problem.

The equation for a mixture problem is obtained by using the following rule:

Basic Idea 3: The sum of the values of the components equals the value of the mixture.

Using the table, the equation is obtained by setting the sum of the entries in the last column above the double line equal to the entry in the last column below the double line.

SAMPLE PROBLEM 1

How many pounds of dried prunes costing 70¢ per lb. should be mixed with 30 lb. of dried apricots costing $1.20 per lb. to obtain a mixture costing $1.00 per lb.?

Component	Unit Value (¢ per pound)	Amount =	Value (in ¢)
Prunes	70	x	
Apricots	120	30	
Mixture	100		

SOLUTION

First, complete the amount column. Using the sum rule, the amount of the mixture is $x + 30$.

Then, complete the last column, using the product rule. The table should look as follows:

Prunes	70	x	$70x$
Apricots	120	30	120 (30)
Mixture	100	$x + 30$	$100(x + 30)$

EQUATION

$100(x + 30) = 70x + 120(30)$

SOLUTION

$100x + 3000 = 70x + 3600$

$30x = 600$

$x = 20$ lb.

SAMPLE PROBLEM 2

How much distilled water should be added to a 12% acid solution to obtain 100 cc. of a solution that is 6% acid?

Component	Unit Value (% as decimal) ·	Amount (in cc.) =	Value (in cc.)
Water	0	x	
12% acid	0.12		
6% acid	0.06	100	

SOLUTION

Completing the amount column by the sum rule, the amount of 12% acid solution is $100 - x$.

Completing the last column by the product rule, the table should look as follows:

Water	0	x	0
12% acid	0.12	$100 - x$	$0.12\,(100 - x)$
6% acid	0.06	100	0.06 (100)

The equation is obtained from the last column.

$$0 + 0.12(100 - x) = 0.06(100)$$

$$12(100 - x) = 6(100)$$

$$100 - x = 50$$

$$x = 50 \text{ cc}$$

EXERCISES 6.7

A.

1. A dealer wants to mix 30 lb. of rice costing 40¢ per lb. with macaroni worth 55¢ per lb. to produce a mixture worth 45¢ per lb. How many pounds of macaroni should he use?

Component	Unit Value (in ¢) ·	Amount (in lb.) =	Value (in ¢)
Rice			
Macaroni			
Mixture			

Equation:

Solution:

2. Tickets to a certain event cost $1.50 for each adult and 50¢ for each child. A total of $545 was paid by 450 persons who attended the event. How many adult tickets were sold?

Component	Unit Value (in ¢) ·	Amount (no. of people) =	Value (in ¢)
Adults			
Children			
Mixture			

Equation:

Solution:

3. $3000 more is invested in mortgages earning 7% than in bonds earning 5%. If the total income from both investments is $630, how much is invested in mortgages?

Component	Unit Value (% as decimal) ·	Amount Invested =	Value (dollars)
Bonds			
Mortgages			
Mixture			

Equation:

Solution:

4. How many quarts of pure methanol should be added to an antifreeze containing 16% methanol to obtain 18 qt. of an antifreeze containing 30% methanol?

Components	Unit Value (% as decimal) ·	Amount = (quarts)	Value (amount of methanol)
Pure methanol			
16% methanol			
30% methanol			

Equation:

Solution:

5. How much water must be evaporated from 60 gal. of a 1% salt solution to obtain a 3% salt solution?

Components	Unit Value ·	Amount =	Value
Water			
3% solution			
1% solution			

Equation:

Solution:

6. How many pounds of walnuts costing 50¢ per lb. should be mixed with 20 lb. of peanuts costing 20¢ per lb. to obtain a mixture costing 30¢ per lb.?

7. How many pounds of caramels worth 45¢ per lb. should be combined with chocolates worth $1.20 per lb. to obtain a 10-lb. mixture of chocolates and caramels worth 60¢ per lb.?

8. A man has twice as much money invested in 7% stocks as he has in 5% bonds. If his total income from these stocks and bonds is $760, how much money does he have invested in bonds and how much money does he have invested in stocks?

B.

1. How many 2¢ stamps should there be in a packet containing 2¢ stamps and 32 eleven-cent stamps if the value of the packet is $4.00?

Component	Unit Value · (in ¢)	Amount = (no. of stamps)	Value (in ¢)
2¢ stamps			
11¢ stamps			
Mixture			

Equation:

Solution:

2. Mrs. Smith bought 3 times as many oranges as apples and found her bill to be $3.04. If the oranges cost 60¢ per doz. and the apples cost 48¢ per doz., how many oranges did she buy?

Components	Unit Value · (in ¢)	Amount = (no. of fruit)	Value (in ¢)
Oranges			
Apples			

Equation:

Solution:

3. A 60-lb. mixture of nuts contains walnuts worth 50¢ per lb., cashews worth $1.00 per lb., and almonds worth 75¢ per lb. If there are twice as many pounds of walnuts as cashews, how many pounds of cashews are in the mixture which is worth 70¢ per lb.?

Components	Unit Value ·	Amount =	Value
Cashews			
Walnuts			
Almonds			
Mixture			

Equation:

Solution:

4. A corporation has $100,000 invested in 6% bonds, 9% stocks, and 7% mortgages. If the amount invested in bonds is the same as the amount invested in stocks and if the total income from the bonds, stocks, and mortgages is $7300, how much money is invested in mortgages?

Components	Unit Value ·	Amount =	Value
Bonds			
Stocks			
Mortgages			
Mixture			

Equation:

Solution:

5. How much pure copper should be combined with 30 lb. of an alloy containing 75% copper to obtain another alloy containing 90% copper?

6. How much cream containing 35% butterfat should be added to milk containing 4% butterfat to produce 155 qt. of half-and-half containing 12% butterfat?

7. How much water (distilled water containing no salt) should be added to brine containing 3% salt to obtain 900 gal. of a brine containing 2% salt?

8. A certain paint product consists of 1 lb. of pigment for every 5 lb. of vehicle. If the pigment costs $20 per lb. and the vehicle costs $1 per lb., how many pounds of pigment are used in making $400 worth of the paint product?

6.8 UNIFORM MOTION, FORMULA

OBJECTIVE: To express in symbols distances and speeds of objects traveling at a uniform rate.

A uniform motion problem is one in which an object is moving at a constant rate of speed. The product rule is again a key concept. If a car travels at 60 m.p.h. for 2 hr., then the distance traveled is the product $60(2) = 120$ mi.

The uniform motion formula states that the distance d traveled is the product of the constant rate r and the time of travel t. Expressed in symbols,

$$d = rt$$

For Sample Problems 1 to 10, express each of the following in symbols.

SAMPLE PROBLEMS	SOLUTIONS
(1-6) A car is traveling at a uniform rate of 50 m.p.h.:	
1. The distance it goes in 4 hr.	**1. 50 (4) = 200 mi.**
2. The distance it goes in x hr.	**2. $50x$ mi.**
3. The distance it goes in 3 hr. more than h hr.	**3. $50(h + 3)$ mi.**
4. The speed of a car going 10 m.p.h. faster.	**4. 50 + 10 = 60 m.p.h.**
5. The speed of a car going r m.p.h. faster.	**5. $50 + r$ m.p.h.**
6. The speed of a car going x m.p.h. slower.	**6. $50 - x$ m.p.h.**
(7-10) A boat is traveling at a uniform rate of x m.p.h.:	
7. The distance the boat goes in 4 hr.	**7. $4x$ mi.**
8. The speed of a boat moving 10 m.p.h. faster.	**8. $x + 10$ m.p.h.**
9. The speed of a boat moving 10 m.p.h. slower.	**9. $x - 10$ m.p.h.**
10. The speed of a boat moving twice as fast.	**10. $2x$ m.p.h.**

EXERCISES 6.8

Basic Idea: distance = (rate) × (time)

Translate into symbols:

A.

1. The distance a car goes in 3 hr. at a uniform rate of 40 m.p.h. — 1. ______
2. The distance a car goes in h hr. at a uniform rate of 40 m.p.h. — 2. ______
3. The distance a car goes in $h + 2$ hr. at a uniform rate of 40 m.p.h. — 3. ______
4. The distance a car goes in 5 hr. at a uniform rate of x m.p.h. — 4. ______
5. The distance a car goes in 5 hr. at a uniform rate of $x - 2$ m.p.h. — 5. ______
6. A train travels at a rate of 35 m.p.h.: the speed of a plane flying at 6 times this speed. — 6. ______

7. A train travels at a rate of x m.p.h.: the speed of a plane flying at 6 times this speed. 7. ______________

8. A train travels at a rate of $x - 2$ m.p.h.: the speed of a plane flying at 4 times this speed. 8. ______________

9. A car is traveling at 30 m.p.h.:
   a. The speed of a car traveling 20 m.p.h. faster. a. ______________
   b. The speed of a car traveling x m.p.h. faster. b. ______________
   c. The speed of a car traveling 5 m.p.h. slower. c. ______________
   d. The speed of a car traveling x m.p.h. slower. d. ______________

10. A car is traveling at x m.p.h.:
   a. The speed of a car traveling 20 m.p.h. faster. a. ______________
   b. The speed of a car traveling 5 m.p.h. slower. b. ______________

B.

(1–5) An airplane is traveling at a uniform rate of 450 m.p.h.:

1. The distance it goes in 3 hr. 1. ______________
2. The distance it goes in x hr. 2. ______________
3. The distance it goes in 2 hr. less than x hr. 3. ______________
4. The speed of a plane going r m.p.h. slower. 4. ______________
5. The speed of a plane going x m.p.h. faster. 5. ______________

(6–10) A ship is traveling at a uniform rate of x knots (nautical m.p.h.).

6. The distance it goes in 6 hr. 6. ______________
7. The distance it goes in 3 hr. more than 6 hr. 7. ______________
8. The speed of a ship moving 12 knots faster. 8. ______________

9. The speed of a ship moving 12 knots slower. 9. ____________

10. The speed of a ship moving 3 times as fast. 10. ____________

6.9 UNIFORM MOTION, OPPOSITE DIRECTIONS

OBJECTIVE: To solve a uniform motion problem for objects traveling in opposite directions.

Basic Ideas:

1. If two moving objects (cars, trains, planes, boats, etc.) start from the same point and travel in opposite directions, then the distance between them is the sum of the distances traveled by the two objects.

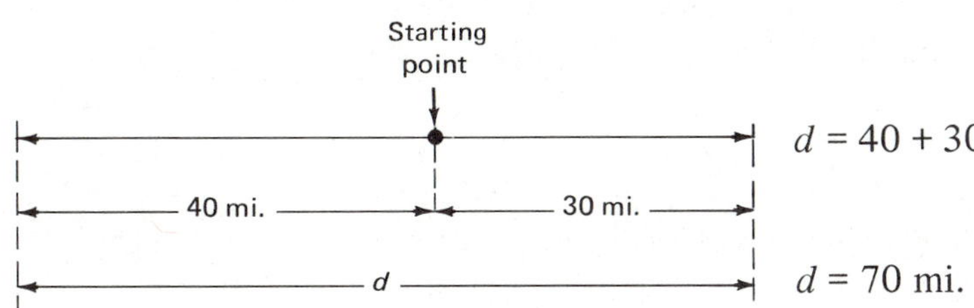

2. If two moving objects (cars, trains, planes, boats, etc.) start from different points and travel toward each other until they meet, then the distance between their starting points is the sum of the distances traveled by the two objects.

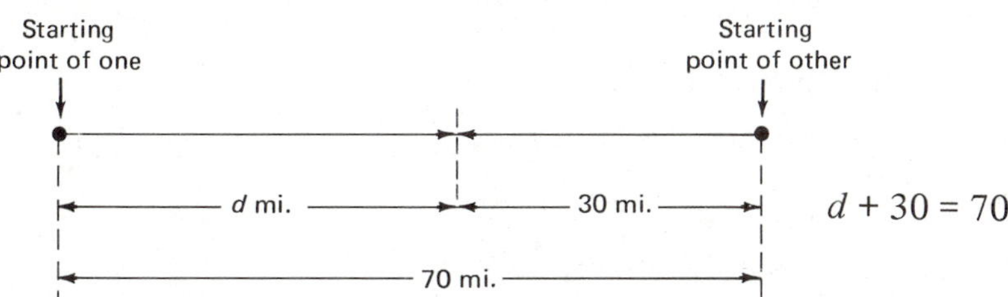

Basic Method

1. Draw a sketch of the motions.
2. Find the values of r (rate) and t (time) for each object.
3. Find the value of d (distance) for each object by using the formula, $d = rt$.
4. Write the equation by noting how the distances are related.
5. Solve the equation and check.

SAMPLE PROBLEM 1

Two cars start at the same time from the same place. One travels north at 40 m.p.h. The other travels south at 60 m.p.h. In how many hours are they 300 mi. apart?

SOLUTION

Let x = the time that each travels.

SKETCH

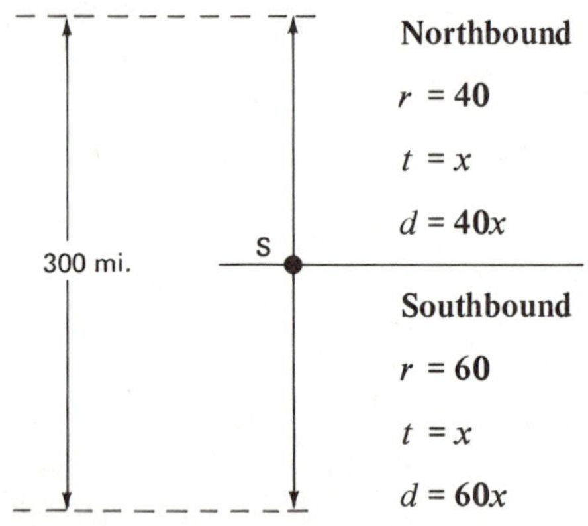

Traveling in opposite directions from same point:

distance of one + distance of other = distance they are apart

EQUATION

$$40x + 60x = 300$$

$$100x = 300$$

$$x = 3 \text{ hr.}$$

SAMPLE PROBLEM 2

Two boats, 81 mi. apart, travel toward each other, one traveling 3 m.p.h. faster than the other. They meet in 3 hr. Find the rate of the slower boat.

SOLUTION

Let x = rate of slower boat.

SKETCH

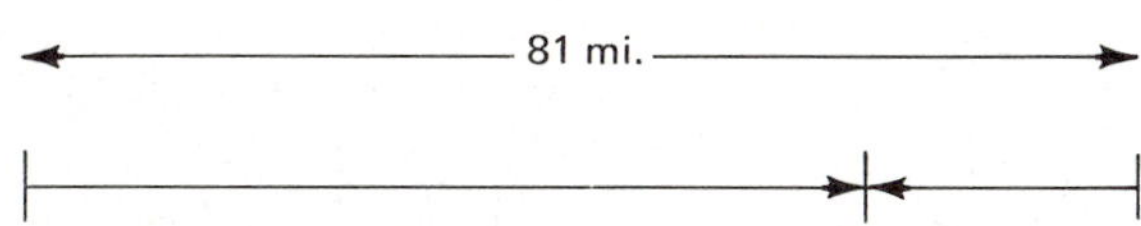

Faster: $r = x + 3$ **Slower:** $r = x$

$t = 3$ $t = 3$

$d = 3\ (x + 3)$ $d = 3x$

Traveling toward each other:

distance of one + distance of other = distance between starting points

EQUATION

$$3x + 3\ (x + 3) = 81$$

$$3x + 3x + 9 = 81$$

$$6x = 72$$

$$x = 12 \text{ m.p.h.}$$

EXERCISES 6.9 Solve each of the following.

A.

1. At 10:00 A.M., two planes leave the same airport, one traveling east at 460 m.p.h. and the other traveling west at 520 m.p.h. In how many hours are they 1470 mi. apart?

Sketch:

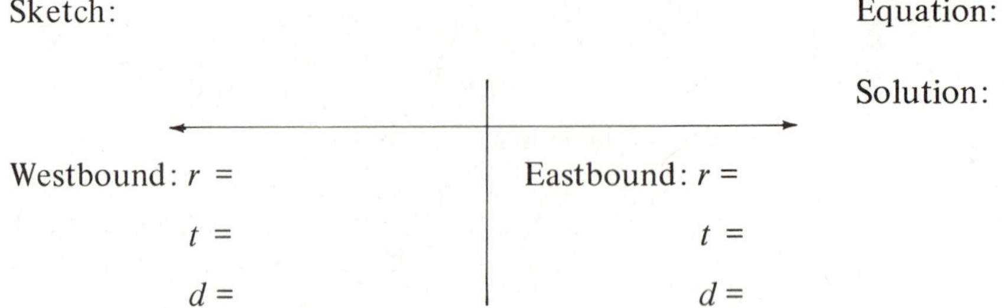

Equation:

Solution:

2. A passenger train and a freight train start from depots 550 mi. apart and travel in opposite directions toward the same station. The freight train takes 6 hr. to reach the station. The passenger train, traveling 50 m.p.h. faster than the freight train, takes 4 hr. to reach the station. Find the rate of each train.

Sketch:

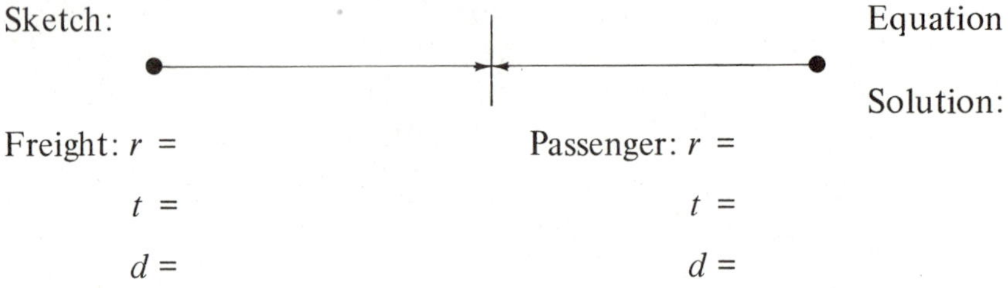

Equation

Solution:

3. At the same time, two ships left two different ports 450 mi. apart and traveled toward each other on parallel routes until they met. If the rate of one was 40 m.p.h. and the rate of the other was 35 m.p.h., how many hours did each travel?

4. At 9:00 A.M. a truck left a warehouse and traveled north. At 11:00 A.M. a second truck left the same warehouse and traveled south. They each traveled at a rate of 55 m.p.h. What was the time of day when they were 330 mi. apart?

B.

1. At 7:00 A.M., two cars leave San Francisco, one traveling north and the other south. The speed of the southbound car is 20 m.p.h. faster than the speed of the northbound car. If they are 440 m. apart at the end of 4 hr., find the speed of each car.

Sketch:

Equation:

Solution:

2. Sacramento and Los Angeles are 400 mi. apart. At 6:00 A.M., a car leaves Sacramento and drives on the freeway to Los Angeles at an average rate of 60 m.p.h. Three hours later, a car leaves Los Angeles and drives on the same freeway toward Sacramento at an average rate of 50 m.p.h. How far from Sacramento do the two cars meet? At what clock time do they meet?

3. At the same time, two airplanes leave the same airport, one traveling east at 650 m.p.h. and the other traveling west at 700 m.p.h. At 12:00 noon they are 4050 mi. apart. What time did they leave the airport?

4. At 10:00 A.M. two trains leave two different stations 380 mi. apart, and travel toward each other on parallel tracks until they meet at 2:00 P.M. If the rate of the faster train was 5 m.p.h. more than the rate of the slower, find the rate of each.

6.10 UNIFORM MOTION, PURSUIT PROBLEMS

OBJECTIVE: To solve a uniform motion problem for one object in pursuit of another.

Starting from the Same Point

Basic Idea: If two moving objects start at the same point, but at different times, and travel in the same direction until one overtakes the other, then the distance of one is the same as the distance of the other.

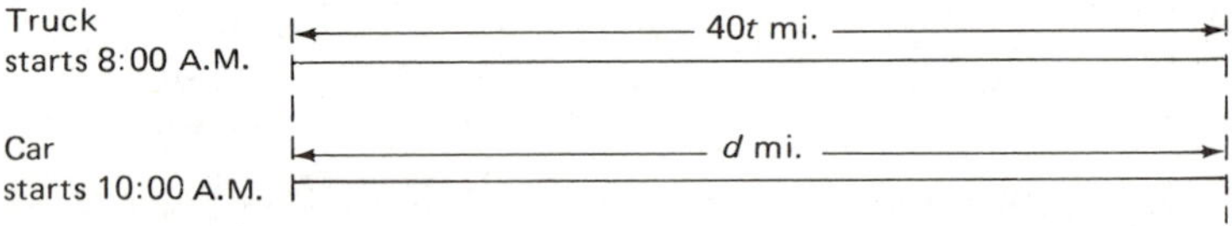

$$\text{distance of truck} = \text{distance of car}$$

$$40t = 60(t - 2)$$

Starting from Different Points

Basic Idea: If two moving objects start at the same time from different points along the same route and one pursues the other, then at the time they meet, the distance of the faster object is the sum of the distance of the slower object and the distance between their starting points.

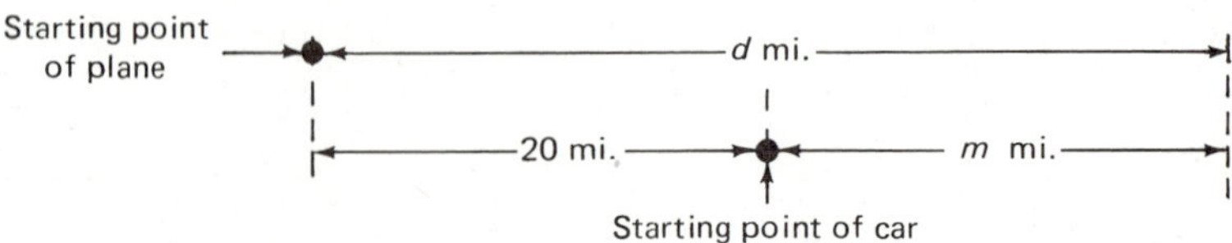

$$\text{distance of faster plane} = 20 + \text{distance of slower car}$$

$$d = 20 + m$$

SAMPLE PROBLEM 1

At 8:00 A.M., a truck leaves a depot, traveling north at 35 m.p.h. Two hours later, a car starts from the same depot and travels at 55 m.p.h. until it overtakes the truck. At what clock time does the car overtake the truck?

SOLUTION

Let x = time the truck travels.

SKETCH

Truck, 8:00 A.M.: $r = 35$

$t = x$

$d = 35x$

Car, 10:00 A.M.: $r = 55$

$d = 55(x - 2)$

One moving object overtakes another, both starting at same point:

distance of one = distance of other

EQUATION

$55(x - 2) = 35x$

$55x - 110 = 35x$

$20x = 110$

$x = \frac{11}{2} = 5\frac{1}{2}$ hr.

Clocktime: $8{:}00 + 5{:}30 = 13{:}30$

$13{:}30 - 12{:}00 = 1{:}30$ P.M.

SAMPLE PROBLEM 2

A helicopter sets out to overtake a train that is 35 mi. ahead. If the rate of the train is 90 m.p.h. and the rate of the helicopter is 160 m.p.h., how long does it take the helicopter to overtake the train?

SOLUTION

Let x = time it takes helicopter.

SKETCH

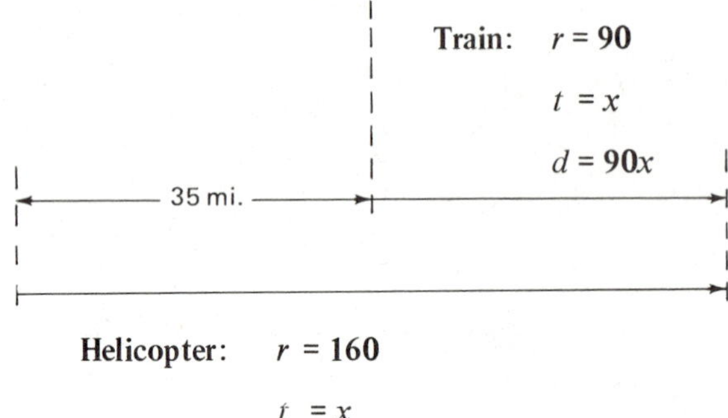

$d = 160x$

One moving object overtakes another, one beginning m mi. away from the other:

distance of faster = m + distance of slower

EQUATION

$160x = 35 + 90x$

$70x = 35$

$x = \frac{1}{2}$ hr. = 30 min.

EXERCISES 6.10

A.

1. At 8:00 A.M., a man left home and walked in the country at a rate of 3 m.p.h. A sudden thunderstorm arose and at 11:00 A.M. his wife left home and started after him, driving at a rate of 39 m.p.h. At what time did the wife overtake her husband?

Sketch:

Equation:

Solution:

Man: r =

t =

d =

Wife: r =

t =

d =

2. A car speeding at 80 m.p.h. is 5 mi. beyond a police car when the police car starts in pursuit at 90 m.p.h. How long does it take the police car to overtake the speeding car?

Sketch:

Equation:

Solution:

3. A truck leaves a depot and travels on a certain road at 30 m.p.h. One hour later, a car leaves the same depot and travels on the same road at 60 m.p.h. How long does it take the car to overtake the truck?

4. At 3:00 P.M. a houseboat on a river is 16 mi. upstream from a motorboat. Both are traveling upstream toward a dock, the houseboat at 9 m.p.h. and the motorboat at 25 m.p.h. If they both reach the dock at the same time, how many hours did each travel?

B.

1. A car leaves a manufacturing plant and travels on a freeway at 55 m.p.h. Two hours later, a helicopter, flying 165 m.p.h., leaves the same plant and flies parallel to the freeway until it overtakes the car. How many hours had the car traveled at this point?

2. City B is 50 mi. due east of city A. A truck leaves city B and travels east at 40 m.p.h. At the same time, a car leaves city A and travels east along the same road at 60 m.p.h. until it overtakes the truck. How many hours did the car travel?

3. At 9:00 A.M. a freight train and an express train leave from stations 120 mi. apart, both traveling in the same direction. The rate of the express train is 75 m.p.h., and the rate of the freight train is 35 m.p.h. At what clock time does the express train overtake the freight train?

4. A coast guard cutter at the shore line wants to overtake a boat that is $\frac{1}{3}$ mi. out at sea, traveling at 15 m.p.h. How fast should he go to overtake the boat within a mile from the shore?

6.11 UNIFORM MOTION, RETURN TRIPS

OBJECTIVE: To solve a uniform motion problem involving return trips.

There are two basic ideas concerning the motion of an object going from one place to another and then returning to the starting place.

1. The distance going is equal to the distance returning.

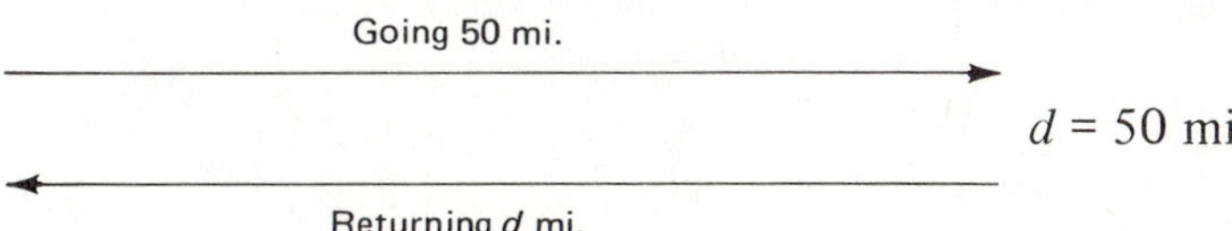

2. The sum of the time going and the time returning is equal to the total time of the trip.

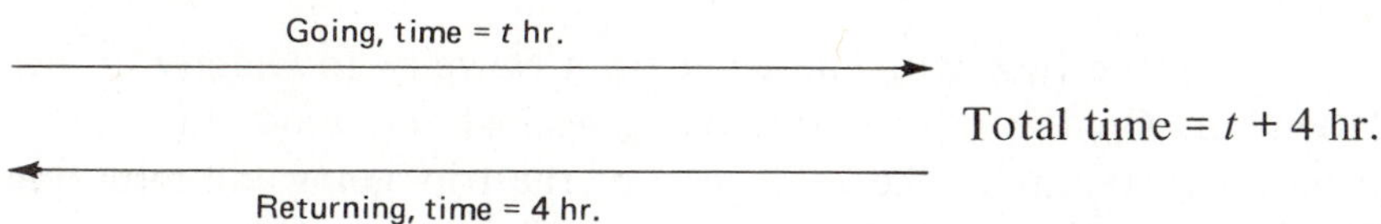

SAMPLE PROBLEM

At 10:00 A.M., a truck left a warehouse to make a delivery to a store. It stayed 1 hr. at the store and then began the return trip to the warehouse, arriving back at 4:00 P.M. Because of heavy traffic, the rate returning was 24 m.p.h., while the rate going was 36 m.p.h. Find the distance from the warehouse to the store.

SOLUTION

Let x = time of trip going.

Total time of travel = 6 − 1 = 5 hours.

Going:		Warehouse ↓↑ Store	Returning:	
r	$= 36$		r	$= 24$
t	$= x$		t	$= 5 - x$
d	$= 36x$		d	$= 24(5 - x)$

For return trips, distance going = distance returning

EQUATION

$36x = 24(5 - x)$

$36x = 120 - 24x$

$60x = 120$

$x = 2$ hr.

$d = 36(2) = 72$ miles

EXERCISES 6.11

A.

1. A plane took 6 hr. to fly to Seattle. The return trip took 1 hr. longer because a strong wind reduced its speed by 25 m.p.h. Find the rate of the plane going to Seattle.

Sketch: Equation:

Solution:

Going		Returning
$r =$	↓ ↑	$r =$
$t =$		$t =$
$d =$		$d =$

2. A plane flew with the wind from Newcity to Oldcity at a rate of 240 m.p.h. On the return trip, flying against the wind, the rate of the plane was 180 m.p.h. The total time of the trip, going and returning, was 7 hr. Find the distance between the two cities.

Sketch: Equation:

Solution:

3. An engineer drove to a construction site at an average speed of 48 m.p.h. When he returned several days later, a detour increased his trip by 18 mi. and increased his time by 1 hr. If his average speed returning was 36 m.p.h., how far did he travel going to the construction site?

Sketch: Equation:

Solution:

B.

1. At 10:00 A.M. a messenger left an office, traveled by car to deliver a message at a construction site, and immediately returned to the office after he delivered the message, arriving back at 12:00 noon. His rate going was 60 m.p.h., but his rate returning was 20 m.p.h. because of very heavy traffic. How far from the office was the construction site?

2. At 9:00 A.M. a man left his home and cycled to a friend's house at a rate of 10 m.p.h. He visited for 1 hr. and then returned home by car, arriving at 3:00 P.M. If the rate of the car was 40 m.p.h., how many hours did he spend cycling?

3. A boat travels downstream for 3 hr. The return trip takes 4 hr. If the rate of the boat downstream was 5 m.p.h. more than the rate upstream, find the rate upstream.

6.12 ASSORTED STATED PROBLEMS

1. The bill for a purchase of dress material and trimming was $15.89, which included 76¢ tax. The cost of the dress material was $1.95 per yd. and the cost of the trimming was 49¢ per yd. If the number of yards of trimming was 5 less than twice the number of yards of material, how much dress material was purchased?

2. Jack's average rate for running a certain race is 8 meters per sec., while John's average rate for the same race is 5 meters per sec. If Jack gives John a 90-meter start, they finish the race in the same time. In what time does Jack run the race?

3. How much distilled water should be added to 500 cc. of a solution containing 25% alcohol to obtain a solution containing 20% alcohol?

4. A man has 48 ft. of fencing for use in enclosing three sides of a rectangular area (1 length and 2 widths). He considers a rectangle, whose length is twice the width, and a square. Which figure would give the greater area and by how many square feet?

5. State College is 45 years younger than City College. Five years from now, the age of City College will be twice that of State College. How many years from now will City College be 100 years old?

6. At 10:00 A.M., an express train leaves City A for City B, a distance of 310 mi. Two hours later, a freight train leaves City B for City A, traveling on tracks parallel to those of the express train. If the express train travels at 90 m.p.h. and if the freight train travels at 40 m.p.h., how far from City A do the trains pass each other?

7. An angle of a triangle is 10° more than the smallest angle and 10° less than the largest angle. Find the angles of the triangle.

8. A messenger drove at the rate of 60 m.p.h. to deliver a message and then returned at the rate of 40 m.p.h. If the total trip took 5 hr., find the distance he drove at 60 m.p.h.

9. How much 40% acid solution should be combined with a 10% acid solution to make 100 cc. of a 16% acid solution?

UNIT 6 POSTTEST

1. Six times the sum of a number and 5 is 10 less than the number. Find the number. (6.1)

2. In 3 years Jay and Kay will celebrate their Golden Wedding Anniversay. Kay would not reveal her age. However, Jay said that he is 4 years older than Kay, and on their Golden Wedding Day, the sum of their ages will be 3 times the sum of their ages on the day that they were married. How old are Jay and Kay? (6.2–6.3)

3. To carpet a rectangular room whose width is 4 ft. less than its length, 45 ft. of tackless stripping are needed. The stripping is placed at all edges of the floor except along the 3-ft. wide doorway. Find the dimensions of the room. (6.4)

4. How many liters of water should be mixed with a 75% alkali solution in order to obtain 60 liters of a 50% alkali solution? (6.5–6.7)

5. A car and a truck, 285 mi. apart, travel toward each other. The car travels 15 m.p.h. faster than the truck. If they meet at the end of 3 hr., find the rate of the car and the rate of the truck. (6.9)

6. A freight train, traveling at 30 m.p.h., leaves Los Angeles for San Francisco. Four hours later, an express train, traveling at 70 m.p.h., also leaves Los Angeles for San Francisco. How far from Los Angeles do the two trains pass? (6.10)

7. A man cycles at the rate of 12 m.p.h. until his bicycle breaks down. He walks home at the rate of 3 m.p.h. If the total trip took 5 hr., how long did he cycle? (6.11)

UNIT 7

OPERATIONS ON POLYNOMIALS

Unit 7
PRETEST

(1–2) Simplify. (7.1)

1. $(-4x^2)(6x^3)$
2. $(-7x^2y^4)(-8x^3y)$

(3–4) Rewrite in descending powers of x. (7.2)

3. $7x - 5 + x^3 - 2x^2$
4. $15 - x - x^3y + x^2y^4$

(5–6) Simplify. (7.3)

5. $(4x^2 - 5 - x) + (8 - 4x - 3x^2)$
6. $(5x - 6y + 4) + (2x + 6y - 9)$

(7–8) Simplify. (7.4)

7. $(x^3 - 8x^2 + 16x) - (4x^2 - 32x + 64)$
8. $(x^2 + 10xy + 25y^2) - (y^2 - 6xy + 9x^2)$

(9–10) Simplify. (7.5)

9. $3x^2(6x^2 - 7x + 1)$
10. $-6xy(4x^2 - 3xy - 5y^2)$

(11–15) Multiply. (7.6)

11. $(2x + 7)(x - 3)$

12. $(5x + 6y)(5x - 6y)$

13. $(3x - 4)(2y - 5)$

14. $(x + 5)(x^2 + 10x + 25)$

15. $(3x - 4y)(9x^2 + 12xy + 16y^2)$

(16–17) Write as a trinomial by using the formula for a perfect square trinomial. (7.8)

16. $(2x + 5)^2$

17. $(7x - 3y)^2$

(18–19) Find the product by using the difference of squares formula. (7.9)

18. $(3x + 7y)(3x - 7y)$

19. $(9x^2 - 4)(9x^2 + 4)$

(20–21) Find the product by using the simple trinomial product formula. (7.10)

20. $(x + 7)(x - 10)$

21. $(y^2 - 9)(y^2 + 16)$

(22–25) Find the product by using the FOIL method. (7.7)

22. $(5x - 2)(4x - 3)$

23. $(3x - 7y)(6x + 5y)$

24. $(x+9)(y-1)$

25. $(x^2 - 36)(x+10)$

7.1 PRODUCT OF MONOMIALS

OBJECTIVE: To simplify an indicated product of monomials.

A **monomial** is a single numeral, a single letter, or an indicated product of factors consisting of one or more letters or numerals.

Examples of monomials are 5, x, $\frac{1}{2}x^3$, $-2xy^2$, and x^2y^3.

Monomials are multiplied by rearranging the factors so that the numerical factors are written first (at the left) and the literal factors are written from left to right in alphabetical order, using the exponential form so that no letter is repeated.

SAMPLE PROBLEM 1

Multiply $(5x^3)(-2x^2)$.

SOLUTION

$$(5x^3)(-2x^2) = (5)(-2)(xxx)(xx)$$
$$= -10x^5$$

SAMPLE PROBLEM 2

Multiply $(3xy^2)(8x^3y^4)$.

SOLUTION

$$(3xy^2)(8x^3y^4) = (3)(8)(x)(x^3)(y^2)(y^4)$$
$$= 24(xxxx)(yy)(yyyy)$$
$$= 24x^4y^6$$

Note that the definition of x^n, where n is a natural number, is used in the multiplication of monomials.

Definition of x^n. If n is a natural number, x^n is defined as

$$x^1 = x \quad \text{and} \quad x^n = x \cdot x \ldots x \ (n \text{ factors})$$

As examples,

$$x^4 = xxxx \quad \text{and} \quad y^5 = yyyyy$$

EXERCISES 7.1

Multiply.

A.

1. $(4x)(5x^3) =$
2. $(-6y^2)(7y^4) =$
3. $(-5y^3)(-8y^2) =$
4. $(-25x^4)(-4x^3) =$
5. $(9x^5)(-4x) =$
6. $(7x^2)(7x^2) =$
7. $(8x^3)(8x^3) =$
8. $(-6y^3)(-6y^3) =$
9. $(-9y^2)(-9y^2) =$
10. $(x^2y)(xy^2) =$
11. $(2xy)(-3x^2y)(4xy^2) =$
12. $(2x)(2x)(2x) =$
13. $(-3x^2)(-3x^2)(-3x^2) =$
14. $-(5xy)^2 =$
15. $(-5xy)^2 =$
16. $(-4x^2)^3 =$
17. $-4(x^2)^3 =$
18. $3x(5x^2)^2 =$
19. $10x^2(2x)^3 =$
20. $-7x(-2x^2)^3 =$

B.

1. $(3x^2)(7x^4) =$
2. $(8y^5)(-2y) =$
3. $(-6t^3)(-5t^5) =$
4. $(-4xy^2)(4x^2y) =$
5. $(9x^2)(9x^2) =$
6. $(10x^2)^2 =$
7. $(-5y^3)(-5y^3) =$
8. $(-12y^3)^2 =$
9. $(5rs)(-2r^2s)(-3rs^2) =$
10. $(-4t^2)(-4t^2)(-4t^2) =$
11. $(-6t^2)^3 =$
12. $-2xy(-2xy)^2 =$
13. $(-3x^4)^2 =$
14. $-3(x^4)^2 =$
15. $2x^2(3x^3)(4x^4) =$
16. $2x^3y^2(5xy^4) =$
17. $-3xy^3(7x^2y^2) =$
18. $6x^5y(-8x^3y^6) =$
19. $(-4x^2y^2)(-6x^4y^6) =$
20. $(9xy^2z^3)(8x^8y^9z) =$

7.2 POLYNOMIALS

OBJECTIVE: To rewrite a given polynomial in descending powers of a variable.

Definitions. A **polynomial** is a monomial or an algebraic sum of monomials.

As examples,

$5x^2 - 2x + 3$ is a polynomial in the variable x.
$2x + 3y - 6$ is a polynomial in two variables: x and y.
$x^2 + 2y^2 - z^2$ is a polynomial in three variables: x, y, and z.

A **binomial** is a polynomial having two terms.
As examples,

$3x + 4$, $2x - 5y$, $x^2 - 9$, and $x^3 - y^3$ are binomials.

A **trinomial** is a polynomial having three terms.
As examples,

$x^2 - 7x + 10$, $2x + 5y - 6$, and $4x^2 + 12xy + 9y^2$ are trinomials.

A polynomial is written in **descending powers** of a variable when the highest power of the variable is written first, at the left, the second highest power is written next, and so on, with the lowest power written last, at the right.

A polynomial is written in **ascending powers** of a variable when it is written in the reverse of the order for descending powers.

As examples,

$5x^3 + 7x^2 - 6x + 3$ is written in descending powers of x.
$8 - 9y - y^5$ is written in ascending powers of y.
$4x^2y + 9xy^3 + y^2$ is written in descending powers of x.

SAMPLE PROBLEM 1

Rewrite $7 - 2x^2 - 5x + 4x^3$ in descending powers of x.

SOLUTION

$4x^3 - 2x^2 - 5x + 7$

SAMPLE PROBLEM 2

Rewrite $35 - 9y^2 + 6y$ in descending powers of y.

SOLUTION

$-9y^2 + 6y + 35$

SAMPLE PROBLEM 3

Rewrite $x^2y^4 + x^4y^3 - 9y^6 - 4xy^5$ in descending powers of x.

SOLUTION

$x^4y^3 + x^2y^4 - 4xy^5 - 9y^6$

SAMPLE PROBLEM 4

Rewrite $-x^2 + 5x + 6$ in ascending powers of x.

SOLUTION

$6 + 5x - x^2$

EXERCISES 7.2

Rewrite each of the following in descending powers of x.

A.

1. $8 - 5x + 3x^2$
2. $x + x^3 - 1 - x^2$
3. $14x^2 - 49 - x^4$
4. $4x + 6x^2 + 7x^3 + 5x^4$
5. $x^2 - y^2 + 2x - 4y - 3$
6. $xy^2 - 8x^3y + y^4$
7. $9y^2 - 4x^2 + 8x - 6y - 10$
8. $1 - x - y - x^3 - y^3$
9. $5 - x^2 + 36y^2 + 6x - 72y$
10. $2x^3y - 7xy^3 + 4x^2y^2$

B.

1. $5 - 2x + 3x^2$
2. $1 + x^3 - 6x$
3. $10x^2 - 25 - x^4$
4. $4x - 2x^3 + x^5 - 10$
5. $x^4 - x - 4x^3 + 4$
6. $y^2 + 6y + x^2 + 4xy$
7. $5x - 2 - x^3$
8. $x^2y^2 + y^4 + x^4$
9. $y^3 - 3xy^2 + 3x^2y - x^3$
10. $y^2 - 3x^2 - 2xy$

7.3 SUM OF POLYNOMIALS

OBJECTIVE: To simplify an indicated sum of polynomials.

A sum of two or more polynomials can often be simplified by first writing each polynomial in descending (or ascending) powers of one variable and then rearranging the terms so that like terms can be combined.

SAMPLE PROBLEM 1

Simplify $(3x^2 - 4 - 6x) + (8x - 7 + 2x^2)$.

SOLUTION

$$(3x^2 - 6x - 4) + (2x^2 + 8x - 7) = (3x^2 + 2x^2) + (-6x + 8x) + (-4 - 7)$$
$$= 5x^2 + 2x - 11$$

SAMPLE PROBLEM 2

Simplify $(6x - 8y + 10) + (8y - 4x - 12)$.

SOLUTION

$$(6x - 8y + 10) + (8y - 4x - 12) = (6x - 4x) + (-8y + 8y) + (10 - 12)$$
$$= 2x + 0 + (-2) = 2x - 2$$

SAMPLE PROBLEM 3

Simplify $(x^3 - 2xy^2 - x^2y) + (x^2y + y^3 - 3xy^2)$.

SOLUTION

$$(x^3 - 2xy^2 - x^2y) + (x^2y + y^3 - 3xy^2) = (x^3 - x^2y - 2xy^2) + (x^2y - 3xy^2 + y^3)$$
$$= x^3 + (-x^2y + x^2y) + (-2xy^2 - 3xy^2) + y^3$$
$$= x^3 + 0 + (-5xy^2) + y^3$$
$$= x^3 - 5xy^2 + y^3$$

Simplify.

EXERCISES 7.3

A.

1. $(5x^2 - 7x + 9) + (3x^2 - 2x - 8)$

2. $(4x - 5y + 8) + (3x + 2y - 5)$

3. $(8 - x^3 - x) + (x^3 + 1 - 4x)$

$(-x^3 - x + 8) + (x^3 - 4x + 1)$

$(-x^3 + x^3) + (4x + 1)$

4. $(x + y - 6) + (y - x - 6)$

5. $(4x - 12y - 8) + (6x + 12y + 6)$

6. $(x^3 - x^2 + x) + (x^2 - x + 1)$

7. $(5x - 4y + 3) + (5y - 5x - 10)$

$= y - 7$

8. $(y + 3z - 4x) + (5z - 6y - 4x)$

9. $(y^4 + y - 3y^2 - 3) + (2y^2 + 7 - y)$

10. $(x^3 - 2x^2y + xy^2) + (2xy^2 - y^3 - x^2y)$

B.

1. $(4x^2 + 5x - 6) + (2x^2 - 4x - 3)$

2. $(1 - 7y - 8y^2) + (5y + 3y^2 - 1)$

3. $(x^3 + 2x^2 + x) + (x^2 + 2x + 1)$

4. $(2x - 3y - 5) + (3x + 3y - 10)$

5. $(4y - 8 - 2y^2) + (y^3 - 4y + 2y^2)$

6. $(y - x + 4) + (x + y - 7)$

7. $(2x - y + 3z - 5) + (2y - 3x - 3z - 4)$

8. $(x^2 + xy + y^2) + (y^2 - yx - 2x^2)$

9. $(x^3 - 3x^2 y + 9xy^2) + (27y^3 - 9xy^2 + 3x^2 y)$

10. $(x^4 + 2x^3 + 2x^2) + (-2x^3 - 4x^2 - 4x) + (2x^2 + 4x + 4)$

7.4 DIFFERENCE OF POLYNOMIALS

OBJECTIVE: To simplify an indicated difference of polynomials.

Since subtraction is defined in terms of addition, that is,

$$a - b = a + (-b),$$

the difference of two polynomials can also be expressed as an addition of two polynomials.

Since $-a = (-1)a$ for all integers, and since a term such as $5x^2y^3$ is an integer if x and y are integers, this idea can be used to find the additive inverse of a polynomial.

As examples,

$$\begin{aligned} -(x^2 + 4y^2) &= (-1)(x^2 + 4y^2) \\ &= (-1)x^2 + (-1)4y^2 \\ &= -x^2 - 4y^2 \end{aligned}$$

$$\begin{aligned} -(2x^2 - 3x + 6) &= (-1)(2x^2 - 3x + 6) \\ &= (-1)(2x^2) + (-1)(-3x) + (-1)(6) \\ &= -2x^2 + 3x - 6 \end{aligned}$$

Informally, subtracting is done by changing the sign of the subtrahend and adding. It is important to remember, however, that when the subtrahend is a polynomial, each term of the subtrahend must have its sign changed.

SAMPLE PROBLEM 1

Simplify $(8x^2 + 5x - 10) - (2x^2 + 7x - 5)$.

SOLUTION

$$\begin{aligned} (8x^2 + 5x - 10) - (2x^2 + 7x - 5) &= (8x^2 + 5x - 10) + (-1)(2x^2 + 7x - 5) \\ &= (8x^2 + 5x - 10) + (-2x^2 - 7x + 5) \\ &= (8x^2 - 2x^2) + (5x - 7x) + (-10 + 5) \\ &= 6x^2 - 2x - 5 \end{aligned}$$

SAMPLE PROBLEM 2

Simplify $(5x - 2y - 7) - (3y + 5x - 8)$.

SOLUTION

$$\begin{aligned} (5x - 2y - 7) - (3y + 5x - 8) &= (5x - 2y - 7) + (-1)(3y + 5x - 8) \\ &= (5x - 2y - 7) + (-3y - 5x + 8) \\ &= (5x - 5x) + (-2y - 3y) + (-7 + 8) \\ &= 0 - 5y + 1 \\ &= -5y + 1 \end{aligned}$$

SAMPLE PROBLEM 3

Simplify $(x^3 - 10x^2y + 25xy^2) - (5x^2y - 50xy^2 + 125y^3)$.

SOLUTION

$$(x^3 - 10x^2y + 25xy^2) + (-5x^2y + 50xy^2 - 125y^3)$$
$$= x^3 + (-10x^2y - 5x^2y) + (25xy^2 + 50xy^2) + (-125y^3)$$
$$= x^3 - 15x^2y + 75xy^2 - 125y^3$$

EXERCISES 7.4

Simplify.

A.

1. $(9x^2 + 6x + 10) - (4x^2 + 2x + 7)$ 1. ______

$9x^2+6x+10-4x^2-2x-7$

$(9x^2-4x^2)(6x-2x)+(10-7)$

$5x^2$ 4 3

2. $(8x^2 - 5x + 4) - (7x^2 - 2x + 8)$ 2. ______

3. $(3x - 6y - 1) - (4x + 3y - 6)$ 3. ______

4. $(2y - 5x + 7) - (5x - 2y - 8)$ 4. ______

$2y-5x+7-5x+2y+8$

$10x+4y+15$

5. $(x^4 - 12x^2 + 36) - (16x^2 - 8x + 1)$ 5. ______

6. $25 - (x^2 + 14x + 49)$ 6. ______

7. $(x + y - z) - (x - y + z)$ 7. ______

8. $(2x^2y - 3x^2y^2 - 6xy^2) - (4xy^2 - 3x^2y - x^2y^2)$ 8. ______

9. $(5x^2 - 6y^2 + 7) - (5x^2 + 6y^2 - 7)$ 9. ______________

10. $(x^2 - xy + x) - (yx - y^2 + y)$ 10. ______________

B.

1. $(4x + 2y - 5z - 10) - (x + 5y - 5z + 15)$ 1. ______________

2. $36x^2 - (y^2 - 10xy + 25x^2)$ 2. ______________

3. $(x^3 + 7x^2 + 49x) - (7x^2 + 49x - 343)$ 3. ______________

4. $(5x^2 - 3xy - 4y^2) - (4x^2 - 2xy + 3y^2)$ 4. ______________

5. $(x^2 + 2xy - y^2) - (y^2 - 2yx + x^2)$ 5. ______________

6. $(x^2y^2 - 18xy + 81) - (36y^2 - 18yx + x^2)$ 6. ______________

7. $(1 - xy - xz) - (yz - tx)$ 7. ______________

8. $(5 - x - y) - (x + y - 5)$ 8. ______________

9. $(x^3 - 6x^2 + 9x) - (3x^2 - 18x + 27)$ 9. ______________

10. $1 - (8 - x - y + z)$ 10. ______________

$1+8+x+y-z$

$8+x+y-z$

$x+y=z+7$

7.5 PRODUCT OF POLYNOMIAL AND MONOMIAL

OBJECTIVE: To simplify an indicated product of a polynomial and a monomial.

The distributive axiom is used to multiply a polynomial by a monomial.

Each term of the polynomial is multiplied by the monomial.

SAMPLE PROBLEM 1

Multiply $5x^2\ (2x^3 - 3x^2 - 4x + 1)$.

SOLUTION

$$5x^2\ (2x^3 - 3x^2 - 4x + 1) = (5x^2)(2x^3) + (5x^2)(-3x^2) + (5x^2)(-4x) + (5x^2)(1)$$
$$= 10x^5 - 15x^4 - 20x^3 + 5x^2$$

SAMPLE PROBLEM 2

Multiply $-4by\ (7y^2 - 8y + 1)$.

SOLUTION

$$-4by\ (7y^2 - 8y + 1) = (-4by)(7y^2) + (-4by)(-8y) + (-4by)(1)$$
$$= -28by^3 + 32by^2 - 4by$$

SAMPLE PROBLEM 3

Multiply $6xy\ (x^2 - 3xy + 5y^2)$.

SOLUTION

$$6xy\ (x^2 - 3xy + 5y^2) = (6xy)(x^2) + (6xy)(-3xy) + (6xy)(5y^2)$$
$$= 6x^3y - 18x^2y^2 + 30xy^3$$

EXERCISES 7.5

Multiply.

A.

1. $2x\ (7x^2 - 3x - 4)$ $14x^3 - 6x^2 - 8x$
2. $-4x\ (5x - 2y + 7)$ $-20x^2 + 8xy - 28x$

3. $5x^2 (x^2 - 2xy + y^2)$

4. $6y^2 (2y^2 - 3xy - x^2)$

5. $6xy (x^2 - 4xy + 3y^2)$

6. $7a (a^2 - 5ab + 1)$

7. $-9b (b^3 - b^2 - 6)$

8. $8x^2 y^2 (4x - 5y - 6)$

9. $-15a^2 b^2 (4a^2 - 4a + 1)$

10. $100x^2 (0.05x^2 + 1.50x - 4.25)$

B.

1. $-12rs (4r^2 - 4rs + 1)$

2. $-15st (t^2 - 5t + 4)$

3. $-r (r^3 - r^2 + r - 1)$

4. $-c^2 d (c^2 + c - d^2 - d)$

5. $4x^3 (2x^3 + x^2 - 2x - 1)$

6. $8y^3 (1 - y - y^2 - y^3)$

7. $8y^2 (1.25y^2 - 0.75y + 0.25)$

8. $-xy (1 - x + y - xy)$

9. $-7st (s + t - s^2 - t^2)$

10. $5ab^2 c^3 (a + 2b - 2c + 5)$

7.6 PRODUCT OF TWO POLYNOMIALS

OBJECTIVE: To simplify an indicated product of two polynomials.

The product of two polynomials can be written as a single polynomial by applying the distributive axiom. Each term of one polynomial must be multiplied by each term of the other polynomial.

SAMPLE PROBLEM 1

Multiply $(2x + 3)(5x - 2)$.

SOLUTION

Horizontal format:

$$\begin{aligned}(2x + 3)(5x - 2) &= 2x(5x - 2) + 3(5x - 2)\\ &= 10x^2 - 4x + 15x - 6\\ &= 10x^2 + 11x - 6\end{aligned}$$

SAMPLE PROBLEM 2

Multiply $(2x + y)(3x^2 - 2xy - 4y^2)$.

SOLUTION

Vertical format (a vertical format is often used if one polynomial has three or more terms):

$$\begin{array}{rrrr} 3x^2 & - 2xy & - 4y^2 & \\ 2x & + y & & \\ \hline 6x^3 & - 4x^2y & - 8xy^2 & \\ & 3x^2y & - 2xy^2 & - 4y^3 \\ \hline 6x^3 & - x^2y & - 10xy^2 & - 4y^3 \end{array}$$

SAMPLE PROBLEM 3

Multiply $(5x - 4)(25x^2 + 20x + 16)$.

SOLUTION

Vertical format:

$$\begin{array}{rrrr} 25x^2 & + 20x & + 16 & \\ 5x & - 4 & & \\ \hline 125x^3 & + 100x^2 & + 80x & \\ & - 100x^2 & - 80x & - 64 \\ \hline 125x^3 & & & - 64 \end{array}$$

SAMPLE PROBLEM 4

Multiply $(3x^3 + x - 5)(x + 2)$.

SOLUTION

Horizontal format:

$$\begin{aligned}(x+2)(3x^3+x-5) &= x(3x^3+x-5)+2(3x^3+x-5)\\ &= 3x^4+x^2-5x+6x^3+2x-10\\ &= 3x^4+6x^3+x^2-3x-10\end{aligned}$$

SAMPLE PROBLEM 5

Multiply $(x + y - 5)(x - y + 5)$.

SOLUTION

Vertical format:

$$\begin{array}{rrrrr} x & +\,y & -\,5 & & \\ x & -\,y & +\,5 & & \\ \hline x^2 & +\,xy & -\,5x & & \\ & -\,xy & & -y^2 & +\,5y \\ & & +\,5x & & +\,5y-25 \\ \hline x^2 & +\,0 & +\,0 & -y^2 & +\,10y-25 \end{array}$$

The product is $x^2 - y^2 + 10y - 25$.

A. **EXERCISES 7.6**

(1–15) Multiply, using a horizontal format.

1. $(x + 4)(x + 2)$ $x(x+2)+4(x+2)$
   $x^2+2x+4x+8$ x^2+6x+8

2. $(x - 5)(x - 1)$

3. $(y + 5)(y - 6)$

4. $(y - 3)(y + 7)$

5. $(2x - 5)(x - 4)$

6. $(3x + 2)(4x - 5)$

7. $(x + 6)(x - 6)$

8. $(2x - 7)(2x + 7)$

9. $(5x + 3y)(5x - 3y)$

10. $(4x + y)(4x + y)$

11. $(6x - 5y)(6x - 5y)$

12. $(x + 2)(y + 5)$

13. $(4x - 6)(5y + 1)$

14. $(x^2 + 1)(x - 1)$

15. $(x^2 - 4)(5x - 2)$

(16–25) Multiply, using a vertical format.

16. $(x + 3)(x^2 + 4x + 5)$

17. $(x - y)(x^2 - xy - 2y^2)$

18. $(x - 6)(x^2 + 6x + 36)$

19. $(y + 2)(y^2 + 4y + 4)$

20. $(x - 1)(x^3 + x^2 + x + 1)$

21. $(x + y + z)(x + y + z)$

22. $(x^2 + 3x + 2)(x^2 - 3x + 2)$

23. $(y^2 - 4y - 5)(y^2 + 4y - 5)$

24. $(y^2 - 1)(y^2 + y - 2)$

25. $(2x + y - 3z)^2$

B.

(1–15) Multiply, using a horizontal format.

1. $(3x - 4)(x + 2)$

2. $(3x + 4)(x - 2)$

3. $(7x - 2y)(7x + 2y)$

4. $(7x - 2y)(7x - 2y)$

5. $(5x + 2)(5x - 2)$

6. $(x^2 + y^2)(x - y)$

7. $(y + 2x)(x - 2y)$

8. $(4y - 5x)(5x - 4y)$

9. $(x + 2y)(2y + x)$

10. $(7x - 6y)(7x - 6y)$

11. $(a^2 - b^2)(a^2 + b^2)$

12. $(x^2 + 4)(x^2 - 4)$

13. $(x^3 + 1)(x^3 - 1)$

14. $(8x^3 - 27)(8x^3 + 27)$

15. $(4x - 7y)(x + 6)$

(16–25) Multiply, using a vertical format.

16. $(y - 2)(2y^2 - 3y + 1)$

17. $(x + 4y)(y^2 - 5xy - x^2)$

18. $(x + 5)(x^2 - 5x + 25)$

19. $(y - 4)(y^2 - 8y + 16)$

20. $(r + 2s - 5)(r + 2s + 5)$

21. $(x^2 - 2x + 2)(x^2 + 2x + 2)$

22. $(y - 5)(4y^3 + 2y - 1)$

23. $(x^2 + 2)(x^2 - 3y - 1)$

24. $(x - y - z)(x - y - z)$

25. $(3a - 5b + c)^2$

7.7 PRODUCT OF BINOMIALS, FOIL METHOD

OBJECTIVE: To write the product of two binomials by using the FOIL method.

There is a special method, called the FOIL method, that is very useful for finding the product of two binomials.

Multiplying, by using the distributive axiom, gives

$$\begin{aligned}(A + B)(C + D) &= A(C + D) + B(C + D)\\ &= AC + AD + BC + BD\end{aligned}$$

Note the following about the four terms of the product polynomial:

(1) The first term, AC, is the product of the first terms of the factors.

(2) The last term, BD, is the product of the last terms of the factors.

(3) The two middle terms form the sum of the inner product and the outer product as shown below.

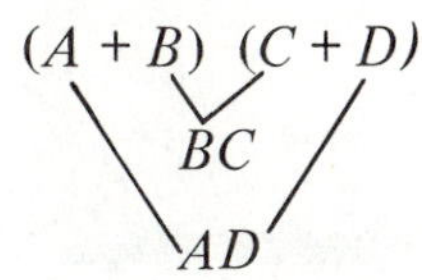

Inner product

Outer product

The word FOIL is a useful device for remembering the terms of this special product: F means first terms product, O means outer product, I means inner product, and L means last terms products.

First term	Last term	First term	Last term	First terms product	Outer product	Inner product	Last terms product
↓	↓	↓	↓	↓	↓	↓	↓

$$(A + B) \quad (C + D) = AC + AD + BC + BD$$

F O I L

In sample problems 1–4 find each indicated product by using the FOIL method.

SAMPLE PROBLEM 1

$(4x + 5)(2x - 3)$

SOLUTION

F O I L

$$(4x + 5)(2x - 3) = 4(2)x^2 + 4(-3)x + 5(2)x + 5(-3)$$
$$= 8x^2 - 12x + 10x - 15$$
$$= 8x^2 - 2x - 15$$

SAMPLE PROBLEM 2

$(3x - 8y)(2x - 5y)$

SOLUTION

F O I L

$$(3x - 8y)(2x - 5y) = (3x)(2x) + (3x)(-5y) + (-8y)(2x) + (-8y)(-5y)$$
$$= 6x^2 - 15xy - 16xy + 40y^2$$
$$= 6x^2 - 31xy + 40y^2$$

SAMPLE PROBLEM 3

$(r - 2t)(s + 4t)$.

SOLUTION

$$\begin{aligned} & \quad\quad\quad\quad\;\; \text{F} \quad\quad\;\; \text{O} \quad\quad\quad\;\; \text{I} \quad\quad\quad\;\; \text{L} \\ (r - 2t)(s + 4t) &= (r)(s) + (r)(4t) + (-2t)(s) + (-2t)(4t) \\ &= rs + 4rt - 2st - 8t^2 \end{aligned}$$

SAMPLE PROBLEM 4

$(y^3 - 8)(y - 5)$.

SOLUTION

$$\begin{aligned} & \quad\quad\quad\; \text{F} \quad\quad\quad\;\; \text{O} \quad\quad\quad\; \text{I} \quad\quad\;\; \text{L} \\ (y^3 - 8)(y - 5) &= y^3(y) + y^3(-5) + (-8)y + (-8)(-5) \\ &= y^4 - 5y^3 - 8y + 40 \end{aligned}$$

EXERCISES 7.7

Find each product by using the FOIL method.

A.

1. $(2x + 5)(3x + 4)$

   $6x^2 + 8x + 15x + 20 \quad 6x^2 + 23x + 20$

2. $(4x - 3)(5x - 1)$

   $20x^2 - 4x - 15x + 3$

   $20x^2 - 19x + 3$

3. $(6y + 7)(2y - 5)$

   $12y^2 - 30y + 14y + 35 \quad 12y^2 - 16y - 35$

   $-16y$

4. $(6y - 7)(2y + 5)$

   $12y^2 + 30y - 14y + 35 \quad 12y^2 + 16y - 35$

5. $(6y - 7)(2y - 5)$

   $12y^2 + 30y - 14y - 35$

6. $(6y + 7)(2y + 5)$

7. $(5x - 2y)(3x + 4y)$

8. $(5x + 4y)(3x - 2y)$

9. $(5x + 2y)(3x - 4y)$

10. $(5x - 4y)(3x + 2y)$

11. $(4x^2 - 1)(x^2 + 1)$

12. $(9y^2 - 4)(4y^2 - 1)$

13. $(x + 3)(y + 2)$

14. $(x - 4)(y - 5)$

15. $(x + 7)(y - 1)$

16. $(x - 6)(y + 4)$

17. $(x^2 - 1)(x - 1)$

18. $(y^2 + 5)(y - 6)$

19. $(x^3 + 1)(x - 5)$

20. $(y^3 - 2)(y^2 + 4)$

B.

1. $(5x + 6)(2x - 5)$

2. $(6x - 7)(3x - 4)$

3. $(2x - 1)(x + 2)$

4. $(8x + y)(x - 3y)$

5. $(12x - 5y)(3x - 4y)$

6. $(9r^2 - 1)(4r^2 - 1)$

7. $(4x^2 - y^2)(x^2 - 9y^2)$

8. $(7 - 2x^2)(5 - 2x^2)$

9. $(3rs + 5)(3rs - 7)$

10. $(4xy - 3)(4xy - 5)$

11. $(x - 7)(y + 6)$

12. $(x - 5)(y - 8)$

13. $(3r + 2s)(r - 5t)$

14. $(a + b)(c + d)$

15. $(x^2 - 4)(x + 6)$

16. $(x - 8)(x^2 - 9)$

17. $(y^2 + 1)(y - 7)$

18. $(y^2 - 1)(y^3 - 1)$

19. $(x^3 + 27)(x^2 - 3)$

20. $(xy + 5)(5x - 2y)$

7.8 SQUARE OF BINOMIAL

OBJECTIVE: To write the square of a binomial as a trinomial by using the perfect square trinomial formula.

The square of a binomial is a polynomial having three terms, called a **perfect square trinomial.**

Using the distributive axiom,

$$\begin{aligned}(A + B)^2 &= (A + B)\ (A + B)\\ &= (A + B)A + (A + B)B\\ &= A^2 + BA + AB + B^2\\ &= A^2 + 2AB + B^2\end{aligned}$$

Therefore, the **perfect square trinomial formula** is

$$(A + B)^2 = A^2 + 2AB + B^2$$

This formula shows that the square of a binomial is obtained by squaring the first term, doubling the product of the two terms, and squaring the last term, and then adding these three products.

SAMPLE PROBLEM 1

Use the formula to obtain the product $(6x + 7)^2$.

SOLUTION

Use $A = 6x$ and $B = 7$.

$$\begin{aligned}(A + B)^2 &= A^2 + 2AB + B^2\\ (6x + 7)^2 &= (6x)^2 + 2(6x)(7) + 7^2\\ &= 36x^2 + 84x + 49\end{aligned}$$

SAMPLE PROBLEM 2

Use the formula to obtain the product $(3x - 5)^2$.

SOLUTION

Use $A = 3x$ and $B = -5$.

$$(A + B)^2 = A^2 + 2AB + B^2$$

$$(3x - 5)^2 = (3x)^2 + 2(3x)(-5) + (-5)^2$$

$$= 9x^2 - 30x + 25$$

SAMPLE PROBLEM 3

Find the product $(4x^2 - 9y^3)^2$ by using the formula.

SOLUTION

Use $A = 4x^2$ and $B = -9y^3$.

$$(A + \quad B)^2 = A^2 \quad + 2AB \quad + B^2$$

$$(4x^2 + (-9y^3))^2 = (4x^2)^2 + 2(4x^2)(-9y^3) + (-9y^3)^2$$

$$= 16x^4 \quad - 72x^2y^3 \quad + 81y^6$$

EXERCISES 7.8

Use the perfect square trinomial formula to obtain each product.

A.

1. $(x + 4)^2$
2. $(x - 3)^2$
3. $(2x - 5)^2$
4. $(5 - 2x)^2$
5. $(5x - 2)^2$
6. $(x - 7y)^2$
7. $(-y + x)^2$
8. $(x^2 + 6)^2$
9. $(4y^3 + 1)^2$
10. $(xy + 8)^2$

B.

1. $(x + 5)^2$
2. $(y - 6)^2$
3. $(10x - 3y)^2$
4. $(xy + 4)^2$
5. $(8x^2 - 1)^2$
6. $(-5x - y)^2$
7. $(9a - 4b)^2$
8. $(4b - 9a)^2$
9. $(-4b - 9a)^2$
10. $(4a - 9b)^2$

7.9 PRODUCT OF SUM AND DIFFERENCE

OBJECTIVE: To write the product of a sum and difference of two terms by using the difference of squares formula.

The product of the sum and the difference of the same two terms is a special product, called the **difference of squares.**

Multiply, using the distributive axiom:

$$(A + B)(A - B) = A(A - B) + B(A - B)$$
$$= A^2 - AB + AB - B^2$$
$$= A^2 - B^2$$

This special product can be obtained rapidly by substituting in the general formula called the **difference of squares formula.**

$$(A + B)(A - B) = A^2 - B^2$$

Expressed in words, when the sum and difference of the same two terms are multiplied, the product is the square of the first term minus the square of the second term.

SAMPLE PROBLEM 1

Use the formula to obtain the product of $(7x + 2)(7x - 2)$.

SOLUTION

Use $A = 7x$ and $B = 2$.

$$(A + B)(A - B) = A^2 - B^2$$
$$(7x + 2)(7x - 2) = (7x)^2 - (2)^2$$
$$= 49x^2 - 4$$

SAMPLE PROBLEM 2

Use the formula to obtain the product of $(4x^2 - 5y)(4x^2 + 5y)$.

SOLUTION

Note that $(4x^2 - 5y)(4x^2 + 5y) = (4x^2 + 5y)(4x^2 - 5y)$ by the commutative axiom for multiplication.

Use $A = 4x^2$ and $B = 5y$.

$$(A + B)(A - B) = A^2 - B^2$$
$$(4x^2 + 5y)(4x^2 - 5y) = (4x^2)^2 - (5y)^2$$
$$= 16x^4 - 25y^2$$

EXERCISES 7.9

Use the formula $(A + B)(A - B) = A^2 - B^2$ to obtain each product.

A.

1. $(x + 3)(x - 3)$
2. $(x + 5)(x - 5)$
3. $(y - 2)(y + 2)$
4. $(y - 4)(y + 4)$
5. $(5x + 1)(5x - 1)$
6. $(4x + 7)(4x - 7)$
7. $(x + y)(x - y)$
8. $(r + s)(s - r)$

9. $(6r - s)(s + 6r)$

10. $(x^2 + 6)(x^2 - 6)$

11. $(x^2 - 8y^2)(x^2 + 8y^2)$

12. $(3y^2 - 10)(3y^2 + 10)$

13. $(a^2 + b^2)(b^2 - a^2)$

14. $(7a - 2b)(2b + 7a)$

15. $(1 + xy)(1 - xy)$

B.

1. $(x + 6y)(x - 6y)$

2. $(2a - 7b)(2a + 7b)$

3. $(x^2 + 4)(4 - x^2)$

4. $(rs - 1)(rs + 1)$

5. $(xy - z)(z + xy)$

6. $(3x^3 - 8)(3x^3 + 8)$

7. $(9r^2 - s^2)(9r^2 + s^2)$

8. $(x + y)(y - x)$

9. $(ax + by)(by - ax)$

10. $(x^2 - y^2)(-y^2 - x^2)$

7.10 PRODUCT OF RELATED BINOMIALS, SIMPLE TRINOMIAL

OBJECTIVE: To write the product of two binomials by using the simple trinomial product formula.

Multiplying $(X + A)(X + B)$, using the distributive axiom, yields the simple trinomial product formula.

$$(X + A)(X + B) = (X + A)(X) + (X + A)(B)$$
$$= X^2 + AX + BX + AB$$
$$= X^2 + (A + B)X + AB$$

Therefore, the **simple trinomial product formula** is

$$(X + A)(X + B) = X^2 + (A + B)X + AB$$

Note that $(A + B)$, the coefficient of X in the trinomial, is the *sum* of the constants A and B in the related binomial factors and that AB, the constant term of the trinomial, is the *product* of the constant terms of the binomial factors.

In sample problems 1–4 find the product by using the simple trinomial product formula.

SAMPLE PROBLEM 1

$(x + 5)(x + 4)$

SOLUTION

$$(X + A)(X + B) = X^2 + (A + B)X + AB$$
$$(x + 5)(x + 4) = x^2 + (5 + 4)x + (5)(4)$$
$$= x^2 + 9x + 20$$

For $A = 5$ and $B = 4$, note that 9, the coefficient of x in the solution, is the sum of 5 and 4. Similarly, the constant 20 in the solution is the product of 5 and 4.

SAMPLE PROBLEM 2

$(x + 5)(x - 6)$

SOLUTION

$$(X + A)(X + B) = X^2 + (A + B)X + AB$$
$$(x + 5)(x - 6) = x^2 + (5 + (-6))x + 5(-6)$$
$$= x^2 + (-1)x + (-30)$$
$$= x^2 - x - 30$$

SAMPLE PROBLEM 3

$(x^2 - 9)(x^2 + 16)$

SOLUTION

Use $X = x^2$, $A = -9$, and $B = 16$.

$$X^2 + (A + B)X + AB = (x^2)^2 + (-9 + 16)x^2 + (-9)(16)$$
$$= x^4 + 7x^2 - 144$$

SAMPLE PROBLEM 4

$(r - 5s)(r - 7s)$

SOLUTION

Use $X = r$, $A = -5s$, and $B = -7s$.

$$X^2 + (A + B)X + AB = r^2 + (-5s - 7s)r + (-5s)(-7s)$$
$$= r^2 - 12rs + 35s^2$$

EXERCISES 7.10

Obtain each product by using the simple trinomial product formula.

$$(X + A)(X + B) = X^2 + (A + B)X + AB$$

A.

1. $(x + 5)(x + 2)$
2. $(x + 4)(x + 7)$
3. $(x - 6)(x - 3)$
4. $(x - 4)(x - 2)$
5. $(x + 7)(x - 6)$
6. $(x + 8)(x - 9)$
7. $(x - 5)(x + 1)$
8. $(x - 2)(x + 8)$

9. $(y + 2)(y + 3)$

10. $(y + 2)(y - 3)$

11. $(x - 2y)(x - 3y)$

12. $(x - 2y)(x + 3y)$

13. $(xy + 1)(xy - 6)$

14. $(xy - 6)(xy + 4)$

15. $(x^2 - 8)(x^2 - 2)$

16. $(x^2 + 9y^2)(x^2 - 4y^2)$

17. $(t^2 - 5)(t^2 + 6)$

18. $(t^2 - 7)(t^2 - 4)$

19. $(y^2 + 5)(y^2 - 4)$

20. $(y^2 - 9)(y^2 + 1)$

B.

1. $(x + 8)(x + 6)$

2. $(x - 8)(x - 6)$

3. $(x + 8)(x - 6)$

4. $(x - 8)(x + 6)$

5. $(x - 2y)(x + 4y)$

6. $(2x + 3y)(2x - 5y)$

7. $(x^2 - 4)(x^2 + 1)$

8. $(x^2 + 16y^2)(x^2 - y^2)$

9. $(r^2 - 9s^2)(r^2 - 4s^2)$

10. $(5 - 2x)(5 - 2y)$

11. $(xy + 10z)(xy - 5z)$

12. $(u^3 - v^3)(u^3 + 2v^3)$

13. $(x^2 - 6)(x^2 + 9)$

14. $(abc + 1)(abc - 2)$

15. $(rx - 3s)(rx - 4s)$

UNIT 7 POSTTEST

(1–2) Simplify. (7.1)

1. $(9x^4)(-7x^3)$

2. $(-25xy^2)(-4x^3y^4)$

(3–4) Rewrite in descending powers of x. (7.2)

3. $18x - x^2 - 81 + x^4$

4. $5x^2y^3 - 2x^3y^2 - 10xy^4$

(5–6) Simplify. (7.3)

5. $(17 - x^2 - 6x) + (2x^2 - 20 + 6x)$

6. $(8x - 5y - 9) + (5y - 3x + 2)$

(7–8) Simplify. (7.4)

7. $(x^4 - 16x^2 + 64) - (9x^2 - 12x + 4)$

8. $(5x^2 + 4y^2 - 7xy) - (3y^2 - 4x^2 - 7xy)$

(9–10) Simplify. (7.5)

9. $-2x^2(7x^3 - x + 3)$

10. $5x^3y^2(2x^2 + 4y^2 - 15xy)$

(11–15) Multiply. (7.6)

11. $(4x - 3)(5x + 2)$

12. $(3x - 7)(6y - 1)$

13. $(9x - 8y)(9x + 8y)$

14. $(x - 4)(x^2 + 4x - 4)$

15. $(6x + 5y)(2x^2 - 3xy - y^2)$

(16–17) Write as a trinomial by using the formula for a perfect square trinomial. (7.8)

16. $(4x + 9y)^2$

17. $(6x - 7)^2$

(18–19) Find the product by using the difference of squares formula. (7.9)

18. $(8x^2 - y^2)(8x^2 + y^2)$

19. $(6x + 5)(6x - 5)$

(20–21) Find the product by using the simple trinomial product formula. (7.10)

20. $(y - 8)(y + 9)$

21. $(x^2 + 4)(x^2 - 25)$

(22–25) Find the product by using the FOIL method. (7.7)

22. $(8x + 9)(2x - 7)$

$16x^2\ 56x + 18x + 63$

$16x^2 - 38x - 63$

23. $(3x - 4y)(6x + 5y)$

$18x^2 + 15xy - 24yx - 20y^2$

$18x^2 + 9xy - 20y^2$

24. $(x - 6)(x^2 + 6)$

$x^3 + 6x - 6x^2 - 36$

25. $(x - 4)(y - 5)$

$xy - 5x - 4y + 20$

UNIT 8

FACTORING

Unit 8 PRETEST

(1–2) Factor. (8.1)

1. $20x^3 + 15x^2$
2. $24x^2y - 12xy^3$

(3–4) Factor. (8.2)

3. $x^2 - 36$
4. $x^4 - 81y^2$

(5–8) Factor. (8.3)

5. $x^2 + 12x + 35$
6. $y^2 - 10y + 24$
7. $x^2 + 3x - 28$
8. $y^2 - 6y - 16$

(9–12) If possible, express as the square of a binomial. If not possible, write "not possible." (8.4)

9. $x^2 - 10x + 25$
10. $y^2 + 4y + 16$
11. $x^2 + 9$
12. $36x^2 + 12xy + y^2$

(13–14) Factor. (8.5)

13. $3x^2 - 19x + 6$
14. $6x^2 - 7xy - 5y^2$

(15–19) Factor completely. (8.6–8.7)

15. $4x^3 - 100x$
16. $3y^2 - 9y - 120$

17. $x^4 - 15x^2 - 16$

18. $50t^2 - 40t + 8$

19. $14y^3 + 2y^2 - 12y$

8.1 MONOMIAL FACTORS

OBJECTIVE: To express a given polynomial as a product of its greatest common monomial factor and a simpler polynomial.

(It may be helpful to review sections 2.2 and 2.3)

For certain algebraic problems, it is desirable to express a polynomial as a product of factors. This process is called **factoring**.

To discuss factoring, we need to talk about the **coefficients** of a polynomial. The numerical factors of the terms of a polynomial are called the coefficients of the polynomial.

For example, for the polynomial $5x^2 - 6x + 4$,

5 is the coefficient of x^2,

-6 is the coefficient of x, and

4 is the coefficient of 1.

The coefficient of the highest power of the variable is also called the **leading coefficient**.

The term not involving the variable is called the **constant term**.

Factors may be found by recognizing the form of a special product and recalling the special product formula.

The first step in factoring a polynomial is to find the common monomial factors, if there are any. The distributive axiom is used for this type of factoring.

Examples of the **distributive axiom**:

$$AB + AC = A\ (B + C)$$

$$AB + AC + AD = A\ (B + C + D)$$

SAMPLE PROBLEM 1

Factor $5x^4 + 20x^2$.

SOLUTION

(1) Examine the numerical coefficients, 5 and 20. The highest common factor is 5.

(2) Examine the literal factors of each term, x^4 and x^2. The highest power of x that divides each is x^2.

(3) The highest common monomial factor is $5x^2$. This is the product of 5 from step (1) and x^2 from step (2).

(4) Using $AB + AC = A(B + C)$,

$$\begin{aligned} 5x^4 + 20x^2 &= (5x^2)(x^2) + (5x^2)(4) \\ &= 5x^2(x^2 + 4) \end{aligned}$$

SAMPLE PROBLEM 2

Factor $3x^2y^3 - 18x^2y^2 + 3xy^2$.

SOLUTION

(1) The common numerical factor is 3.

(2) The common power of x is x.

(3) The common power of y is y^2.

(4) The common monomial factor is $3xy^2$, the product of 3, x, and y^2.

$$\begin{aligned} 3x^2y^3 - 18x^2y^2 + 3xy^2 &= 3xy^2(xy) - 3xy^2(6x) + 3xy^2(1) \\ &= 3xy^2(xy - 6x + 1) \end{aligned}$$

SAMPLE PROBLEM 3

Factor $-24t^5 - 36t^4 + 48t^3$.

SOLUTION

(1) The greatest common numerical factor of 24, 36, and 48 is 12.

(2) The highest power of t that divides t^5, t^4, and t^3 is t^3.

(3) The product of all common factors of the terms of the polynomial is $12t^3$.

When the leading coefficient of a polynomial is negative, it is useful to include the minus sign in the monomial factor. This makes the polynomial factor simpler, as a rule.

Therefore, $-12t^3$ is taken as the common monomial factor.

$$\begin{aligned} -24t^5 - 36t^4 + 48t^3 &= (-12t^3)(2t^2) + (-12t^3)(3t) + (-12t^3)(-4) \\ &= -12t^3(2t^2 + 3t - 4) \end{aligned}$$

EXERCISES 8.1

Factor.

A.

1. $12x^3 + 6x^2$
2. $18y^4 - 12y^3$
3. $10xy + 15x^2y^2$
4. $14xy^2 - 21x^2y$
5. $16x^3 + 8x$
6. $-18x^4 - 9x^2$
7. $16x^3y^2 - 40x^2y^3$
8. $54x^3y^3 + 9x^2y^5$
9. $-15at^4 - 75at^2$
10. $24c^2s^2 - 42c^3s$
11. $7x + 7y + 7z$
12. $4x - 4y + 4$
13. $15ax - 45ay - 15a$
14. $4a^2x^2 + 8a^3x - 4a^2$
15. $-9y^4 - 9y^3 + 27y^2$
16. $150y^3 - 200y^2 - 250y$
17. $70t^4 - 105t^3 + 140t^2$
18. $a^2b^2c + ab^2c^2 + a^2bc^2$
19. $r^4s^2t^2 - r^2s^4t^2 - r^2s^2t^4$
20. $24x^3yz + 48xy^3z - 72xyz^3$

B.

1. $40x^2y - 50xy^2$
2. $25x^4 - 100x^3$
3. $4x^6 + 64x^4$
4. $-15x^2y^2 - 15y^4$
5. $42x^3 + 6x^2$
6. $-40x^2y^2 + 32xy^3$
7. $21x - 42y + 21z$
8. $x^4y^2 - x^2y^4 - x^2y^2$

9. $48r - 84s + 24$

10. $6x^2 - 18x + 30$

11. $4x^3 + 16x^2 - 4x$

12. $10x^3y - 5x^2y^2 - 10xy^3$

13. $-24y^4 + 40y^3 - 32y^2$

14. $r^2st + rs^2t + rst^2$

15. $-3a^4b^2c^2 - 6a^2b^4c^2 + 3a^2b^2c^4$

8.2 DIFFERENCE OF SQUARES

OBJECTIVE: To express a difference of squares as a product of a sum and a difference.

A polynomial having the form $A^2 - B^2$ can be expressed as the product of a sum and a difference by using the special product for a sum and a difference.

Difference of squares = Product of sum and difference

$$A^2 - B^2 = (A + B)(A - B)$$

SAMPLE PROBLEM 1

Factor $4x^2 - 9$.

SOLUTION

(1) Recognize difference of squares: $4x^2 - 9 = (2x)^2 - 3^2$.

(2) Use the formula $A^2 - B^2 = (A + B)(A - B)$, replacing A by $2x$ and replacing B by 3.

(3) $$(2x)^2 - (3)^2 = (2x + 3)(2x - 3)$$

SAMPLE PROBLEM 2

Factor $36y^2 - 1$.

SOLUTION

$$A^2 - B^2 = (A + B)(A - B)$$

$$36y^2 - 1 = (6y)^2 - (1)^2 = (6y + 1)(6y - 1)$$

SAMPLE PROBLEM 3

Factor $25x^6 - 16y^4$

SOLUTION

$25x^6 = (5)(5)(xxx)(xxx) = (5x^3)(5x^3) = (5x^3)^2$

$16y^4 = (4)(4)(yy)(yy) = (4y^2)(4y^2) = (4y^2)^2$

Use $A^2 - B^2 = (A + B)(A - B)$.

$$25x^6 - 16y^4 = (5x^3)^2 - (4y^2)^2$$
$$= (5x^3 + 4y^2)(5x^3 - 4y^2)$$

EXERCISES 8.2

Factor.

A.

1. $x^2 - 25$
2. $x^2 - 16$
3. $y^2 - 36$
4. $y^2 - 49$
5. $64x^2 - 1$
6. $9y^2 - 1$
7. $81x^2 - 25$
8. $100y^2 - 49$
9. $25x^2 - 64y^2$
10. $4x^2 - 9y^2$
11. $16x^4 - 25y^2$
12. $36y^2 - x^4$
13. $x^2y^4 - z^2$
14. $400\,a^2b^2 - c^2$
15. $x^6 - y^2$

B.

1. $x^2 - 1$
2. $16x^2 - y^2$

3. $36x^2 - 49$
4. $64y^2 - 25$
5. $x^2 - 100y^2$
6. $9x^2y^2 - 4$
7. $81 - k^2$
8. $x^2y^2 - z^4$
9. $900a^4 - 1$
10. $a^6 - b^2c^2$
11. $4 - 81y^6$
12. $c^6 - a^4$
13. $121x^6 - y^4$
14. $y^6 - 144x^4$
15. $x^6 - 169y^6$

8.3 SIMPLE TRINOMIALS

OBJECTIVE: To express a simple trinomial as a product of binomial factors, if possible.

Simple trinomial product:

$$X^2 + (A + B)X + AB = (X + A)(X + B)$$

To factor a simple trinomial product, it is necessary to find two integers whose product is the constant term and whose sum is the coefficient of the variable.

SAMPLE PROBLEM 1

Factor $x^2 + 9x + 20$.

SOLUTION

(1) Find the pairs of factors of 20: (1)(20); (2)(10); (4)(5).

(2) Select those factors whose sum is 9: 4 and 5.

(3) Substitute into the simple trinomial product formula, using $A = 4$ and $B = 5$.

$$X^2 + (A + B)X + AB = (X + A)(X + B)$$

$$x^2 + 9x + 20 = x^2 + (4 + 5)x + (4)(5) = (x + 4)(x + 5)$$

SAMPLE PROBLEM 2

Factor $y^2 - 7y + 12$.

SOLUTION

(1) Factors of 12: (1)(12); (2)(6); (3)(4).

(2) Since the product is positive and the sum is negative, the factors of 12 must both be negative.

Factors whose sum is -7 are -3 and -4.

(3) Use $A = -3$ and $B = -4$.

$$\begin{aligned} y^2 - 7y + 12 &= y^2 + ((-3) + (-4))\,y + (-3)(-4) \\ &= (y + (-3))\,(y + (-4)) \\ &= (y - 3)(y - 4) \end{aligned}$$

SAMPLE PROBLEM 3

Factor $x^2 - 3x - 10$.

SOLUTION

(1) Factors of 10: (1)(10); (2)(5).

(2) Since the product, -10, is negative, the factors of -10 must be selected with opposite sign and the sum must be -3. The combination desired is $+2$ and -5.

(3) Use $A = -5$ and $B = 2$.

$$X^2 + (A + B)\,X + AB = (X + A)\,(X + B)$$

$$x^2 - 3x - 10 = x^2 + (-5 + 2)\,x + (-5)\,(2) = (x - 5)\,(x + 2)$$

SAMPLE PROBLEM 4

Factor $x^2 + 7xy - 18y^2$.

SOLUTION

(1) Factors of $18y^2$: $(y)\,(18y)$; $(2y)\,(9y)$; $(3y)\,(6y)$.

(2) Since the product, $-18y^2$, is negative, the factors of $-18y^2$ must be selected with opposite sign and the sum must be $+7y$.

The combination desired is $-2y$ and $+9y$.

(3) Use $A = -2y$ and $B = 9y$.

$$\begin{aligned} x^2 + 7xy - 18y^2 &= x^2 + (-2y + 9y)\,x + (-2y)\,(9y) \\ &= (x - 2y)\,(x + 9y) \end{aligned}$$

SAMPLE PROBLEM 5

Factor $x^2 + 4x + 2$, if possible.

SOLUTION

The only pairs of factors of 2 are (1) (2) and (−1) (−2). Since neither combination has a sum of 4, the polynomial can not be factored using integers for A and B. Such a polynomial is said to be *prime* with respect to the integers.

The following table shows the factorization of some simple trinomials. Note the sign patterns that are typical in this type of factoring: when the product AB is positive, A and B have the same sign, the sign of the coefficient of X. When the product AB is negative, A and B have opposite signs.

Trinomials and Factors

$$X^2 + (A + B)X + AB = (X + A)(X + B)$$

$$x^2 + 8x + 15 = (x + 5)(x + 3)$$

$$x^2 - 8x + 15 = (x - 5)(x - 3)$$

$$x^2 + 2x - 15 = (x + 5)(x - 3)$$

$$x^2 - 2x - 15 = (x - 5)(x + 3)$$

EXERCISES 8.3

Factor, if possible. If the polynomial is prime, write "prime."

A.

1. $x^2 + 9x + 14$
2. $x^2 - 9x + 14$
3. $x^2 - 5x - 14$
4. $x^2 + 5x - 14$
5. $x^2 - 15x + 14$
6. $x^2 - 13x - 14$
7. $x^2 + 13x - 14$
8. $x^2 + 11x + 30$
9. $x^2 - 5x + 4$
10. $x^2 - 5x + 6$

B.

1. $x^2 + 6x + 8$
2. $x^2 - 11x + 30$
3. $x^2 - 2x - 24$
4. $x^2 + 2x - 8$
5. $x^2 + 2x + 4$
6. $y^2 - 4y - 5$
7. $y^2 + 4y - 5$
8. $y^2 - 5y + 4$
9. $y^2 - 5y - 4$
10. $x^2 + 11xy + 18y^2$

11. $x^2 - 4x - 21$
12. $x^2 + 4x - 21$
13. $y^2 + y - 20$
14. $y^2 - y - 20$
15. $y^2 - 14y - 15$
16. $y^2 - 16y + 8$
17. $x^2 + 15x + 26$
18. $x^2 + 11x - 26$
19. $x^2 - 11x - 26$
20. $t^2 - 15t + 26$
21. $x^2 + 4xy - 12y^2$
22. $x^2 - 4xy - 12y^2$
23. $r^2 - 11rs - 12s^2$
24. $s^2 - 13st + 12t^2$
25. $u^2 - 3uv - 5v^2$

11. $x^2 + 5xy - 24y^2$
12. $x^2 - 2xy - 35y^2$
13. $x^2 + 4xy - 3y^2$
14. $u^2 - uv - 2v^2$
15. $u^2 + uv + v^2$
16. $u^2 + uv - 56v^2$
17. $r^2 - 3rs - 54s^2$
18. $x^4 - 2x^2y^2 - 3y^4$
19. $x^4 - 15x^2 + 50$
20. $x^4 - 16x^2 + 64$

8.4 PERFECT SQUARE TRINOMIALS

OBJECTIVE: To recognize and express a perfect square trinomial as the square of a binomial.

A **perfect square trinomial** is a trinomial having the form

$$A^2 + 2AB + B^2$$

A perfect square trinomial is factored by using the formula for the perfect square trinomial product.

Perfect square trinomial product:

$$A^2 + 2AB + B^2 = (A + B)^2$$

$$A^2 - 2AB + B^2 = (A - B)^2$$

To recognize a perfect square trinomial:

(1) Two terms must be perfect square monomials: $(A)^2$ and $(B)^2$.

(2) The third term must be *twice* the product of A and B, where A^2 and B^2 are the perfect square monomials.

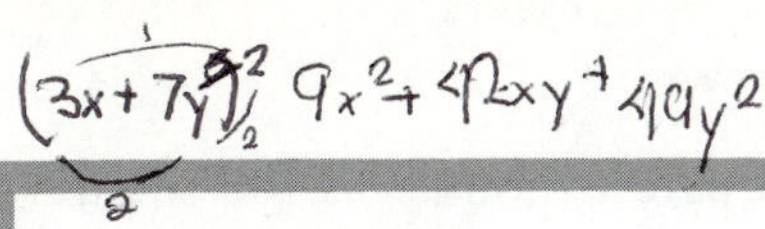

SAMPLE PROBLEM 1

Factor $9x^2 + 30xy + 25y^2$.

SOLUTION

(1) The perfect square monomials are $9x^2$ and $25y^2$; that is, $9x^2 = (3x)^2$ and $25y^2 = (5y)^2$.

(2) Use the formula: $A^2 + 2AB + B^2 = (A + B)^2$.

$$(3x)^2 + 30xy + (5y)^2 = (3x + 5y)^2$$

(3) Check that $30xy = 2AB$, where $A = 3x$ and $B = 5y$.

$$\begin{aligned} 2AB &= 2\,(3x)\,(5y) \\ &= 2\,(15xy) \\ &= 30xy \end{aligned}$$

Therefore, $9x^2 + 30xy + 25y^2 = (3x + 5y)^2$.

SAMPLE PROBLEM 2

Factor $x^4 - 14x^2 + 49$.

SOLUTION

(1) Recognize the perfect square monomials $x^4 = (x^2)^2$ and $49 = 7^2$.

(2) Use the formula: $A^2 - 2AB + B^2 = (A - B)^2$.

$$(x^2)^2 - 2\,(x^2)\,(7) + 7^2 = (x^2 - 7)^2$$

(3) Check the middle term: $2\,(x^2)\,(7) = 14x^2$.

Therefore, $x^4 - 14x^2 + 49 = (x^2 - 7)^2$.

SAMPLE PROBLEM 3

Show that each of the following is not a perfect square trinomial: (a) $4x^2 + 9$, (b) $4x^2 + 6x + 9$, and (c) $4x^2 + 12x - 9$.

SOLUTION

Noting that $(2x + 3)^2 = 4x^2 + 2\,(2x)\,(3) + 9 = 4x^2 + 12x + 9$,

(a) $4x^2 + 9$ is not a perfect square since the third term is missing.

(b) $4x^2 + 6x + 9$ is not a perfect square since the term $6x$ is *not twice* the product of $2x$ and 3.

(c) $4x^2 + 12x - 9$ is not a perfect square since -9 is not the square of a monomial. (Note that $(-3)\,(-3) = +9$.)

EXERCISES 8.4

Express each perfect square trinomial as the square of a binomial. If the trinomial is not a perfect square, write "not a perfect square."

A.

1. $x^2 + 8x + 16$
2. $y^2 + 2y + 1$
3. $y^2 - 4y + 4$
4. $y^2 - 6y + 9$
5. $x^2 - 6x + 36$
6. $x^2 + 2x + 4$
7. $4x^2 + 20x + 25$
8. $4x^2 - 20x + 25$
9. $4x^2 - 20x - 25$
10. $4x^2 + 10x + 25$
11. $x^2 + 25$
12. $x^2 - 16xy + 64y^2$
13. $x^2 + 18xy + 81y^2$
14. $x^2 + 20xy - 100y^2$
15. $x^2 + 100y^2 - 20xy$

B.

1. $9x^2 + 49y^2 - 42xy$
2. $x^4 - 10x^2 + 25$
3. $y^4 + 4y^2 + 4$
4. $25x^4 + 60x^2y^2 + 36y^4$
5. $16x^4 + 20x^2y^2 + 25y^4$
6. $49x^2 + 100y^2$

7. $25x^2 + 10x + 1$

8. $49y^2 - 14y + 1$

9. $36x^2 + 25$

10. $36x^2 - 12x + 1$

11. $100t^4 + 20t^2 + 1$

12. $64t^4 - 16t^2 + 1$

13. $4x^2 + y^2 - 4xy$

14. $4x^4 + 1 - 2x^2$

15. $9x^4 + 1 - 6x^2$

8.5 GENERAL TRINOMIALS

OBJECTIVE: To express a general trinomial as a product of binomial factors, if possible.

A trinomial having the form $AX^2 + BX + C$ can sometimes be factored by using the **general trinomial product formula.**

$$acx^2 + (bc + ad)\,x + bd = (ax + b)\,(cx + d)$$

Note the following in the general trinomial product formula:

(1) The coefficients of x in the binomial factors are factors of the coefficient of x^2 in the product.

(2) The constant terms in the binomial factors are factors of the constant term of the product.

(3) The sum of the inner product and the outer product of the binomial factors produces the coefficient of x in the product.

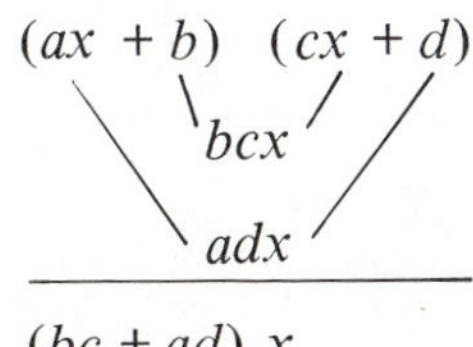

Inner product

Outer product

Sum of inner and outer products

SAMPLE PROBLEM 1

Factor $3x^2 + 11x + 10$.

SOLUTION

(1) Note that the factored form will be $(ax + b)(cx + d)$.

(2) Factor the coefficient of x^2: $3 = (1)(3)$. Thus, the factored form will be $(x + b)(3x + d)$.

(3) Factor the constant term: $10 = (1)(10)$ or $(2)(5)$. Thus, the possible factored forms are $(x + 1)(3x + 10)$, $(x + 10)(3x + 1)$, $(x + 2)(3x + 5)$, or $(x + 5)(3x + 2)$.

(4) To select the correct set of factors, find the sum of the inner and outer products for each set.

$(x + 1)(3x + 10)$	$(x + 10)(3x + 1)$
$3x$	$30x$
$10x$	x
$13x$	$31x$
$(x + 2)(3x + 5)$	$(x + 5)(3x + 2)$
$6x$	$15x$
$5x$	$2x$
$11x$	$17x$

(5) Select the set of factors whose sum of inner and outer products is the middle term of the given trinomial, in this case $11x$.

Therefore, $3x^2 + 11x + 10 = (x + 2)(3x + 5)$.

SAMPLE PROBLEM 2

Factor $5y^2 + 3y - 14$.

SOLUTION

(1) Using $acy^2 + (ad + bc)y + bd = (ay + b)(cy + d)$, note that $ac = 5$ and $5 = (5)(1)$. The factors will have the form $(5y + b)(y + d)$.

(2) Since $bd = -14$, the possible pairs for b and d are $(-1)(14)$, $(1)(-14)$, $(2)(-7)$, and $(-2)(7)$.

(3) Try each possibility to obtain the correct middle term:

$(5y - 1)(y + 14)$	and	$-y + 70y = 69y$
$(5y + 14)(y - 1)$	and	$14y - 5y = 9y$
$(5y + 1)(y - 14)$	and	$y - 70y = -69y$
$(5y - 14)(y + 1)$	and	$-14y + 5y = -9y$
$(5y + 2)(y - 7)$	and	$2y - 35y = -33y$
$(5y - 7)(y + 2)$	and	$-7y + 10y = 3y$
$(5y - 2)(y + 7)$	and	$-2y + 35y = 33y$
$(5y + 7)(y - 2)$	and	$7y - 10y = -3y$

(4) Select the set of factors whose sum of inner and outer products is $3y$.

$$5y^2 + 3y - 14 = (5y - 7)(y + 2)$$

EXERCISES 8.5

(1–10) Supply the missing factor for each of the following.

A.

1. $6x^2 + 23x + 20 = (2x + 5)(\quad)$
2. $8x^2 + 18x + 7 = (4x + 7)(\quad)$
3. $5y^2 - 47y + 18 = (5y - 2)(\quad)$
4. $6y^2 - 13y + 6 = (3y - 2)(\quad)$
5. $10x^2 - 11x - 6 = (5x + 2)(\quad)$
6. $4x^2 + 5x - 6 = (4x - 3)(\quad)$
7. $2x^2 - 19x - 10 = (2x + 1)(\quad)$
8. $5x^2 - 48x + 27 = (x - 9)(\quad)$
9. $6t^2 - t - 15 = (3t - 5)(2t + 3)$
10. $6t^2 - t - 7 = (t + 1)(\quad)$

B.

1. $3r^2 - 5r + 2 = (r - 1)(\quad)$
2. $18r^2 + 3r - 28 = (3r + 4)(6r - 7)$
3. $52x^2 + xy - 30y^2 = (4x - 3y)(13x + 10y)$

4. $2x^2 - 5xy + 2y^2 = (x - 2y)(\qquad)$

5. $15x^2 - xy - 2y^2 = (3x + y)(\qquad)$

6. $10x^2 + 13xy - 3y^2 = (5x - y)(\qquad)$

7. $4x^2 - 15xy - 4y^2 = (4x + y)(\qquad)$

8. $6x^2 - 25x + 14 = (2x - 7)(\qquad)$

9. $6x^2 - 17x - 14 = (2x - 7)(\qquad)$

10. $6x^2 + 17x - 14 = (2x + 7)(\qquad)$

(11–18) Supply the missing signs for each pair of factors.

A.

11. $6x^2 - x - 7 = (6x \quad 7)(x \quad 1)$

12. $6x^2 + x - 7 = (6x \quad 7)(x \quad 1)$

13. $6x^2 - 13x + 7 = (6x \quad 7)(x \quad 1)$

14. $6x^2 + 13x + 7 = (6x \quad 7)(x \quad 1)$

15. $9y^2 + 9y + 2 = (3y \quad 1)(3y \quad 2)$

16. $9y^2 + 3y - 2 = (3y \quad 1)(3y \quad 2)$

17. $9y^2 - 3y - 2 = (3y \quad 1)(3y \quad 2)$

18. $9y^2 - 9y + 2 = (3y \quad 1)(3y \quad 2)$

B.

11. $8s^2 - 6st - 5t^2 = (4s \quad 5t)(2s \quad t)$

12. $40s^2 - 17st - 12t^2 = (8s \quad 3t)(5s \quad 4t)$

13. $12x^2 - 97xy + 8y^2 = (12x \quad y)(x \quad 8y)$

14. $20x^2 - 109xy - 100y^2 = (4x \quad 25y)(5x \quad 4y)$

15. $10a^2 + 31ab - 14b^2 = (2a \quad 7b)(5a \quad 2b)$

16. $30x^2 + 31xy - 21y^2 = (2x \quad 3y)(15x \quad 7y)$

17. $21a^2 - 26ab + 8b^2 = (7a \quad 4b)(3a \quad 2b)$

18. $21a^2 + 2ab - 8b^2 = (7a \quad 4b)(3a \quad 2b)$

(19–34) Factor completely, if possible. If the polynomial is prime, write "prime."

A.

19. $5x^2 + 11x + 2$

20. $3x^2 - 22x + 7$

21. $2x^2 - 5x - 3$

22. $7x^2 + 13x - 2$

23. $3y^2 + 2y - 5$

24. $2y^2 - 3y - 5$

25. $7t^2 - 2t - 5$

26. $5t^2 + 6t + 2$

27. $11x^2 + 9xy - 2y^2$

28. $33x^2 - 8xy - y^2$

29. $2x^2 - 7xy + 6y^2$

30. $2x^2 - xy - 6y^2$

31. $5y^2 + 6y - 8$

32. $5y^2 - 6y - 8$

33. $6x^2 - 17x + 5$

34. $6x^2 + 15x - 5$

B.

19. $15y^2 - y - 2$

20. $15y^2 + y - 2$

21. $15x^2 - 11xy + 2y^2$

22. $15x^2 - 17xy + 2y^2$

23. $15x^2 + 29x - 2$

24. $5t^2 - 8t + 7$

25. $6t^2 + t - 2$

26. $27a^2 + 48a + 5$

27. $10b^2 + b - 2$

28. $8x^2 - 37x - 15$

29. $8x^2 + 14x - 15$

30. $15x^2 - 22xy + 8y^2$

31. $6x^2 - 2xy - 5y^2$

32. $18r^2 + 3rs - 28s^2$

33. $18r^2 - 3rs - 28s^2$

34. $18r^2 + 55rs - 28s^2$

8.6 ONE STEP PROBLEMS

OBJECTIVE: To factor a given polynomial in one step, if possible, by using factoring formulas.

The factoring formulas we have studied so far are listed below for convenience.

1. Monomial factor: $AX + AY = A\,(X + Y)$

   $AX + AY + AZ = A\,(X + Y + Z)$

2. Difference of squares: $A^2 - B^2 = (A - B)\,(A + B)$

3. Perfect square trinomial: $A^2 + 2AB + B^2 = (A + B)^2$

   $A^2 - 2AB + B^2 = (A - B)^2$

4. Simple trinomial: $X^2 + (a + b)\,X + ab = (X + a)\,(X + b)$

5. General trinomial:

$$acX^2 + (ad + bc)\,X + bd = (aX + b)\,(cX + d)$$

The first step in factoring a polynomial is to recognize the polynomial as having one of the forms on the left side of the above equations. (This means that the formulas must be memorized.) Then the polynomial is written in its factored form, the form on the right side of one of the above formulas.

Unless we recognize the form immediately, we test the above five formulas in the order in which they are listed: monomial factor first, then difference of squares, then perfect square trinomial, and so on.

SAMPLE PROBLEM 1

Factor $9x^2 - 25$.

SOLUTION

We first check to see if there is a monomial factor common to each term. There is none in this case.

Next we test if the polynomial is a difference of squares:

$$9x^2 - 25 = (3x)^2 - 5^2 = A^2 - B^2$$

Recalling that $A^2 - B^2 = (A - B)\,(A + B)$,

$$9x^2 - 25 = (3x)^2 - 5^2 = (3x - 5)\,(3x + 5)$$

SAMPLE PROBLEM 2

Factor $x^2 - 4x - 21$.

SOLUTION

Checking, we find there is no common monomial factor. We do not have a difference of squares since we have three terms and not two. We do not have a perfect square trinomial since two terms are not perfect squares. So we try the simple trinomial formula: $X^2 + (a + b) X + ab = (X + a)(X + b)$.

We need to find 2 numbers, a and b, whose product is -21 and whose sum is -4. The numbers are -7 and 3. Therefore,

$$x^2 - 4x - 21 = (x - 7)(x + 3)$$

SAMPLE PROBLEM 3

Factor $36x^2 + 36x + 36$.

SOLUTION

Checking for a common monomial factor first, we note that 36 divides each term. Therefore,

$$36x^2 + 36x + 36 = 36(x^2 + x + 1)$$

Since no two integers have both a sum of 1 and a product of 1, our factoring is completed.

Note how many trials it would have taken if we had tried to use the general trinomial formula first. It always pays to check for a common monomial factor first.

SAMPLE PROBLEM 4

Factor $36x^2 - 60x + 25$

SOLUTION

Checking for the common monomial factor first, we note that there is none. Noting that $36x^2$ and $+25$ are perfect squares, we test the perfect square trinomial formula:

If we have the form $A^2 - 2AB = B^2$, then $A = 6x$ and $B = 5$. We form $(A - B)^2$ and multiply.

$$(6x - 5)^2 = 36x^2 - 2(6x)(5) + 25 = 36x^2 - 60x + 25$$

Therefore, $36x^2 - 60x + 25 = (6x - 5)^2$

Note how many trials were saved by not trying the general trinomial formula first.

EXERCISES 8.6

A.

(1–8) Match each polynomial in Column 1 with its factored form in column 2.

Column 1	Column 2
1. $4x^2 + 36$	a. $(2x + 3)^2$
2. $4x^2 - 9$	b. $(2x + 3)(2x - 3)$
3. $4x^2 - 9x$	c. $(4x + 1)(x - 3)$
4. $4x^2 + 9$	d. $(4x + 1)(x + 3)$
5. $4x^2 + 12x + 9$	e. $4(x^2 + 9)$
6. $4x^2 + 4x + 4$	f. $x(4x - 9)$
7. $4x^2 + 13x + 9$	g. $4(x^2 + x + 1)$
8. $4x^2 - 11x - 3$	h. not factorable

(9–26) Factor if possible. If not possible, write "prime."

9. $9x^2 - 25$
10. $9x^2 - 25x$
11. $9x^2 + 36$
12. $9x^2 + 25$
13. $x^2 - 12x + 36$
14. $9x^2 - 30x + 25$
15. $x^2 - 5x - 36$
16. $9x^2 + 19x + 2$
17. $9x^2 + 18x + 27$
18. $49x^2 - 1$
19. $x^2 + 4x - 32$
20. $5x^2 - 2x - 3$
21. $x^4 - 100$
22. $x^4 - 100x^3$
23. $x^4 + 20x^2 + 100$
24. $x^4 - x^3 + x^2$
25. $x^4 + x^2y - 2y^2$
26. $x^4 + 9x^2 + 20$

B.

EXERCISES 8.6

(1–8) Each polynomial in Column 1 has a factor listed in Column 2. Find this factor for each polynomial.

Column 1	Column 2
1. $25x^2 - 4$	a. $x + 1$
2. $25x^2 + x^4$	b. $x + 5$
3. $25x^2 + 20x$	c. $x - 5$
4. $25x^2 + 25$	d. $5x + 1$
5. $x^2 - 5x - 50$	e. $5x + 2$
6. $x^2 + 26x + 25$	f. $5x + 4$
7. $25x^2 + 10x + 1$	g. $x^2 + 25$
8. $5x^2 - 26x + 5$	h. $x^2 + 1$

(9–20) Factor if possible. If not possible, write "prime."

9. $x^2 - 36$
10. $x^4 - 36$
11. $x^4 + 36x^2$
12. $x^2 - 14x + 49$
13. $x^4 - 14x^2 + 49$
14. $x^2 + 6x - 16$
15. $x^4 + 6x^2 - 16$
16. $7x^2 + 4x - 3$
17. $7x^4 + 4x^2 - 3$
18. $x^4 - 64y^2$
19. $x^4 + 64x^2$
20. $36x^4 + 12x^2 + 1$

8.7 MULTIPLE STEP PROBLEMS

OBJECTIVE: To completely factor a given polynomial.

A polynomial whose coefficients are integers is said to be **completely factored** if

(1) The polynomial is expressed as a product of polynomial factors whose coefficients are integers.

(2) None of these factors can be expressed as a product of two different polynomials whose coefficients are integers.

After a polynomial has been factored by using one of the formulas listed in section 8.6, each polynomial factor must be checked to see if it can be factored.

SAMPLE PROBLEM 1

Completely factor $2x^5 - 162x$.

Check by multiplication.

SOLUTION

Always find the monomial factor first, if there is one.

$2x^5 - 162x = 2x(x^4 - 81)$ **Distributive axiom**

$= 2x(x^2 - 9)(x^2 + 9)$ **Difference of squares**

$= 2x(x - 3)(x + 3)(x^2 + 9)$ **Difference of squares**

CHECK

$$2x(x^2 - 9)(x^2 + 9) = 2x(x^4 - 81)$$

$$= 2x(x^4) - 2x(81)$$

$$= 2x^5 - 162x$$

SAMPLE PROBLEM 2

Completely factor $18x^2 - 66x - 24$.

SOLUTION

$18x^2 - 66x - 24 = 6(3x^2 - 11x - 4)$ **Removing the monomial factor**

$= 6(3x + 1)(x - 4)$ **Using the general trinomial product formula**

CHECK

Check by multiplication.

$$6(3x + 1)(x - 4) = 6[3x(x - 4) + (1)(x - 4)]$$

$$= 6(3x^2 - 12x + x - 4)$$

$$= 6(3x^2 - 11x - 4)$$

$$= 18x^2 - 66x - 24$$

SAMPLE PROBLEM 3

Completely factor $x^4 + 4x^2 - 32$.

SOLUTION

$x^4 + 4x^2 - 32 = (x^2 - 4)(x^2 + 8)$	General trinomial product
$= (x - 2)(x + 2)(x^2 + 8)$	Difference of squares

EXERCISES 8.7
EXERCISES 8.7

Completely factor each of the following. Check by multiplication. sad

A.

1. $5x^3 - 35x^2 + 50x$
2. $16x^4 - 81y^4$
3. $100x^2 - 4y^2$
4. $100x^2 - 4x$
5. $16y^4 + 64y^2$
6. $6x^4 - 24x^3 - 126x^2$
7. $9x^4 - 72x^3 + 144x^2$
8. $25x^2 - 50x + 100$
9. $256x^4 - 625y^4$
10. $15x^2 + 60x + 60$
11. $12x^2 + 60x - 72$
12. $36x^4 - 144x^2$
13. $21x^3 - 21x^2 - 21x$
14. $3x^3 - 8x^2 + 5x$
15. $42x^2 - 9x - 6$
16. $72x^2 - 60x - 28$

17. $x^4 - 10x^2 + 9$

18. $16x^4 - 68x^2 + 16$

19. $625x^4 + 600x^2 - 25$

20. $72x^4 + 178x^2 - 5$

B.

1. $9x^2 - 9x - 18$

2. $9x^2y + 9xy - 18y$

3. $3x^3 - 17x^2 - 6x$

4. $144x^3y - 100xy^3$

5. $324x^4 - 4y^4$

6. $27t^3 + 36t^2 + 63t$

7. $49x - x^3$

8. $y^8 - 16y^4$

9. $5x^2 - 15x - 50$

10. $4y^2 - 28y + 49$

11. $6x^2 - 20x + 16$

12. $2c^2 - 20c^3 + 50c^4$

13. $y^5 - y$

14. $6x^2 - 54x + 84$

15. $-3x^3 - 27x$

16. $5a^5 + 60a^3 - 320a$

17. $6b^4 + b^3 - 12b^2$

18. $x^4 - 16x^2$

19. $x^4 - 13x^2 + 36$

20. $4a^4 - 16b^4$

21. $2x^4y^4 - 32$

22. $t^6 + 5t^4 - 36t^2$

23. $4x^4 - 48x^3 + 144x^2$

24. $x^4 - 24x^2 - 25$

25. $16u^4 - 36v^4$

26. $-x^4 + 4x^3 + 5x^2$

UNIT 8 POSTTEST

(1–2) Factor. (8.1)

1. $18ax^3 + 3ax^2$
2. $9x^3 - 18x^2 + 36x$

(3–4) Factor. (8.2)

3. $49t^2 - 36$
4. $25x^4 - 64y^2$

(5–8) Factor. (8.3)

5. $x^2 + 7x + 12$
6. $x^2 - 9xy + 20y^2$
7. $y^2 + 6y - 27$
8. $x^2 - 2xy - 15y^2$

(9–12) If possible, express as the square of a binomial. If not possible, write "not possible." (8.4)

9. $x^2 - 14x + 49$
10. $y^2 - 10y - 25$
11. $y^4 + 16y^2 + 64$
12. $x^2 + 4$

(13–14) Factor. (8.5)

13. $2x^2 - 3x + 1$
14. $10x^2 + 13x - 14$

(15–19) Factor completely. (8.6–8.7)

15. $x^4 - 16$
16. $3x^2 - 6x - 105$
17. $360x^3 - 600x^2 + 250x$
18. $2x^4 - 7x^3 - 15x^2$
19. $y^4 - 18y^2 + 81$

UNIT 9

ALGEBRAIC FRACTIONS

Unit 9
PRETEST

(1–5) Simplify. (9.1)

1. $\dfrac{21xy}{14y^2}$

2. $\dfrac{2x + 6}{4x + 12}$

3. $\dfrac{y^2 + 3y}{y^2 - 3y}$

4. $\dfrac{x^2 - 4x + 4}{x^2 + x - 6}$

5. $\dfrac{12x^2 + 4x}{3x^2 - 5x - 2}$

(6–8) Simplify. (9.2)

6. $\dfrac{10xy^3}{-35x^2y}$

7. $\dfrac{x - 8}{8 - x}$

8. $\dfrac{x - x^2}{x^2 - 1}$

(9–12) Find the missing numerator. (9.3)

9. $\dfrac{8}{3x} = \dfrac{\quad}{15x^2y^2}$

10. $\dfrac{1}{2 - x} = \dfrac{\quad}{x - 2}$

11. $\dfrac{3x}{x - 3} = \dfrac{\quad}{x^2 - 9}$

12. $\dfrac{x - 6}{x - 1} = \dfrac{\quad}{x^2 + 4x - 5}$

(13–16) Express as a single simplified fraction. (9.4)

13. $\frac{1}{10x} + \frac{4}{15x}$

14. $\frac{n-1}{4n+8} - \frac{n+1}{5n+10}$

15. $\frac{1}{y} + \frac{1}{5} + \frac{1}{y+5}$

16. $\frac{x+2}{x^2-4x-5} - \frac{x-2}{x^2-6x+5}$

(17–18) Express each product as a single simplified fraction. (9.5)

17. $\frac{5x^3}{9y^3} \cdot \frac{27y^2}{35x^2}$

18. $\frac{y^2-4}{y^2-y} \cdot \frac{y^3-y^2}{y^2+2y}$

(19–20) Express each quotient as a single simplified fraction. (9.6)

19. $\frac{8r^4t^6}{13x^2y} \div \frac{40r^3t^5}{26xy}$

20. $\frac{y^2-y-12}{2y^2-32} \div \frac{3y^2-27}{12y^2-36y}$

(21–22) Simplify. (9.7)

21. $\frac{\frac{1}{x}+\frac{1}{2}}{\frac{x}{2}-\frac{2}{x}}$

22. $\frac{x+3-\frac{10}{x}}{x-\frac{25}{x}}$

(23–24) Divide and check. (9.8)

23. $\dfrac{x^3 - 7x^2 + 17x - 8}{x - 3}$

$x - 3\,\overline{)\,x^3 - 7x^2 + 17x - 8}$

Check:

24. $\dfrac{x^4 - 29x^2 + 4}{x^2 + 5x - 2}$

$x^2 + 5x - 2\,\overline{)\,x^4 \quad - 29x^2 \quad + 4}$

Check:

THE SET OF RATIONAL NUMBERS

While the set of integers contains the answer to every addition, subtraction, and multiplication problem, it does not contain the answer to every division problem. For example,

$2 + (-6)$ is an integer,
$2 - 6$ is an integer,
$\dfrac{-6}{2}$ is an integer, but
$\dfrac{2}{6}$ and $\dfrac{2}{-6}$ are not integers.

So that every division by a nonzero divisor will have a solution, the set of integers is extended to include the quotients of counting numbers and their negatives. This new set of numbers is called the **set of rational numbers,** denoted by Q (for quotient).

Definition of the Set of Rationals, Q

The set of rational numbers is the set of numbers that can be written in the form $\dfrac{x}{y}$ where x and y are integers and $y \neq 0$.

It is understood that $\dfrac{5}{1}$ and 5 are forms of the same number. Similarly, $\dfrac{-5}{1}$ and $\dfrac{5}{-1}$ and -5 are forms of the same number. In general, for any integer n, $\dfrac{n}{1} = n$. Thus, every integer is a rational number.

The axioms for the set of integers also hold for the set of rationals. However, there is now a new axiom, the inverse axiom for multiplication.

Inverse axiom for multiplication	For each nonzero rational number x, there is exactly one rational number $\frac{1}{x}$, called the reciprocal or multiplicative inverse of x, so that $x\left(\frac{1}{x}\right) = 1$ and $\left(\frac{1}{x}\right)x = 1$.

The set of rational numbers has the property that it is closed with respect to the operations of addition, subtraction, multiplication, and division. In other words, every sum, difference, product, and quotient (excluding zero as a divisor) of two rational numbers is a rational number.

9.1 SIMPLIFICATION

OBJECTIVE: To reduce a quotient of polynomials to lowest terms.

The simplification of an algebraic fraction is similar to the process of reducing an arithmetic fraction to lowest terms.

An algebraic fraction of the form $\frac{P}{Q}$ where P and Q are polynomials is said to be in **simplified form** if the numerator P and the denominator Q have no common factor other than 1.

To simplify an algebraic fraction,

(1) Factor the numerator and the denominator.

(2) Use the fundamental principle of fractions

$$\frac{nk}{dk} = \frac{n}{d}$$

so that the resulting fraction is in simplified form.

SAMPLE PROBLEM 1

Reduce $\frac{30}{42}$ to lowest terms.

SOLUTION

(1) Factor the numerator and the denominator:

$$\frac{30}{42} = \frac{(2)(3)(5)}{(2)(3)(7)} = \frac{(5)(6)}{(7)(6)}$$

(2) Use the fundamental principle of fractions:

$$\frac{nk}{dk} = \frac{n}{d}$$

$$\frac{30}{42} = \frac{(5)(6)}{(7)(6)} = \frac{5}{7}$$

SAMPLE PROBLEM 2

Simplify $\frac{36x^3y}{60x^2y^3}$

SOLUTION

(1) **Factor the numerator and the denominator:**

$$\frac{2\cdot2\cdot3\cdot3xxxy}{2\cdot2\cdot3\cdot5xxyyy}$$

(2) **Line out the common factors of the numerator and the denominator:**

$$\frac{3\,(\cancel{12})\,x\,(\cancel{xx})\,(\cancel{y})}{5(\cancel{12})\,(\cancel{xx})\,yy\,(\cancel{y})}$$

(3) **Write the simplified result:**

$$\frac{3x}{5y^2}$$

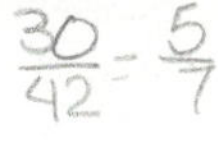

SAMPLE PROBLEM 3

Simplify $\frac{5x^2 - 45}{5x^2 + 10x - 75}$

SOLUTION

(1) $5x^2 - 45 = 5\,(x^2 - 9) = 5\,(x + 3)\,(x - 3)$

$5x^2 + 10x - 75 = 5\,(x^2 + 2x - 15) = 5\,(x + 5)\,(x - 3)$

(2) $\frac{5x^2 - 45}{5x^2 + 10x - 75} = \frac{(x + 3)\,(5)\,(x - 3)}{(x + 5)\,(5)\,(x - 3)} = \frac{x + 3}{x + 5}$

SAMPLE PROBLEM 4

Simplify $\frac{x^4 - 4}{x^4 + 4x^2 + 4}$

SOLUTION

$$\frac{x^4 - 4}{x^4 + 4x^2 + 4} = \frac{(x^2 - 2)\,(x^2 + 2)}{(x^2 + 2)\,(x^2 + 2)} = \frac{x^2 - 2}{x^2 + 2}$$

Note that $\frac{x^2 - 2}{x^2 + 2}$ is in simplified form. It would be incorrect to line out either the two x^2 terms or the 2's. Terms of a sum can not be lined out. Only a common *factor* of the numerator and the denominator can be omitted.

EXERCISES 9.1 Simplify.

A.

1. $\frac{10}{15}$
2. $\frac{56}{64}$
3. $\frac{6}{54}$
4. $\frac{4}{100}$
5. $\frac{6}{10}$
6. $\frac{125}{1000}$
7. $\frac{6x}{9x^2}$
8. $\frac{20xy}{45x^2y^2}$
9. $\frac{28x^3y^4}{7x^2y^4}$
10. $\frac{x^2 - 4}{x^2 + 3x + 2}$
11. $\frac{5x^2 - 5}{5x^2 + 5}$
12. $\frac{x^2 + 5x}{x^2 - 25}$
13. $\frac{x^2 + 6x + 5}{x^2 + 7x + 10}$
14. $\frac{4x^2 - 100x}{x^2 - 24x - 25}$
15. $\frac{x^2 + x - 42}{x^2 - 2x - 24}$

B.

1. $\frac{21}{28}$
2. $\frac{8}{32}$
3. $\frac{36}{64}$
4. $\frac{25}{100}$
5. $\frac{8}{10}$
6. $\frac{16}{1000}$
7. $\frac{12x^2y}{18xy^2}$
8. $\frac{14x^3}{21x^2}$
9. $\frac{25x^4y^2}{125x^4y^3}$
10. $\frac{4x + 8}{x^2 - 4}$
11. $\frac{2x^2 + 18}{4x^2 - 36}$
12. $\frac{x^2 - 36}{x^2 - 6x}$
13. $\frac{x^2 - 9x + 14}{x^2 - 10x + 21}$
14. $\frac{9x^2 - 9x - 108}{9x^2 - 36x}$
15. $\frac{x^2 + 3x - 40}{x^2 - 3x - 88}$

A.

16. $\dfrac{2x^2 - 9x - 5}{x^2 - 3x - 10}$

17. $\dfrac{x^2 - 49}{x^2 + 14x + 49}$

18. $\dfrac{4y^2 - 9}{4y^2 - 12y + 9}$

B.

16. $\dfrac{3x^2 + 8x - 3}{3x^2 - 10x + 3}$

17. $\dfrac{y^2 - 36}{y^2 - 12y + 36}$

18. $\dfrac{25x^2 + 20xy + 4y^2}{25x^2 - 4y^2}$

9.2 NEGATIVE FRACTIONS

OBJECTIVES: To express a negative fraction in simplified form.

To simplify fractions containing the forms $a - b$ and $b - a$.

Since the quotient of a positive number and a negative number is negative, there are three ways to express such a quotient. For example,

$$\frac{-3}{4} = \frac{3}{-4} = -\frac{3}{4}$$

In general,

$$\frac{-a}{b} = \frac{a}{-b} = -\frac{a}{b}$$

The preferred form is $\frac{-a}{b}$, with the minus sign in the numerator. The binomial $5 - 9$ is the negative of the binomial $9 - 5$; that is,

$$5 - 9 = -(9 - 5)$$

In general,

$$b - a = -(a - b)$$

Just as $\frac{5 - 9}{9 - 5} = \frac{-4}{4} = -1$, in general,

$$\frac{b - a}{a - b} = \frac{-(a - b)}{(a - b)} = -1$$

SAMPLE PROBLEM 1

Simplify $\frac{4x^2}{-8x}$

SOLUTION

$$\frac{4x^2}{-8x} = \frac{-4x^2}{8x} = \frac{(-x)(4x)}{(2)(4x)} = \frac{-x}{2}$$

SAMPLE PROBLEM 2

Simplify $\frac{x-4}{4-x}$

SOLUTION

$$\frac{x-4}{4-x} = \frac{-(4-x)}{(4-x)} = -1$$

SAMPLE PROBLEM 3

Simplify $\frac{(5-x)(7-y)}{(x-5)(y-7)}$

SOLUTION

$$\frac{(5-x)(7-y)}{(x-5)(y-7)} = \frac{(-1)(x-5)(-1)(y-7)}{(x-5)(y-7)} = (-1)(-1) = 1$$

EXERCISES 9.2

Simplify.

A.

1. $\frac{5x}{-15x^2}$

2. $\frac{-12x^2y}{54xy^2}$

3. $\frac{21x^2y^2}{-63x^2y^2}$

B.

1. $\frac{8xy}{-20x^2y}$

2. $\frac{-56x^3}{84x^4}$

3. $\frac{42xy^3}{-70x^3y}$

A.

4. $\dfrac{x-7}{7-x}$

5. $\dfrac{5-y}{y-5}$

6. $\dfrac{5x^2-10x}{10x-5x^2}$

7. $\dfrac{3x-9}{3x-x^2}$

8. $\dfrac{-5+x}{5-x}$

9. $\dfrac{(3-x)(2-x)}{(x-3)(x-2)}$

10. $\dfrac{(x-y)(y-z)}{(z-y)(y-x)}$

B.

4. $\dfrac{y-4}{4-y}$

5. $\dfrac{6-x}{x-6}$

6. $\dfrac{x^2-x^3}{x^3-x^2}$

7. $\dfrac{25-5x}{x^2-5x}$

8. $\dfrac{4-x}{-4+x}$

9. $\dfrac{(6-y)(5-y)}{(y-5)(y-6)}$

10. $\dfrac{(a-b)(b-c)}{(c-a)(b-a)}$

9.3 RAISING TO HIGHER TERMS

OBJECTIVE: To raise an algebraic fraction to higher terms.

Just as an arithmetic fraction is raised to higher terms by multiplying the numerator and denominator by a common factor, an algebraic fraction is also raised to higher terms by multiplying the numerator and denominator by a common factor.

Basic idea: the fundamental principle of fractions

$$\frac{n}{d} = \frac{nk}{dk} \text{ where } dk \neq 0$$

SAMPLE PROBLEM 1

Raise to higher terms as indicated: $\frac{5}{x} = \frac{?}{4x^3}$

SOLUTION

(1) Factor the new denominator to find the multiplier: $4x^3 = 4xxx = x\,(4x^2)$

(2) Apply the fundamental principle of fractions by multiplying the numerator and the denominator by the multiplier found in step (1). $\frac{5}{x} = \frac{(5)(4x^2)}{(x)(4x^2)} = \frac{20x^2}{4x^3}$

SAMPLE PROBLEM 2

Raise to higher terms as indicated: $\frac{x}{x+7} = \frac{?}{x^2 + 2x - 35}$

SOLUTION

(1) $x^2 + 2x - 35 = (x + 7)(x - 5)$

(2) $\frac{x}{x+7} = \frac{(x)(x-5)}{(x+7)(x-5)} = \frac{x^2 - 5x}{x^2 + 2x - 35}$

SAMPLE PROBLEM 3

Find the missing numerator in $\frac{3}{4-y} = \frac{\quad}{y-4}$

SOLUTION

$$\frac{3}{4-y} = \frac{3}{-(y-4)} = \frac{-3}{y-4}$$

EXERCISES 9.3

Find the missing numerator.

A.

1. $\frac{3}{4} = \frac{15}{20}$

B.

1. $\frac{2}{3} = \frac{12}{18}$

2. $\frac{7}{8} = \frac{\quad}{32}$

3. $4 = \frac{\quad}{15}$

4. $\frac{5}{2} = \frac{\quad}{16}$

5. $\frac{1}{8} = \frac{\quad}{200}$

6. $1 = \frac{\quad}{64}$

7. $\frac{4}{25} = \frac{\quad}{100}$

8. $\frac{2}{x} = \frac{\quad}{5x^2}$

9. $\frac{x}{x+1} = \frac{\quad}{5x+5}$

10. $\frac{4}{-x} = \frac{\quad}{x^2}$

11. $\frac{x}{-5} = \frac{\quad}{5}$

12. $\frac{1}{5-x} = \frac{\quad}{x-5}$

13. $\frac{4}{y-7} = \frac{\quad}{7-y}$

14. $\frac{-2}{4-x} = \frac{\quad}{x^2-16}$

15. $\frac{6}{x+3} = \frac{\quad}{x^2-9}$

16. $\frac{4}{x-5} = \frac{\quad}{(x-5)^2}$

17. $\frac{5}{x+4} = \frac{\quad}{2x^2-32}$

2. $\frac{5}{9} = \frac{\quad}{36}$

3. $6 = \frac{\quad}{8}$

4. $\frac{7}{5} = \frac{\quad}{25}$

5. $\frac{5}{4} = \frac{\quad}{100}$

6. $\frac{1}{7} = \frac{\quad}{56}$

7. $1 = \frac{\quad}{16}$

8. $\frac{7}{x} = \frac{\quad}{9x^3}$

9. $\frac{5}{x-5} = \frac{\quad}{2x-10}$

10. $\frac{8}{-y} = \frac{\quad}{y^2}$

11. $\frac{y}{-7} = \frac{\quad}{7}$

12. $\frac{2}{y-6} = \frac{\quad}{6-y}$

13. $\frac{1}{8-5x} = \frac{\quad}{5x-8}$

14. $\frac{-5}{6-x} = \frac{\quad}{x^2-36}$

15. $\frac{3}{x-4} = \frac{\quad}{x^2-16}$

16. $\frac{2}{x+7} = \frac{\quad}{(x+7)^2}$

17. $\frac{8}{x-2} = \frac{\quad}{5x^2-20}$

A.	B.
18. $\dfrac{x+1}{x-1}=\dfrac{}{x^2-1}$	18. $\dfrac{x-5}{x+5}=\dfrac{}{x^2-25}$
19. $\dfrac{x-4}{4x}=\dfrac{}{4x^2+16x}$	19. $\dfrac{x-3}{9x}=\dfrac{}{9x^2+27x}$
20. $\dfrac{x+4}{x+6}=\dfrac{}{x^2+5x-6}$	20. $\dfrac{x-2}{x+4}=\dfrac{}{x^2+2x-8}$
21. $\dfrac{5x+1}{5x-1}=\dfrac{}{5x^2+24x-5}$	21. $\dfrac{4x-3}{4x-5}=\dfrac{}{4x^2-x-5}$
22. $\dfrac{x}{(x-8)^2}=\dfrac{}{(x^2-64)^2}$	22. $\dfrac{5}{(x+9)^2}=\dfrac{}{(x^2-81)^2}$

9.4 ADDITION AND SUBTRACTION

OBJECTIVE: To express a sum or a difference of algebraic fractions as a single simplified fraction.

Basic Idea

Addition of fractions:

$$\frac{a}{d}+\frac{b}{d}=\frac{a+b}{d} \quad \text{where } d \neq 0$$

Subtraction of fractions:

$$\frac{a}{d}-\frac{b}{d}=\frac{a-b}{d} \quad \text{where } d \neq 0$$

The basic idea states that an indicated sum (or difference) of two fractions having the *same* denominator can be expressed as a single fraction whose numerator is the sum (or difference) of the numerators of the terms and whose denominator is the *common* denominator.

An indicated sum (or difference) of two fractions having *different* denominators can be expressed as a single fraction by first renaming the given fractions so that the resulting fractions have a common denominator. The work is simplest when the **least common denominator (LCD)** is used.

The **LCD (least common denominator)** of two or more given fractions is the LCM (least common multiple) of their denominators.

In sample problems 1–3 express each of the following as a single simplified fraction.

SAMPLE PROBLEM 1

$$\frac{1}{3x} + \frac{5}{3x}$$

SOLUTION

Since the denominators are the same, the fractions do not have to be raised to higher terms.

(1) Add the numerators and place the sum over the common denominator:

$$\frac{1+5}{3x} = \frac{6}{3x}$$

(2) Factor the numerator and the denominator:

$$\frac{2\,(3)}{x\,(3)}$$

(3) Line out the factor common to the numerator and denominator. Write the result as a single simplified fraction:

$$\frac{2}{x}$$

SAMPLE PROBLEM 2

$$\frac{7}{10} + \frac{8}{15}$$

SOLUTION

Note that the denominators are not the same. Therefore, the numerators cannot be added until the fractions are raised to higher terms, each having the same common denominator.

(1) Factor each denominator and hold the place of the build-up multiplier with open parens:

$$\frac{7\,(\ )}{2\,(5)\,(\ )} + \frac{8\,(\ )}{3\,(5)\,(\ )}$$

(2) Insert the build-up factors by finding the smallest number to make the denominators the same:

$$\frac{7\,(3)}{2\,(5)\,(3)} + \frac{8\,(2)}{3\,(5)\,(2)}$$

(3) Write the sum of the numerators over the common denominator (if you have found the smallest common denominator possible, then this is the least common denominator [LCD]):

$$\frac{7\,(3) + 8\,(2)}{2\,(3)\,(5)}$$

(4) Do the indicated operations in the numerator and the denominator:

$$\frac{37}{30}$$

Since the numerator and the denominator do not have a factor in common, the problem is finished.

SAMPLE PROBLEM 3

$$\frac{4}{x^2 + 4x} + \frac{4}{x^2 - 4x}$$

SOLUTION

(1) Factor each denominator and insert open parens:

$$\frac{4\,(\quad)}{x\,(x + 4)\,(\quad)} + \frac{4\,(\quad)}{x\,(x - 4)\,(\quad)}$$

(2) Insert the least number of prime polynomial build-up factors to make the denominators the same:

$$\frac{4\,(x - 4)}{x\,(x + 4)\,(x - 4)} + \frac{4\,(x + 4)}{x\,(x - 4)\,(x + 4)}$$

(3) Write the simplified sum of the numerators over the (least) common denominator:

$$\frac{8x}{x\,(x + 4)\,(x - 4)}$$

(4) Line out factors common to the numerator and denominator. Write the result as a single simplified fraction:

$$\frac{8}{(x + 4)\,(x - 4)}$$

or

$$\frac{8}{x^2 - 16}$$

Either of the fractions shown in step (4) is accepted as the final form for the fraction.

SAMPLE PROBLEM 4

$$\frac{y + 2}{2y + 10} - \frac{y + 3}{3y + 15}$$

SOLUTION

(1) Use the definition of subtraction, $a - b = a + (-1)b$:

$$\frac{y + 2}{2y + 10} + \frac{(-1)\,(y + 3)}{3y + 15}$$

(2) Factor each denominator and insert open parens:

$$\frac{(y + 2)\,(\;)}{2\,(y + 5)\,(\;)} + \frac{(-1)\,(y + 3)\,(\;)}{3\,(y + 5)\,(\;)}$$

(3) Insert the build-up factors to form the LCD:

$$\frac{(y+2)(3)}{2(y+5)(3)} + \frac{(-1)(y+3)(2)}{3(y+5)(2)}$$

(4) Do the multiplications in each numerator and then place the sum of the numerators over the LCD:

$$\frac{3y+6-2y-6}{6(y+5)}$$

(5) Simplify the numerator:

$$\frac{y}{6(y+5)} \text{ or } \frac{y}{6y+30}$$

Since the numerator and the denominator do not have a common factor, this is the final result.

EXERCISES 9.4

Express each of the following as a single simplified fraction.

A.

1. $\frac{1}{3}+\frac{1}{9}$

2. $\frac{2}{9}-\frac{1}{18}$

3. $\frac{5}{6}+\frac{7}{8}$

4. $\frac{1}{2}-\frac{1}{3}$

5. $2-\frac{1}{3}$

B.

1. $\frac{1}{4}+\frac{5}{12}$

2. $\frac{3}{14}-\frac{2}{7}$

3. $\frac{13}{16}-\frac{5}{8}$

4. $\frac{1}{3}+\frac{1}{4}$

5. $1+\frac{2}{5}$

6. $\frac{1}{2}+\frac{1}{3}+\frac{1}{6}$

6. $\frac{1}{5}+\frac{1}{12}+\frac{1}{20}$

7. $\frac{1}{42}+\frac{1}{30}-\frac{1}{7}$

7. $\frac{1}{6}+\frac{1}{4}-\frac{1}{2}$

8. $\frac{1}{8}-\frac{1}{14}-\frac{1}{56}$

8. $\frac{1}{5}-\frac{1}{15}-\frac{1}{30}$

9. $\frac{3}{4x}+\frac{5}{4x}$

9. $\frac{5}{12x}-\frac{1}{12x}$

10. $\frac{x}{x+3}+\frac{6}{x+3}$

10. $\frac{x}{x-5}+\frac{5}{x-5}$

11. $\frac{y}{y-8}-\frac{8}{y-8}$

11. $\frac{x}{x-7}-\frac{7}{x-7}$

12. $\frac{x-5}{x-9}-\frac{x+1}{x-9}$

12. $\frac{y+2}{y-6}-\frac{y-4}{y-6}$

13. $\frac{1}{3-x}+\frac{1}{x-3}$

13. $\frac{1}{x-10}-\frac{1}{10-x}$

14. $\frac{2}{x^2}+\frac{5}{3x}$

14. $2+\frac{4}{y}$

15. $1-\frac{5}{y^2}$

15. $\frac{4}{5x}-\frac{3}{2x^2}$

16. $\dfrac{7}{3x^2} - \dfrac{2}{9x}$

16. $\dfrac{5}{6x^2} - \dfrac{7}{8x^2}$

17. $\dfrac{4}{x - 7} + \dfrac{x - 20}{x^2 - 49}$

17. $\dfrac{10}{x^2 - 25} + \dfrac{2}{x + 5}$

18. $\dfrac{4}{(x + 4)^2} - \dfrac{1}{(x + 4)}$

18. $\dfrac{3}{(x - 6)^2} - \dfrac{2}{(x - 6)}$

19. $\dfrac{x}{x - 7} + \dfrac{x}{x + 7}$

19. $\dfrac{5}{x - 5} - \dfrac{5}{x + 5}$

20. $\dfrac{1}{(x + 4)^2} - \dfrac{1}{x^2 - 16}$

20. $\dfrac{6}{25x^2 - 9} + \dfrac{6}{(5x - 3)^2}$

$25x^2 - 9 = (5x + 3)(5x - 3)$

21. $\dfrac{1}{x} + \dfrac{1}{x + 2}$

21. $\dfrac{1}{5} + \dfrac{1}{y + 5}$

22. $\dfrac{1}{6y - 18} + \dfrac{1}{10y - 30}$

22. $\dfrac{1}{y^2 + 3y} + \dfrac{1}{3y + 9}$

23. $\dfrac{t}{t + 1} - \dfrac{t}{t - 1}$

23. $\dfrac{2}{2k - 1} - \dfrac{2}{2k + 1}$

24. $\dfrac{n + 1}{2n + 10} - \dfrac{n - 1}{3n + 15}$

24. $\dfrac{n - 2}{n - 3} - \dfrac{n + 2}{2n - 6}$

25. $\dfrac{3}{x^2 - 12x + 36} + \dfrac{4}{x^2 - 36}$

$\dfrac{3}{(x-6)^2} \quad \dfrac{4}{(x+6)(x-6)}$

$\dfrac{3(x+6)}{(x-6)^2 (x+6)} + \dfrac{4(x-6)}{(x-6)^2(x+6)}$

25. $\dfrac{4}{4x^2 - 1} - \dfrac{4}{4x^2 - 4x + 1}$

9.5 MULTIPLICATION

OBJECTIVE: To express a product of algebraic fractions as a single simplified fraction.

The product of two fractions is defined as a single fraction whose numerator is the product of the numerators of the two fractions being multiplied and whose denominator is the product of the denominators of the two fractions being multiplied.

Multiplication of Fractions

$$\frac{a}{b} \cdot \frac{c}{d} = \frac{ac}{bd} \quad \text{where } bd \neq 0$$

To multiply two or more fractions:

(1) Factor each numerator and denominator and reduce each fraction, if possible.

(2) Multiply the numerators to obtain the new numerator.

(3) Multiply the denominators to obtain the new denominator.

(4) Reduce the product fraction to lowest terms.

SAMPLE PROBLEM 1

Simplify $\left(\frac{2}{5}\right)\left(\frac{3}{7}\right)$.

SOLUTION

$$\left(\frac{2}{5}\right)\left(\frac{3}{7}\right) = \frac{(2)(3)}{(5)(7)} = \frac{6}{35}$$

SAMPLE PROBLEM 2

Simplify $\left(\frac{5}{12}\right)\left(\frac{-18}{35}\right)$.

SOLUTION

$$\left(\frac{5}{12}\right)\left(\frac{-18}{35}\right) = \frac{(-1)(5)(18)}{(12)(35)}$$

$$= \frac{(-1)(5)(2)(3)(3)}{(2)(2)(3)(5)(7)}$$

Factor in order to reduce the product to lowest terms.

$$= \frac{-3}{14}$$

SAMPLE PROBLEM 3

Multiply $\frac{15}{x^2} \cdot \frac{x^3}{20} \cdot \frac{12}{x^4}$

SOLUTION

$$\frac{15}{x^2} \cdot \frac{x^3}{20} \cdot \frac{12}{x^4} = \frac{(15)\,(12)\,x^3}{(20)\,x^2\,x^4}$$

$$= \frac{(3)\,(5)\,(3)\,(4)\,x^3}{(5)\,(4)\,x^2 x x^3}$$

$$= \frac{9}{x^3}$$

SAMPLE PROBLEM 4

Multiply $\frac{4x + 4y}{x^2 - y^2} \cdot \frac{(x - y)^2}{8x - 8y}$

SOLUTION

$$\frac{4\,(x + y)}{(x + y)\,(x - y)} \cdot \frac{(x - y)\,(x - y)}{8\,(x - y)} = \frac{4\,(x - y)}{(x - y)\,8}$$

$$= \frac{1}{2}$$

SAMPLE PROBLEM 5

Multiply $\frac{x^2 + 3x - 10}{2x^2 - x - 10} \cdot \frac{2x^2 - 5x}{x^3 - 4x^2 + 4x}$

SOLUTION

$$\frac{(x + 5)\,(x - 2)}{(2x - 5)\,(x + 2)} \cdot \frac{x\,(2x - 5)}{x\,(x - 2)\,(x - 2)} = \frac{(x + 5)\,(x - 2)\,(2x - 5)}{(2x - 5)\,(x + 2)\,(x - 2)\,(x - 2)}$$

$$= \frac{x + 5}{(x + 2)\,(x - 2)}$$

$$= \frac{x + 5}{x^2 - 4}$$

SAMPLE PROBLEM 6

Multiply $\frac{t^2 - 9}{2t^2 + 2t} \cdot \frac{t^2 + 4t + 3}{t^2 + 6t + 9}$

SOLUTION

$$\frac{(t + 3)\,(t - 3)}{2t\,(t + 1)} \cdot \frac{(t + 1)\,(t + 3)}{(t + 3)\,(t + 3)} = \frac{(t + 3)\,(t - 3)\,(t + 1)}{2t\,(t + 1)\,(t + 3)}$$

$$= \frac{t - 3}{2t}$$

EXERCISES 9.5 Multiply.

A.

1. $\left(\frac{1}{3}\right)\left(\frac{3}{5}\right)$

2. $\left(\frac{5}{16}\right)\left(\frac{-32}{35}\right)$

3. $\left(\frac{2}{9}\right)(54)$

4. $\left(\frac{-5}{14}\right)\left(\frac{-21}{15}\right)$

5. $(-12)\left(\frac{5}{6}\right)$

6. $\left(\frac{-7}{9}\right)(0)$

7. $\frac{36x^2}{60x^3} \cdot \frac{10x^4}{21x}$

8. $\frac{3x}{2y^2} \cdot \frac{8y^3}{9x}$

9. $10x^2\left(\frac{-7}{5x}\right)$

B.

1. $\left(\frac{4}{7}\right)\left(\frac{14}{15}\right)$

2. $\left(\frac{35}{36}\right)\left(\frac{-12}{55}\right)$

3. $\left(\frac{3}{5}\right)(45)$

4. $\left(-\frac{7}{9}\right)\left(-\frac{36}{14}\right)$

5. $\left(-\frac{3}{18}\right)(27)$

6. $\left(\frac{8}{3}\right)\left(\frac{3}{8}\right)$

7. $\frac{xy}{xz} \cdot \frac{yz}{4}$

8. $\left(\frac{-2x}{y}\right) \cdot \left(\frac{-y}{2x}\right)$

9. $14xy\left(\frac{2}{7y}\right)$

A.

10. $\left(\frac{2a}{3b}\right)\left(\frac{-5b}{4c}\right)\left(\frac{-6c}{35a^2}\right)$

11. $\left(\frac{9}{t^2-9}\right)\left(\frac{t-3}{3}\right)$

12. $\frac{4}{3x+3y}\cdot\frac{6x+6y}{2x+3y}$

13. $\frac{6x^2-42x}{45x-90}\cdot\frac{5x^2-10x}{2x^2-28x+98}$

14. $\frac{x^2-x}{x^2+2x}\cdot\frac{8x+8}{4x^3-4x^2}$

15. $\frac{ab+a}{b-ab}\cdot\frac{a^2b-ab}{a^2b+a^2}$

16. $\frac{2t^2+3t-2}{2t^2-t-1}\cdot\frac{2t^2-3t-2}{2t^2+t-1}$

B.

10. $(x^2-y^2)\left(\frac{1}{x+y}\right)$

11. $\left(\frac{6}{(x+8)^2}\right)\left(\frac{x+8}{12}\right)$

12. $\left(\frac{35}{x-10}\right)\left(\frac{(x-10)^2}{45}\right)$

13. $\frac{6x+12y}{2x-6y}\cdot\frac{4x^2-36y^2}{9x^2-36y^2}$

14. $\frac{xy-x}{xy+y}\cdot\frac{x^2y^2-y^2}{x^2y^2-x^2}$

15. $\frac{b^2-3b+2}{b^2-b-2}\cdot\frac{b^2-2b-3}{b^2+b-2}$

16. $\frac{9r^4s^2-r^2s^4}{36r^4-108r^2s}\cdot\frac{9r^2-27s}{3r^2s^2-10rs^3+3s^4}$

9.6 DIVISION

OBJECTIVE: To express a quotient of algebraic fractions as a single simplified fraction.

Definition of division:

$$\frac{a}{b}=(a)\left(\frac{1}{b}\right),\ \text{where } b\neq 0$$

Division of fractions

$$\frac{\frac{a}{b}}{\frac{c}{d}} = \frac{a}{b} \div \frac{c}{d} = \frac{a}{b} \cdot \frac{d}{c} \quad \text{where } bcd \neq 0$$

The same rule used for arithmetic fractions also applies; that is, "invert the divisor and multiply."

SAMPLE PROBLEM 1

Simplify $\dfrac{\left(\frac{2}{3}\right)}{\left(\frac{4}{15}\right)}$

SOLUTION

$$\frac{\frac{2}{3}}{\frac{4}{15}} = \left(\frac{2}{3}\right)\left(\frac{15}{4}\right) = \frac{(2)(3)(5)}{(2)(3)(2)} = \frac{5}{2}$$

SAMPLE PROBLEM 2

Simplify $\dfrac{12}{\frac{3}{4}}$

SOLUTION

$$\frac{12}{\frac{3}{4}} = \left(\frac{12}{1}\right)\left(\frac{4}{3}\right) = \frac{(4)(4)(3)}{(1)(3)} = \frac{16}{1} = 16$$

SAMPLE PROBLEM 3

Divide $\dfrac{15}{x^2} \div \dfrac{10}{x^3}$

SOLUTION

$$\frac{15}{x^2} \div \frac{10}{x^3} = \frac{15}{x^2} \cdot \frac{x^3}{10}$$

$$= \frac{(3)(5)(x)(x)(x)}{(2)(5)(x)(x)}$$

$$= \frac{3x}{2}$$

SAMPLE PROBLEM 4

Divide $\dfrac{\dfrac{6x}{7y^2}}{\dfrac{3x^2}{14y}}$

SOLUTION

$$\frac{6x}{7y^2} \div \frac{3x^2}{14y} = \frac{6x}{7y^2} \cdot \frac{14y}{3x^2}$$

$$= \frac{6x\ (14y)}{3x^2\ (7y^2)}$$

$$= \frac{2\ (2)}{x\ (y)}$$

$$= \frac{4}{xy}$$

SAMPLE PROBLEM 5

Divide $\dfrac{\dfrac{x+y}{x-y}}{x^2 - y^2}$

SOLUTION

$$\frac{x+y}{x-y} \div x^2 - y^2 = \frac{x+y}{x-y} \div \frac{x^2 - y^2}{1}$$

$$= \frac{x+y}{x-y} \cdot \frac{1}{(x+y)\,(x-y)}$$

$$= \frac{1}{(x-y)^2}$$

SAMPLE PROBLEM 6

Divide $\dfrac{r^3 - 36r}{12r^2 + 2r} \div \dfrac{r^3 - 6r^2}{36r + 6}$

SOLUTION

$$\frac{r\ (r+6)\ (r-6)}{2r\ (6r+1)} \cdot \frac{6\ (6r+1)}{r^2\ (r-6)} = \frac{6\ (r+6)}{2r^2}$$

$$= \frac{3\ (r+6)}{r^2}$$

$$= \frac{3r + 18}{r^2}$$

EXERCISES 9.6 Divide.

A.

1. $\dfrac{2}{x} \div \dfrac{5}{y}$

2. $\dfrac{\dfrac{24}{x^2}}{\dfrac{18}{x^3}}$

3. $6t \div \dfrac{18}{t^2}$

4. $\dfrac{\dfrac{18a^2 b}{5c^2}}{60a^2 b^2 c^2}$

5. $\dfrac{3x + 3y}{10x - 10y} \div \dfrac{6x - 6y}{5x + 5y}$

B.

1. $\dfrac{6}{x} \div \dfrac{18}{xy}$

2. $\dfrac{\dfrac{25x^2}{y^2}}{\dfrac{5x}{y^2}}$

3. $35r^2 s \div \dfrac{42r^2}{s^2}$

4. $\dfrac{\dfrac{4a}{5t}}{20at}$

5. $\dfrac{a^2 - 4b^2}{4a + 2b} \div \dfrac{a^2 - 4ab + 4b^2}{8a + 4b}$

A.

6. $\dfrac{-5x}{x^2+5} \div \dfrac{-5x}{x^2+5}$

7. $(25a^2 - 36b^2) \div \dfrac{5a-6b}{5a+6b}$

8. $\dfrac{\dfrac{(x-4)^2}{3x^2-12x}}{\dfrac{20-5x}{10}}$

9. $\dfrac{4a^2-4a}{10a-20} \div \dfrac{2a^2+2a}{5a-10}$

10.

$\dfrac{15x^2y}{x^2-8xy+12y^2} \div \dfrac{75xy^2}{x^2-4xy-12y^2}$

B.

6. $\dfrac{y^2-36}{y^2+36} \div \dfrac{36-y^2}{36+y^2}$

7. $\dfrac{7r+2s}{7r-2s} \div (49r^2-4s^2)$

8. $\dfrac{\dfrac{10y}{y^2-36}}{\dfrac{5y^2}{(6-y)^2}}$

9. $\dfrac{10t^2}{t^2-7t+10} \div \dfrac{25t^2}{t^2+3t-10}$

10.

$\dfrac{b^2x^2-b^4}{a^2x^2+2a^2bx+a^2b^2} \div \dfrac{ax-ab}{bx+b^2}$

9.7 COMBINED OPERATIONS

OBJECTIVE: To express a sequence of combined operations on fractions as a single simplified fraction.

Combined operations on fractions are done in a similar way to the method used for integers, as explained in sections 1.3 and 1.4.

Operations within the innermost set of grouping symbols are done first. It is important to remember, when dealing with fractions, that the bar is a grouping symbol.

Unless grouping symbols indicate otherwise, the operations are done in the following order:

1. Squaring and/or cubing as read from left to right.
2. Multiplication and/or division as read from left to right.
3. Addition and/or subtraction as read from left to right.

SAMPLE PROBLEM 1

Simplify $\dfrac{\dfrac{2}{5x^3}}{\dfrac{3}{10x^2}}$

SOLUTION

$$\frac{2}{5x^3} \div \frac{3}{10x^2} = \frac{2}{5x^3} \cdot \frac{10x^2}{3} = \frac{4}{3x}$$

SAMPLE PROBLEM 2

Simplify $\dfrac{\dfrac{1}{2} - \dfrac{1}{3}}{1 + \dfrac{1}{6}}$

SOLUTION

(1) Numerator: $\dfrac{1}{2} - \dfrac{1}{3} = \dfrac{3}{6} - \dfrac{2}{6} = \dfrac{1}{6}$

(2) Denominator: $1 + \dfrac{1}{6} = \dfrac{6}{6} + \dfrac{1}{6} = \dfrac{7}{6}$

(3) $\dfrac{\dfrac{1}{6}}{\dfrac{7}{6}} = \dfrac{1}{6} \div \dfrac{7}{6} = \dfrac{1}{6} \cdot \dfrac{6}{7} = \dfrac{1}{7}$

SAMPLE PROBLEM 3

Simplify $\dfrac{\dfrac{1}{2}-\dfrac{1}{x}}{\dfrac{1}{4}-\dfrac{1}{x^2}}$

SOLUTION

(1) **Numerator:** $\dfrac{1}{2}-\dfrac{1}{x}=\dfrac{x}{2x}-\dfrac{2}{2x}=\dfrac{x-2}{2x}$

(2) **Denominator:** $\dfrac{1}{4}-\dfrac{1}{x^2}=\dfrac{x^2}{4x^2}-\dfrac{4}{4x^2}=\dfrac{x^2-4}{4x^2}$

(3) $\dfrac{x-2}{2x}\div\dfrac{x^2-4}{4x^2}=\dfrac{x-2}{2x}\cdot\dfrac{(2x)(2x)}{(x-2)(x+2)}=\dfrac{2x}{x+2}$

EXERCISES 9.7

Simplify.

A.

1. $\dfrac{\dfrac{3}{5}}{\dfrac{6}{20}}$

2. $\dfrac{5+\dfrac{1}{2}}{4}$

3. $\dfrac{1-\dfrac{1}{x}}{x}$

4. $\dfrac{x-5}{\dfrac{1}{2}}$

B.

1. $\dfrac{\dfrac{4x^2}{9y^2}}{\dfrac{6x}{y}}$

2. $\dfrac{1-\dfrac{x}{y}}{2}$

3. $\dfrac{2x+14}{\dfrac{2}{7}}$

4. $\dfrac{1-\dfrac{1}{x}}{1+\dfrac{1}{x}}$

A.

5. $\dfrac{\frac{1}{6}+\frac{1}{9}}{\frac{2}{3}}$

6. $\dfrac{2x}{\frac{1}{x}+\frac{1}{5}}$

7. $\dfrac{\frac{1}{3}-\frac{1}{x}}{\frac{1}{9}-\frac{1}{x^2}}$

8. $\dfrac{\frac{1}{4x}+\frac{1}{6x}}{\frac{5}{3x}}$

9. $\dfrac{1+\frac{5}{t}}{1+\frac{25}{t^2}}$

10. $\dfrac{1-\frac{6}{x}-\frac{7}{x^2}}{1-\frac{8}{x}+\frac{7}{x^2}}$

B.

5. $\dfrac{\frac{1}{2}+\frac{1}{3}}{1-\frac{1}{3}}$

6. $\dfrac{\frac{8}{y}}{\frac{1}{4}-\frac{1}{y}}$

7. $\dfrac{\frac{x}{5}+\frac{5}{x}}{\frac{x}{5}-\frac{5}{x}}$

8. $\dfrac{2d}{\frac{d}{x}+\frac{d}{y}}$

9. $\dfrac{\frac{7}{t}+1}{\frac{49}{t^2}-1}$

10. $\dfrac{1+\frac{1}{x}-\frac{20}{x^2}}{1+\frac{9}{x}+\frac{20}{x^2}}$

A.

11. The average speed, r, of a car that travels x miles at 40 m.p.h. and then returns the same distance at 60 m.p.h. is obtained by dividing the total distance traveled by the total travel time. In symbols, r is expressed as

$$r = \frac{2x}{\frac{x}{40} + \frac{x}{60}}$$

Simplify the expression for r and find the average speed.

12. The following expression occurs in trigonometry and analytic geometry for finding the angle between two lines. Express m as a simplified single fraction.

$$m = \frac{\frac{1}{3} - \frac{1}{x}}{1 + \frac{1}{3x}}$$

13. a, b, and c are reciprocal triplets if and only if $\frac{1}{a} + \frac{1}{b} = \frac{1}{c}$; that is,

$$c = \frac{1}{\frac{1}{a} + \frac{1}{b}}$$

Show that if $a = r\,(r + s)$ and $b = s\,(r + s)$, where r and s are positive integers, then c is also a positive integer.

B.

11. The total resistance, R, in ohms due to a resistor of x ohms and a resistor of 40 ohms connected in parallel is given by

$$R = \frac{1}{\frac{1}{x} + \frac{1}{40}}$$

a. Express R as a single fraction.
b. Evaluate R for $x = 24$, using the result of a.

12. A problem from differential calculus:

a. Simplify $\dfrac{\frac{1}{x+h} - \frac{1}{x}}{h}$.

b. Evaluate the expression obtained above in a for $h = 0$. (This result is called the derivative of $\frac{1}{x}$.)

13. Simplify the expression below which occurs in the theory of magnetism.

$$\frac{k}{a^2 + b^2} - \frac{2ka^2}{(a^2 + b^2)^2}$$

9.8 DIVISION OF POLYNOMIAL BY POLYNOMIAL

OBJECTIVE: To express the quotient of two polynomials as the sum of a polynomial and an algebraic fraction, the degree of whose numerator is less than the degree of the divisor.

To divide a polynomial by a polynomial:

(1) Arrange each polynomial in descending powers of a variable, if necessary.

(2) Write the polynomials in the long division format used in arithmetic, leaving space for any missing powers of a variable.

(3) Obtain the first term of the quotient by dividing the first term of the dividend by the first term of the divisor.

(4) Multiply the term of the quotient by each term of the divisor.

(5) Subtract this product from the dividend.

(6) Repeat steps (3) through (5) until the degree of the remainder is less than the degree of the divisor.

(7) Add the remainder to the quotient.

To check, multiply the quotient polynomial by the divisor and add the remainder to this product. The result should be the dividend.

SAMPLE PROBLEM 1

Divide and check $\dfrac{2x - 66 - 23x^2 + 5x^3}{x - 5}$

SOLUTION

(1) $\dfrac{5x^3 - 23x^2 + 2x - 66}{x - 5}$ — Arranging each polynomial in descending powers of x

(2) $x - 5\,\overline{)\,5x^3 - 23x^2 + 2x - 66}$ — Using the long division format

(3) Dividing $5x^3$ by x to obtain $5x^2$

$$\begin{array}{r} 5x^2 \\ x - 5\,\overline{)\,5x^3 - 23x^2 + 2x - 66} \end{array}$$

(4) Multiplying $x - 5$ by $5x^2$

$$\begin{array}{r} 5x^2 \\ x - 5\,\overline{)\,5x^3 - 23x^2 + 2x - 66} \\ 5x^3 - 25x^2 \end{array}$$

(5) Subtracting

$$\begin{array}{r} 5x^2 \\ x - 5\,\overline{)\,5x^3 - 23x^2 + 2x - 66} \\ \underline{5x^3 - 25x^2} \\ 2x^2 + 2x - 66 \end{array}$$

(6) Repeating steps (3) to (5)

$$\begin{array}{r} 5x^2 + 2x + 12 \\ x - 5\,\overline{)\,5x^3 - 23x^2 + 2x - 66} \\ \underline{5x^3 - 25x^2} \\ 2x^2 + 2x - 66 \\ \underline{2x^2 - 10x} \\ 12x - 66 \\ \underline{12x - 60} \\ -6 \end{array}$$

(7) The quotient is $5x^2 + 2x + 12 + \dfrac{-6}{x - 5}$

CHECK

$$(x - 5)\left(5x^2 + 2x + 12 + \frac{-6}{x - 5}\right) = 5x^3 - 25x^2 + 2x^2 - 10x + 12x - 60 - 6$$

$$= 5x^3 - 23x^2 + 2x - 66$$

SAMPLE PROBLEM 2

Divide and check $\frac{x^2 - 4x + 15}{x - 5}$

SOLUTION

$$\begin{array}{r} x + 1 \\ x - 5 \overline{) x^2 - 4x + 15} \\ \underline{x^2 - 5x } \\ x + 15 \\ \underline{x - 5} \\ 20 \end{array}$$

$$\frac{x^2 - 4x + 15}{x - 5} = x + 1 + \frac{20}{x - 5} \text{ (answer)}$$

CHECK

$(x - 5)(x + 1) + 20 = (x^2 - 4x - 5) + 20$

$= x^2 - 4x + 15$

SAMPLE PROBLEM 3

Divide $15xy + 14y^2 + 4x^2$ by $2y + x$ and check.

SOLUTION

Rewriting in descending powers of x.

$$\begin{array}{r} 4x + 7y \\ x + 2y \overline{) 4x^2 + 15xy + 14y^2} \\ \underline{4x^2 + 8xy } \\ 7xy + 14y^2 \\ \underline{7xy + 14y^2} \end{array}$$

$$\frac{4x^2 + 15xy + 14y^2}{x + 2y} = 4x + 7y \text{ (answer)}$$

CHECK

$(x + 2y)(4x + 7y) = 4x^2 + 8xy + 7xy + 14y^2$

$= 4x^2 + 15xy + 14y^2$

SAMPLE PROBLEM 4

Divide and check $\dfrac{x^3 - 5x - 4}{x - 2}$

SOLUTION

$$\begin{array}{r}
x^2 + 2x - 1 \\
x - 2\,\overline{)\,x^3 \qquad - 5x - 4} \\
\underline{x^3 - 2x^2} \qquad\quad \\
2x^2 - 5x - 4 \\
\underline{2x^2 - 4x} \quad \\
-x - 4 \\
\underline{-x + 2} \\
-6
\end{array}$$

$$\frac{x^3 - 5x - 4}{x - 2} = x^2 + 2x - 1 + \frac{-6}{x - 2} \text{ (answer)}$$

CHECK

$$\begin{aligned}
(x - 2)(x^2 + 2x - 1) + (-6) &= (x^3 + 2x^2 - x) + (-2x^2 - 4x + 2) - 6 \\
&= x^3 - 5x - 4
\end{aligned}$$

SAMPLE PROBLEM 5

Divide and check $\dfrac{x^3 + 8}{x + 2}$

SOLUTION

$$\begin{array}{r}
x^2 - 2x + 4 \\
x + 2\,\overline{)\,x^3 \qquad\qquad + 8} \\
\underline{x^3 + 2x^2} \qquad\qquad \\
-2x^2 \qquad + 8 \\
\underline{-2x^2 - 4x} \quad \\
4x + 8 \\
\underline{4x + 8}
\end{array}$$

CHECK

$$\begin{aligned}
(x + 2)(x^2 - 2x + 4) &= x(x^2 - 2x + 4) + 2(x^2 - 2x + 4) \\
&= (x^3 - 2x^2 + 4x) + (2x^2 - 4x + 8) \\
&= x^3 + (-2x^2 + 2x^2) + (4x - 4x) + 8 \\
&= x^3 + 8
\end{aligned}$$

Divide and check.

EXERCISES 9.8

A.

1. $\dfrac{x^2 - 7x + 10}{x + 2}$

2. $\dfrac{y^2 - 6y + 5}{y - 5}$

3. $\dfrac{6x^2 - 5x - 4}{2x - 1}$

4. $\dfrac{12x^2 + 11x - 15}{3x + 5}$

B.

1. $\dfrac{x^2 + 4x - 8}{x + 3}$

2. $\dfrac{y^2 - 7y + 10}{y - 2}$

3. $\dfrac{8x^2 + 6x - 9}{4x + 3}$

4. $\dfrac{5x^2 - 22x + 8}{5x - 2}$

A.

5. $\dfrac{4y^2 - y + 2y^3 - 5}{1 + y}$

6. $\dfrac{x^3 - 10x - 24}{x - 4}$

7. $\dfrac{y^3 + 125}{y + 5}$

8. $\dfrac{4x^2 + 9}{2x + 3}$

B.

5. $\dfrac{5y + 7 - 8y^3 - 2y^2}{1 - 2y}$

6. $\dfrac{x^3 + 5x^2 - 12}{x + 2}$

7. $\dfrac{y^3 - 27}{y - 3}$

8. $\dfrac{25x^2 + 16y^2}{4y + 5x}$

UNIT 9 POSTTEST

(1-5) Simplify. (9.1)

1. $\dfrac{15x^2y^3}{40x^3y^3}$

2. $\dfrac{4y^2 - 12y}{5y^3 - 15y^2}$

3. $\dfrac{x^3 - 4x}{x^2 - 2x}$

4. $\dfrac{x^2 - x - 20}{x^2 - 10x + 25}$

5. $\dfrac{2x^2 - 3x - 35}{2x^2 - 10x}$

(6-8) Simplify. (9.2)

6. $\dfrac{12x^2y}{-8xy^2}$

7. $\dfrac{10x - y}{y - 10x}$

8. $\dfrac{4 - x^2}{4x - 8}$

(9-12) Find the missing numerator. (9.3)

9. $\dfrac{5x}{4y} = \dfrac{\quad}{32x^2y^2}$

10. $\dfrac{1}{x - 8} = \dfrac{\quad}{8 - x}$

11. $\dfrac{6t}{t - 6} = \dfrac{\quad}{t^3 - 36t}$

12. $\dfrac{x + 4}{x + 6} = \dfrac{\quad}{x^2 + 3x - 18}$

(13–16) Express each sum as a single simplified fraction. (9.4)

13. $\dfrac{5}{x} + \dfrac{3}{x + 3}$

14. $\frac{3}{n - 6} - \frac{2}{n + 4}$

15. $1 - \frac{6}{x + 3} - \frac{1}{(x + 3)^2}$

16. $\frac{2}{4x^2 - 1} - \frac{2}{2x - 1} + \frac{1}{2x + 1}$

(17–18) Express each product as a single simplified fraction. (9.5)

17. $\frac{2x - 2}{15} \cdot \frac{18x}{3x - 3}$

18. $\frac{x^2 - 6x + 9}{x^2 - 4x + 3} \cdot \frac{2x^2 - 2x}{x^2 - 9}$

(19–20) Express each quotient as a single simplified fraction. (9.6)

19. $\frac{x^4 y^6}{x^2 - 2x - 8} \div \frac{x^5 y^3}{x^2 - 8x + 16}$

20. $\frac{x^4 y - 4x^3 y}{9x^2 - 4} \div \frac{x^3 y - 4x^2 y}{3x^2 + 10x - 8}$

(21–22) Simplify. (9.7)

21. $\dfrac{1 - \dfrac{1}{x+1}}{1 + \dfrac{1}{x-1}}$

22. $\dfrac{1}{\dfrac{1}{x} + \dfrac{1}{y}}$

(23–24) Divide and check. (9.8)

23. $\dfrac{8x^3 - 60x^2 + 150x - 125}{2x - 5}$

24. $\dfrac{x^4 + x^3 - 3x - 3}{x + 3}$

$2x - 5 \overline{)\, 8x^3 - 60x^2 + 150x - 125}$

$x + 3 \overline{)\, x^4 + x^3 - 3x - 3}$

Check:

Check:

UNIT 10

FRACTIONAL EQUATIONS, APPLICATIONS

Unit 10
PRETEST

(1–4) Solve and check. (10.1)

1. $\dfrac{x+2}{x+3}=2$

2. $\dfrac{1}{4x}+\dfrac{1}{6x}=\dfrac{5}{36}$

3. $\dfrac{x-1}{x-4}=\dfrac{3}{x-4}$

4. $\dfrac{320}{60+x}=\dfrac{280}{60-x}$

(5–10) Solve and check. (10.2)

5. $\dfrac{2}{x+1}+\dfrac{1}{x-1}=\dfrac{5}{x^2-1}$

6. $\dfrac{x+3}{2x+8}-\dfrac{x-5}{3x+12}=1$

7. $\frac{2}{x - 6} + \frac{5}{x - 4} = \frac{18}{x^2 - 10x + 24}$

8. $\frac{1}{2x^2 - x - 10} = \frac{1}{x^2 - 3x - 10}$

9. $\frac{2}{x^2 - 2x} + \frac{3}{x^2 + 2x} = \frac{4}{x^2 - 4}$

10. $\frac{3}{x + 3} + \frac{4}{x - 4} = \frac{7x}{x^2 - x - 12}$

(11–12) Solve. (10.3)

11. The ratio of a speed in miles per hour to a speed in kilometers per hour is about 5 to 8. If the speed limit on a highway in Mexico is 100 km. per hour, what is the speed in miles per hour?

12. A gardener has 234 ft. of fencing to enclose a rectangular area. He wants the ratio of the width of the rectangle to its length to be 5 to 8. How long and how wide should he make the garden?

13. A plane with a tailwind completes a 3200-mi. trip in the same time that a plane going in the opposite direction completes a trip of 2800 mi. Find the speed of the wind if each plane has a speed in still air of 600 m.p.h. (10.4)

14. One thousand checks can be processed by an old machine in 40 min. and by a new machine in 24 min. How long would it take to process these checks if both machines worked together? (10.5)

10.1 FRACTIONAL EQUATIONS, ELEMENTARY TYPES

OBJECTIVE: To solve a simple fractional equation.

A **fractional equation** is an equation that has one or more terms that are algebraic fractions.

The basic idea in solving a fractional equation is to eliminate the fractions. This is done by multiplying each side of the equation by the same number, namely the LCD.

Theorem. If $C \neq 0$, then $A = B$ and $AC = BC$ have the same solution set.

For example, $\frac{6}{x} = \frac{3x}{x}$ and $6 = 3x$ have the same solution set since $6 = 3x$ is obtained by multiplying each side of $\frac{6}{x} = \frac{3x}{x}$ by x and since $x \neq 0$ ($x = 2$ in this case).

Division by 0 is undefined, and a variable may not equal a value that makes a denominator 0. For example, consider

$$\frac{x-2}{x-5} + \frac{5x}{x+4} = 7.$$

If $x = 5$, then $x - 5 = 5 - 5 = 0$.

If $x = -4$, then $x + 4 = -4 + 4 = 0$.

Therefore, for this equation, the restrictions on x are $x \neq 5$ and $x \neq -4$.

To solve a fractional equation:

(1) Factor each denominator to find the LCD (least common denominator) of all the fractions in the equation. State the restrictions on the variable.

(2) Raise fractions to higher terms so that the denominator of each fraction is the LCD.

(3) Multiply each side of the equation by the LCD (each term of the equation must be multiplied by the LCD).

(4) Remove parentheses, if necessary, by using the distributive axiom.

(5) Combine like terms.

(6) Solve the resulting equation.

(7) Check the solution in the original equation.

SAMPLE PROBLEM 1

Solve and check $\frac{x}{10} + 2 = \frac{x}{2}$

SOLUTION

LCD = 10; no restrictions on x.

$$\frac{x}{2 \cdot 5} + \frac{2\,(10)}{10} = \frac{5x}{5 \cdot 2}$$

$$x + 20 = 5x$$

$$5x = x + 20$$

$$4x = 20$$

$$x = 5$$

CHECK

$$\frac{5}{10} + 2 = \frac{5}{2}$$

$$\frac{1}{2} + \frac{4}{2} = \frac{5}{2}$$

$$\frac{5}{2} = \frac{5}{2} \text{ (check)}$$

SAMPLE PROBLEM 2

Solve and check $\frac{2}{3x} - \frac{1}{x^2} = \frac{1}{2x}$

SOLUTION

LCD = $6x^2$**. Restriction:** $x \neq 0$**.**

$$\frac{2\,(2x)}{3x\,(2x)} - \frac{1\,(6)}{xx\,(6)} = \frac{1\,(3x)}{2x\,(3x)}$$

$$\frac{4x - 6}{6x^2} = \frac{3x}{6x^2}$$

$$4x - 6 = 3x$$

$$4x = 3x + 6$$

$$x = 6$$

CHECK

$$\frac{2}{3\,(6)} - \frac{1}{6^2} = \frac{1}{2\,(6)}$$

$$\frac{2}{18} - \frac{1}{36} = \frac{1}{12}$$

$$\frac{4}{36} - \frac{1}{36} = \frac{1}{12}$$

$$\frac{3}{36} = \frac{1}{12}$$

$$\frac{1}{12} = \frac{1}{12} \quad \text{(check)}$$

SAMPLE PROBLEM 3

Solve and check $\frac{1}{x+1} = \frac{3}{x-5}$

SOLUTION

LCD = $(x + 1)(x - 5)$. Restrictions: $x \neq -1, x \neq 5$.

$$\frac{1(x-5)}{(x+1)(x-5)} = \frac{3(x+1)}{(x-5)(x+1)}$$

$$x - 5 = 3(x + 1)$$

$$x - 5 = 3x + 3$$

$$3x + 3 = x - 5$$

$$2x + 3 = -5$$

$$2x = -8$$

$$x = -4$$

CHECK

$$\frac{1}{-4+1} = \frac{3}{-4-5}$$

$$\frac{1}{-3} = \frac{3}{-9}$$

$$-\frac{1}{3} = -\frac{1}{3} \text{ (check)}$$

SAMPLE PROBLEM 4

Solve $\frac{x}{x-6} = \frac{6}{x-6}$

SOLUTION

LCD = $x - 6$. Restriction: $x \neq 6$.

$$(x-6)\left(\frac{x}{x-6}\right) = (x-6)\left(\frac{6}{x-6}\right)$$

$$x = 6$$

Since x is restricted so that $x \neq 6$, 6 can not be a solution. In this case the original equation has no solution. The solution set is the empty set, $\emptyset$.

SAMPLE PROBLEM 5

Solve $\frac{1}{2x} + \frac{1}{3x} = \frac{5}{6x}$

SOLUTION

LCD = 6x. Restriction: $x \neq 0$.

$$\frac{1\,(3)}{2x\,(3)} + \frac{1\,(2)}{3x\,(2)} = \frac{5}{6x}$$

$$3 + 2 = 5$$

$$5 = 5$$

Since 5 = 5 is true no matter what value x has, the original equation is true for all values of x except $x = 0$, the restricted value.

EXERCISES 10.1

Solve and check.

A.

1. $\frac{x}{3} + 6 = x$

2. $5 - \frac{x}{2} = 1$

3. $\frac{y}{6} + \frac{y}{3} = 12$

B.

1. $x + \frac{x}{7} = 16$

2. $\frac{3x}{5} - 2 = \frac{2x}{5}$

3. $\frac{y}{4} - \frac{y}{8} = 5$

A.

4. $\frac{t}{15} + 1 = \frac{t}{10}$

5. $\frac{1}{2x} + \frac{1}{5x} = \frac{1}{10}$

6. $\frac{4}{5x} - \frac{8}{x} = \frac{9}{5}$

7. $\frac{1}{2x} + \frac{1}{4x} + \frac{1}{8x} = \frac{7}{8x}$

8. $\frac{1}{6} + \frac{2}{x} = \frac{x^2 + 12}{6x^2}$

9. $\frac{4}{x - 4} = \frac{x}{x - 4}$

B.

4. $\frac{t}{12} - 2 = \frac{t}{9}$

5. $\frac{3}{x} + \frac{1}{2} = \frac{2}{x}$

6. $\frac{1}{2x} + \frac{1}{3x} + \frac{1}{6x} = 1$

7. $\frac{2}{x} + \frac{5}{x^2} = \frac{4}{x^2}$

8. $\frac{x}{x - 5} = \frac{5}{x - 5}$

9. $\frac{5x}{2x + 7} = 3$

A.

10. $\frac{7}{y+1} = \frac{5}{y-1}$

11. $\frac{3}{t} + \frac{5}{t} = \frac{8}{2t}$

12. $\frac{1}{x} - \frac{1}{2} = \frac{1}{6} - \frac{1}{x}$

13. $\frac{2x+3}{x+5} = \frac{2x-7}{x-1}$

B.

10. $\frac{1}{3y-4} = \frac{1}{4y+5}$

11. $\frac{1}{x} + \frac{1}{5} = \frac{1}{4}$

12. $\frac{x+2}{x+5} = \frac{x}{x+4}$

13. $\frac{3x-2}{3x-1} = \frac{3x-1}{3x+2}$

10.2 FRACTIONAL EQUATIONS, GENERAL TYPES

OBJECTIVE: To solve a general fractional equation.

The general fractional equation is solved by the same method used for the elementary types. However, to find the LCD, it is often necessary to factor one or more denominators first. Moreover, it is important to bear in mind that the fraction bar is a grouping symbol. Many errors can be avoided by enclosing a binomial numerator or denominator within parentheses.

SAMPLE PROBLEM 1

Solve and check $\frac{5}{x+1} - \frac{3}{x-1} = \frac{8}{x^2-1}$

SOLUTION

Factors of $x^2 - 1$ are $(x + 1)(x - 1)$, and $x \neq -1, x \neq 1$.

$$\frac{5}{(x+1)} - \frac{3}{(x-1)} = \frac{8}{(x+1)(x-1)}$$

Rename fractions so each denominator is the LCD, $(x + 1)(x - 1)$.

$$\frac{5(x-1)}{(x+1)(x-1)} - \frac{3(x+1)}{(x-1)(x+1)} = \frac{8}{(x+1)(x-1)}$$

Multiply each side by the LCD.

$$5(x-1) - 3(x+1) = 8$$

Remove parentheses and solve.

$$5x - 5 - 3x - 3 = 8$$

$$2x - 8 = 8$$

$$2x = 16$$

$$x = 8$$

CHECK

$$\frac{5}{8+1} - \frac{3}{8-1} = \frac{8}{8^2-1}$$

$$\frac{5}{9} - \frac{3}{7} = \frac{8}{64-1}$$

$$\frac{5(7)}{9(7)} - \frac{3(9)}{7(9)} = \frac{8}{63}$$

$$\frac{35-27}{63} = \frac{8}{63}$$

$$\frac{8}{63} = \frac{8}{63} \quad \text{(check)}$$

SAMPLE PROBLEM 2

Solve and check $\frac{x+6}{4x+20} - \frac{x-5}{6x+30} = 2.$

SOLUTION

Factoring each denominator,

$$\frac{(x+6)}{2 \cdot 2\,(x+5)} - \frac{(x-5)}{2 \cdot 3\,(x+5)} = 2$$

LCD = $12\,(x+5)$ and $x \neq -5$.

Raising fractions to higher terms,

$$\frac{3\,(x+6)}{3 \cdot 4\,(x+5)} - \frac{2\,(x-5)}{2 \cdot 6\,(x+5)} = \frac{2\,(12)\,(x+5)}{12\,(x+5)}$$

Multiplying by the LCD,

$$3\,(x+6) - 2\,(x-5) = 24x + 120$$

Solving,

$$\begin{aligned} 3x + 18 - 2x + 10 &= 24x + 120 \\ x + 28 &= 24x + 120 \\ 24x + 120 &= x + 28 \\ 23x &= -92 \\ x &= -4 \end{aligned}$$

CHECK

$$\frac{-4+6}{4\,(-4)+20} - \frac{-4-5}{6\,(-4)+30} = 2$$

$$\frac{2}{4} - \frac{-9}{6} = 2$$

$$\frac{1}{2} + \frac{3}{2} = 2$$

$$\frac{4}{2} = 2$$

$2 = 2$ (check)

SAMPLE PROBLEM 3

Solve and check $\frac{x}{x-8}+\frac{9}{8-x}=\frac{1}{2}$

SOLUTION

Since $8-x=-x+8=(-1)(x-8)$, $\frac{9}{8-x}=\frac{9}{(-1)(x-8)}=\frac{-9}{x-8}$

The equation can then be written as $\frac{x}{x-8}+\frac{-9}{x-8}=\frac{1}{2}$. The LCD = $2(x-8)$ and $x \neq 8$.

Raising to higher terms, $\frac{2x}{2(x-8)}+\frac{2(-9)}{2(x-8)}=\frac{x-8}{2(x-8)}$

Multiplying each side by the LCD,

$$2x-18=x-8$$

$$2x=x+10$$

$$x=10$$

CHECK

$$\frac{10}{10-8}+\frac{9}{8-10}=\frac{1}{2}$$

$$\frac{10}{2}+\frac{9}{-2}=\frac{1}{2}$$

$$\frac{10}{2}-\frac{9}{2}=\frac{1}{2}$$

$$\frac{1}{2}=\frac{1}{2}\text{(check)}$$

EXERCISES 10.2

Solve and check.

A.

1. $\frac{x}{2}-\frac{x-1}{3}=2$

B.

1. $\frac{y}{4}-\frac{y+1}{5}=1$

A.

2. $\frac{5}{x-3} - \frac{3}{x} = \frac{3}{x^2 - 3x}$

3. $\frac{3}{y^2 - 4} - \frac{5}{y-2} = \frac{12}{y+2}$

4. $\frac{x+1}{x-2} + \frac{x+2}{x-1} = \frac{2x^2 - 7x + 9}{x^2 - 3x + 2}$

5. $\frac{4}{x-5} + \frac{3}{5-x} = \frac{1}{4}$

6. $\frac{5x+2}{3x+2} = \frac{4}{3x+2} + 1$

7. $\frac{1}{5y-3} + \frac{5y}{50y^2 - 18} = \frac{1}{20y + 12}$

B.

2. $\frac{7}{x+5} - \frac{6}{x} + \frac{20}{x^2 + 5x} = 0$

3. $\frac{6}{y+3} = \frac{30}{y^2 - 9} - \frac{5}{y-3}$

4. $\frac{x-4}{x+6} - \frac{x-6}{x+4} = \frac{4x - 8}{x^2 + 10x + 24}$

5. $\frac{x}{7-x} + \frac{2x}{x-7} = \frac{3}{10}$

6. $\frac{5}{x^2 - 7x + 10} = \frac{6}{x^2 - 9x + 20}$

7. $\frac{8}{x+3} - \frac{7}{x-3} = \frac{1}{x}$

A.

8. $\frac{5}{x^2 - 2x - 8} - \frac{3}{x^2 - 6x + 8} = 0$

9. $\frac{1}{3x^2 - x - 2} = \frac{1}{5x^2 - 7x + 2}$

10. $\frac{x}{(x-6)^2} = \frac{12}{(x+6)^2} + \frac{x}{x^2 - 36}$

B.

8. $2 - \frac{8x}{4x - 7} = \frac{7}{4x^2 - 3x - 7}$

9. $\frac{2}{2x^2 + 5x - 3} + \frac{-3}{2x^2 + 3x - 2} = 0$

10. $\frac{x^2 - 7}{x^2 - 4x} + \frac{x^2 - 1}{x^2 + 4x} = 2$

10.3 RATIO AND PROPORTION

OBJECTIVE: To solve a stated problem concerning ratio and proportion.

Ratio and Proportion

A **ratio** is a fraction.

For example, the ratio of 4 to 5 is the fraction $\frac{4}{5}$. The ratio of a to b is also written as $a{:}b$. The ratio 4:5 (read "four to five") is the fraction $\frac{4}{5}$.

A **proportion** is an equation stating that two ratios are equal; thus, a proportion has the form

$$\frac{a}{b} = \frac{c}{d}.$$

The property of equal fractions is very useful in solving proportions. **The property of equal fractions:**

$$\frac{a}{b} = \frac{c}{d}, \text{ if and only if } ad = bc$$

SAMPLE PROBLEM 1

Two cousins are to divide an inheritance of \$36,000 in the ratio 3:5. How much does each one receive?

SOLUTION

Let x = the amount one receives.

Then 36,000 − x = the amount the other receives.

$$\frac{x}{36{,}000 - x} = \frac{3}{5}$$

$$5x = 3\,(36{,}000 - x)$$

$$5x = 108{,}000 - 3x$$

$$8x = 108{,}000$$

$$x = 13{,}500 \text{ dollars}$$

$$36{,}000 - x = 22{,}500 \text{ dollars}$$

CHECK

$$\frac{13{,}500}{22{,}500} = \frac{135}{225} = \frac{3\,(45)}{5\,(45)} = \frac{3}{5}$$

If two numbers have the ratio $\frac{a}{b}$, then the numbers may be expressed as ax and bx, since $\frac{ax}{bx} = \frac{a}{b}$, where x is not zero.

SAMPLE PROBLEM 1: ALTERNATE SOLUTION

Since $\frac{3x}{5x} = \frac{3}{5}$, the problem can also be worked as follows.

Let $3x$ = the amount that one receives; then $5x$ = the amount the other receives.

$$3x + 5x = 36{,}000$$

$$8x = 36{,}000$$

$$x = 4{,}500$$

$$3x = 13{,}500 \text{ dollars}$$

$$5x = 22{,}500 \text{ dollars}$$

SAMPLE PROBLEM 2

A certain city code requires that for a certain room the ratio of the total area of the windows to the area of the floor of the room must be at least 1 to 8. Find the total area of the windows that will just meet this requirement for a room with the dimensions of 14 ft. by 12 ft.

SOLUTION

Let x = the total area of the window.

The area of the floor = (14) (12) = 168 sq. ft.

Equate the ratios.

$$\frac{x}{168} = \frac{1}{8}$$

$$8x = 168$$

$$x = 21 \text{ sq. ft.}$$

If three numbers are in the ratio $a{:}b{:}c$, then the numbers may be expressed as ax, bx, and cx, where $x \neq 0$.

SAMPLE PROBLEM 3

Three cousins are to divide an inheritance of \$36,000 in the ratio 3:5:8. How much does each receive?

SOLUTION

Let $3x$, $5x$, and $8x$ be the amounts they receive.

Then

$$3x + 5x + 8x = 36{,}000$$

$$16x = 36{,}000$$

$$x = 2{,}250$$

$$3x = 6{,}750 \text{ dollars}$$

$$5x = 11{,}250 \text{ dollars}$$

$$8x = 18{,}000 \text{ dollars}$$

EXERCISES 10.3

A.

1. A piece of wire 18 in. long is to be divided into 2 pieces whose lengths have the ratio 5 to 7. Find the length of each piece.

2. The three angles of a triangle are in the ratio 2:3:4. Find the three angles. (The sum of the angles in any triangle is 180°.)

3. A profit of $25,000 on the sale of a certain house was split between the owner and the sales agent in the ratio of 17 to 3. How much did each receive?

4. The scale on a certain blueprint reads "1 in. = 4 ft." How many inches on the paper are needed to draw the length of a house 64 ft. long?

5. In a certain community the ratio of eligible voters under 21 to the eligible voters 21 or over is 1 to 6. If the number of eligible voters 21 or over is 7200, how many eligible voters are under 21?

6. A speed of 30 m.p.h. is the same as a speed of 44 ft. per second.
   a. What speed in feet per second is equal to a speed of 45 m.p.h.?
   b. What speed in miles per hour is equal to a speed of 26,400 ft. per second, the reported speed of the first man-made satellite?

7. A pattern requires $3\frac{1}{2}$ yd. of material. If the material is sold by the meter, and if 100 meters is about 109 yd., how many meters of material should be bought?

8. If 100 liters equal approximately 106 qt., how many gallons of gasoline is 18 liters? (1 gal. = 4 qt.)

9. How many moles of hydrogen are in 0.42 mole of water if water consists of 2 moles of hydrogen for every mole of oxygen?

10. If the school tax rate is $35.20 per $1000 of assessed valuation, what will be the tax on property assessed at $16,000?

B.

1. The ratio of the length of 1 of the 2 equal sides of an isosceles triangle to the length of the third unequal side is 2 to 3. The perimeter of the triangle is 84 in. Find the lengths of the sides of the triangle.

2. A certain paint is a mixture of pigment and vehicle, in the ratio of 1 to 5. How much pigment and how much vehicle are needed to make 300 lb. of this paint?

3. The ratio of A's to B's in a certain class was 2:3. If there were 12 B's, how many A's were there?

4. On a certain day, the ratio of Swiss francs to U.S. dollars was 6 to 25. What was the cost in U.S. dollars of an article costing 12 francs?

5. The ratio by weight of one large egg to one medium egg is 3 to 2. If medium eggs cost 63¢ per dozen and if large eggs cost 72¢ per dozen, which is the better buy?

6. In a certain electrical circuit, the ratio of the voltage to the amperage is 40 to 1.

   a. Find the voltage if the amperage is 5.5.
   b. Find the amperage if the voltage is 110 volts.

7. The pitch of a roof is the ratio of its height to its half-span. Find the height of a roof whose half-span is 24 ft. and whose pitch is 1 to 3.

8. The ratio of the width of a book to its length is to be 1 to 1.618 (the Golden Ratio considered by the ancient Greeks to be the most pleasing visually). Find the length if the width is 25 cm.

9. Two and a half palas of saffron are purchased for $\frac{3}{7}$ of a niska. How many will be purchased for 9 niskas? (From the work of the Hindu Bhaskara, about 1150.)

10. A certain king sent 30 men into his orchard to plant trees. If they could set out 1000 trees in 9 days, in how many days would 36 men set out 4400 trees? (From the *Liber Abaci,* written in 1202 by the Italian mathematician Leonardo Fibonacci.)

10.4 UNIFORM MOTION PROBLEMS

OBJECTIVE: To solve a stated uniform motion problem involving fractions.

Solving $d = rt$ for t and for r,

$$t = \frac{d}{r} \quad \text{and} \quad r = \frac{d}{t}.$$

(Recall that $d = rt$ states that the distance d is the product of the uniform rate r and the time t.)

If a problem states a relationship between times or rates, a fractional equation may be needed to describe this situation.

Suggested Procedure

(1) Let a variable be an unknown rate or time.

(2) Make a table comparing the rates, times, and distances of the trips:

	one trip	other trip
distance		
rate		
time		

(3) Use the information from the problem to fill in the entries for the distances and the rates (or the times).

(4) Use the formula $t = \frac{d}{r}\left(\text{or } r = \frac{d}{t}\right)$ to fill in the entry for the times (or the rates).

(5) Form an equation using the statement in the problem that relates the times (or the rates).

(6) Solve the equation.

After some practice, and according to the difficulty of the problem, the table can often be omitted.

SAMPLE PROBLEM 1

John can walk twice as fast on level road as he can walk uphill. One day it took him a total of 3 hr. to walk 5 mi. on level road and 2 mi. uphill. Find his rate of walking on the level road and his rate of walking uphill.

SOLUTION

(1) Let x = rate of walking uphill, then $2x$ = rate of walking on level road.

		uphill	level
(2)			
(3)	distance	2	5
	rate	x	$2x$
(4)	time, $\frac{d}{r}$	$\frac{2}{x}$	$\frac{5}{2x}$

(5) **time uphill + time on level road = total time**

$$\frac{2}{x} + \frac{5}{2x} = 3$$

(6)

$$(2x)\left(\frac{2}{x}\right) + (2x)\left(\frac{5}{2x}\right) = (2x)(3)$$

$$4 + 5 = 6x$$

$$x = \frac{9}{6} = \frac{3}{2} = 1\frac{1}{2} \text{ m.p.h.}$$

$$2x = 3 \text{ m.p.h.}$$

SAMPLE PROBLEM 2

A train took 2 hr. longer to travel 375 mi. than it took on a trip of 250 mi. If the train traveled at the same rate of speed on both trips, find the time for each trip.

SOLUTION

(1) Let x = time for trip of 250 miles; then $x + 2$ = time for trip of 375 miles.

		375-mi. trip	250-mi. trip
(2)			
(3)	distance	375	250
	time	$x + 2$	x
(4)	rate	$\frac{375}{x+2}$	$\frac{250}{x}$

(5) rate for 375 mi. = rate for 250 mi.

$$\frac{375}{x+2} = \frac{250}{x}$$

(6)

$$375x = 250\,(x + 2)$$

$$375x = 250x + 500$$

$$125x = 500$$

$$x = 4 \text{ hr.}$$

$$x + 2 = 6 \text{ hr.}$$

If an object travels in water or in air where there is a water current or a wind current, then the speed of the current affects the speed of the object.

If r = the speed of an object in still water or still air and s = the speed of the current, then $r + s$ = the speed with the current (downstream or with a tailwind), and $r - s$ = the speed against the current (upstream or with a headwind.)

SAMPLE PROBLEM 3

A boat has a maximum speed in still water of 18 m.p.h. One day, traveling at full speed, the boat took twice as long to go 42 mi. upstream as it did to go 33 mi. downstream. Find the rate of the current.

SOLUTION

(1) Let x = the rate of the current; then $18 - x$ = the rate upstream and $18 + x$ = the rate downstream.

(2)		upstream	downstream
(3)	distance	42	33
	rate	$18 - x$	$18 + x$
(4)	time, $\frac{d}{r}$	$\frac{42}{18 - x}$	$\frac{33}{18 + x}$

(5) time upstream = 2 (time downstream)

$$\frac{42}{18 - x} = \frac{2\,(33)}{18 + x}$$

(6)

$$42\,(18 + x) = 66\,(18 - x)$$

$$66x + 42x = 66\,(18) - 42\,(18)$$

$$108x = 18\,(66 - 42) = 18\,(24) = 432$$

$$x = \frac{432}{108}$$

$$x = 4 \text{ m.p.h.}$$

CHECK

Time upstream $= \frac{42}{18 - 4} = \frac{42}{14} = 3$ hr.

Time downstream $= \frac{33}{18 + 4} = \frac{33}{22} = 1\frac{1}{2}$ hr.

Time upstream should be twice the time downstream: $2\left(1\frac{1}{2}\right) = 3$.

A.

EXERCISES 10.4

1. At 1 P.M. a man left home and walked 14 mi. to his friend's house. He visited with his friend for 1 hr., and then his friend drove him home, where he arrived at 6 P.M. If the rate at which his friend drove was 7 times the rate at which he walked, how much time did it take him to walk to his friend's house?

2. It took a man a total of 6 hr. to row 18 mi. downstream and then return. If his rate of rowing downstream was 3 times his rate of rowing upstream, find his upstream rate.

3. A man traveled 144 mi. at a certain uniform speed. Because of a detour, he then traveled 80 mi. further at $\frac{2}{3}$ his original speed. If the entire trip took $5\frac{1}{2}$ hr., find his two speeds.

4. A family on a vacation drove 165 mi. on a freeway and then 45 mi. on a winding road on which their speed was reduced by 25 m.p.h. If the time spent on the freeway was twice the time spent on the winding road, find their speeds for each part of the trip.

5. A certain helicopter flies 180 m.p.h. in still air. One day it took the same time to fly 250 mi. with a wind as it did to return a distance of 200 mi. against the same wind. Find the rate of the wind.

6. Traveling at full speed against a current of 5 m.p.h., a motorboat went 14 mi. Traveling at half speed with the same current, the motorboat went 10 mi. If both trips took the same time, find the maximum speed of the boat in still water.

7. A motorist has a choice of two routes between two cities: highway 12 and highway 48. The distance along highway 12 is 180 mi., while the distance along highway 48 is 120 mi. However, since he can travel twice as fast on highway 12, it takes him 1 hr. less time using this route. Find the time it takes him to make the trip along highway 12.

B.

1. A boat traveling at maximum speed takes $2\frac{1}{2}$ hr. longer to travel a distance of 120 mi. than to travel a distance of 30 mi. at the same maximum speed. Find the maximum speed of the boat.

2. Two trains left the same station and traveled in opposite directions. At the end of 2 hr. they were 210 mi. apart. If the ratio of their rates was 2:5, find the rate of each.

3. A certain airplane can fly 160 m.p.h. in still air. It can fly 290 mi. against a certain wind in the same time that it can fly 350 mi. with the wind. Find the rate of the wind.

4. A salesman traveled 900 mi. by airplane and 45 mi. by bus. The total time of the trip was $3\frac{1}{2}$ hr. If the rate of the plane was 8 times the rate of the bus, find the time he traveled by plane.

5. A truck left a certain warehouse and traveled at the rate of 45 m.p.h. One hour and 20 min. later, a car traveling at 60 m.p.h. left the warehouse in order to overtake the truck. For how long a time does the car travel before it overtakes the truck?

6. On a nonstop flight from Portland to Sacramento (a distance of 600 mi.), a commercial jet plane flies against a wind current of 25 m.p.h. A smaller private plane whose speed in still air is $\frac{1}{3}$ the speed of the jet in still air takes $3\frac{1}{3}$ times as long to make the same flight against the same wind current. Find the speed of each plane in still air.

7. Cyclist A covers 6 laps of an oval race track in the same time that cyclist B covers 5 laps. In a 20-mi. race, A gives B a 14-min. head start and still beats B by 1 mi. Find A's speed in miles per hour.

10.5 WORK PROBLEMS

OBJECTIVE: To solve a stated work problem.

If a bricklayer can build a certain fireplace in 4 hr., then he builds $\frac{1}{4}$ of the fireplace in 1 hr. The fraction $\frac{1}{4}$ is the hourly rate at which the bricklayer works. The rate of work plays an important role in work problems.

Basic Idea 1. Let T = the time to do a certain job. Then $r = \frac{1}{T}$ is the rate at which the job is done.

Example 1. A painter paints a certain room in 50 min. What is his rate of working?

Solution. Rate $= \frac{1}{50}$ of the room per minute.

Basic Idea 2. The amount of work done is the product of the rate and the time spent working: $W = rt$.

Example 2. John can build a certain type of fence in 5 hr. If he works for 3 hr., how much of the fence does he build?

Solution. Using $r = \frac{1}{T}$, where T = time for the complete job, $r = \frac{1}{5}$.

Using $W = rt$, $W = \frac{1}{5}(3) = \frac{3}{5}$ of the fence.

Basic Idea 3. The total work done is the sum of the amounts of work done by those persons (or machines) working.

Example 3. In 1 hr., machine A can process $\frac{1}{10}$ while machine B can process $\frac{1}{15}$ of a certain amount of data. How much data is processed if A works for 3 hr. and B works for 4 hr.?

Solution. Total work = (work of A) + (work of B).

$$W = 3\left(\frac{1}{10}\right) + 4\left(\frac{1}{15}\right) = \frac{3}{10} + \frac{4}{15} = \frac{17}{30}$$

Therefore, $\frac{17}{30}$ of the data is processed.

Basic Idea 4. If the whole job is completed, the amount of work done is 1; therefore, $W = 1$.

Example 4. If John does $\frac{1}{30}$ of a job in 1 min. and Bill does $\frac{1}{50}$ of the same job in 1 min., how many minutes will it take them working together to do the whole job?

Solution. John does $\frac{x}{30}$ of the job, and Bill does $\frac{x}{50}$ of the job. Together, $\frac{x}{30} + \frac{x}{50} = 1$.

$$150\left(\frac{x}{30}\right) + 150\left(\frac{x}{50}\right) = 150$$

$$5x + 3x = 150$$

$$8x = 150$$

$$x = \frac{75}{4} = 18\frac{3}{4} \text{ hr.}$$

Questions on Basic Ideas

1. Bill takes 40 min. to wash his car. What is his rate of working?

____

2. A pipe takes 5 hr. to fill a certain tank with water. What is the rate at which the pipe fills the tank? ______________

3. A man requires x hr. to brick a certain fireplace. What is the rate at which he works? _______________

4. A machine takes $x + 20$ min. to make a certain article. What is the rate at which the machine works? _______________

5. A roofing crew takes 10 days to roof a certain building. How much of the work does the crew do in 6 days? _______________

6. An outlet pipe takes x hr. to empty a swimming pool. How much of the pool is emptied in 2 hr.? _______________

7. One pipe takes 36 min. to fill a pool. Another pipe takes 45 min. to fill the same pool. If both pipes are open, how much of the pool do they fill in x min.? _______________

8. If John and Bill work together and John does $\frac{1}{4}$ of the work, how much of the work does Bill do if together they do all of the work? _______________

9. Two machines work together and process a certain amount of data. If one machine processes $\frac{1}{3}$ of the data, how much of the data does the other machine process? _______________

10. Two chutes working together fill a certain granary with wheat. If one chute fills $\frac{1}{5}$ of the granary and if the other chute fills $\frac{1}{x}$ of the granary, then $\frac{1}{x} + \frac{1}{5} =$ _______________

SAMPLE PROBLEM 1

A new machine can process a certain amount of data in 30 min., while an old machine can process the same amount of data in 60 min. How long would it take both machines working together to process the same amount of data?

SOLUTION

Let x = the time for both working together.

$\frac{1}{30}$ = the rate of the new machine.

$\frac{1}{60}$ = the rate of the old machine.

Use the formula $RT = W$; total work = 1, since the whole job is done.

$$\frac{x}{30} + \frac{x}{60} = 1$$

$$2x + x = 60$$

$$3x = 60$$

$$x = 20 \text{ min.}$$

SAMPLE PROBLEM 2

A man can build a certain type of fireplace in 4 hr. If his son helps him, then working together they can build the fireplace in 3 hr. How long would it take the son working alone to build the fireplace?

SOLUTION

Let x = the time of the son, $\frac{1}{x}$ = the rate of the son, and $\frac{1}{4}$ = the rate of the father.

Use $RT = W$ and total work = 1.

$$3\left(\frac{1}{4}\right) + 3\left(\frac{1}{x}\right) = 1$$

$$\frac{3}{4} + \frac{3}{x} = 1$$

$$3x + 12 = 4x$$

$$x = 12 \text{ hr.}$$

SAMPLE PROBLEM 3

One pipe takes 10 min. to fill a certain tank full of water. A second pipe takes 15 min. to fill the tank, and a third pipe takes 30 min. to fill the tank. If all 3 pipes are open, how long would it take to fill $\frac{3}{4}$ of the tank?

SOLUTION

Let x = the time to fill $\frac{3}{4}$ of the tank. Then $\frac{x}{10} + \frac{x}{15} + \frac{x}{30} = \frac{3}{4}$.

$$60\left(\frac{x}{10} + \frac{x}{15} + \frac{x}{30}\right) = 60\left(\frac{3}{4}\right)$$

$$6x + 4x + 2x = 45$$

$$12x = 45$$

$$x = \frac{15}{4}$$

$$x = 3\frac{3}{4} \text{ min.}$$

SAMPLE PROBLEM 4

It takes an outlet pipe 4 times as long to empty a tank as it takes an inlet pipe to fill the tank. One day both pipes were open, and it took 4 hr. to fill the tank. How long does it take the inlet pipe to fill the tank when the outlet pipe is closed?

SOLUTION

Let x = time of inlet pipe.

Then $4x$ = time of outlet pipe.

$\frac{4}{x} + \frac{-4}{4x} = 1$ Since the outlet pipe does negative work toward filling the tank, the work done by this pipe is $-4\left(\frac{1}{4x}\right)$.

$$\frac{4}{x} - \frac{1}{x} = 1$$

$$\frac{3}{x} = 1$$

$$x = 3 \text{ hr.}$$

SAMPLE PROBLEM 5

A man can mow his lawn in 35 min., and his son can mow the lawn in 50 min. One day the man started to mow the lawn alone. At the end of 18 min., his son joined him, and they completed the job. How long did the son work?

SOLUTION

Let x = the time the son worked.

	rate	time	work
father	$\frac{1}{35}$	$x + 18$	$\frac{x+18}{35}$
son	$\frac{1}{50}$	x	$\frac{x}{50}$

$$\frac{x+18}{35} + \frac{x}{50} = 1$$

$$350\left(\frac{x+18}{35} + \frac{x}{50}\right) = 350\,(1)$$

$$10\,(x + 18) + 7x = 350$$

$$17x = 170$$

$$x = 10 \text{ min.}$$

EXERCISES 10.5

A.

1. A painter working alone could paint a certain house in 30 hr., while it would take his helper 70 hr. working alone. How long would it take them to paint the house if they worked together?

2. One company could build a certain bridge in 24 days. To rush the work, another company is called in; and both companies working together build the bridge in 15 days. How long would it have taken the second company working alone to build the bridge?

3. Three machines, each working alone, can do a certain job in 10 min., 15 min., and 30 min., respectively. How long would it take to complete the job if all 3 machines were in operation at the same time?

4. An inlet pipe can fill a tank in 24 min. An outlet pipe can empty the tank in 20 min. If the tank is $\frac{3}{4}$ full, how long would it take to empty the tank if both pipes were open?

5. A cabinet maker can make a certain set of cabinets in 21 days. His assistant working alone would take 28 days to make the same set of cabinets. The cabinet maker started to make the cabinets. After he had worked for 7 days alone, his assistant then helped him to complete the job. How long did his assistant work?

B.

1. One machine can produce 100 articles in 20 min., while it takes another machine 1 hr. to produce 100 of the same type of article. How long would it take to produce 100 articles of the same type if both machines were in operation at the same time?

2. One inlet pipe can fill a certain tank in 75 min. It is desired to install a second inlet pipe so that when both pipes are open, the tank will be filled in 30 min. How many minutes should it take the second inlet pipe alone to fill the tank?

3. When 3 machines are in operation together, they can do a certain job in 12 hr. The second machine alone takes twice as long as the first machine alone to do the job. The third machine alone takes 3 times as long as the first machine alone to do the job. Find the times that it takes each of the 3 machines alone to do the job.

4. It takes an old machine 5 times as long to disc a certain field as it does a new one. Both machines started to disc the field. At the end of $2\frac{1}{2}$ hr., the old machine broke down. It took $1\frac{1}{2}$ hr. more for the new machine to complete the job. Find the time it would have taken the new machine if it had done the whole job alone.

5. A mechanic can do a certain job in 12 hr. He works alone on the job for 8 hr. but then is called away on an emergency. His helper completes the job in 10 more hr. How long would it have taken the helper to do the entire job alone?

UNIT 10 POSTTEST

(1–4) Solve and check. (10.1)

1. $\dfrac{1}{36}+\dfrac{1}{45}=\dfrac{1}{x}$

2. $\dfrac{x}{54-x}=\dfrac{1}{8}$

3. $\dfrac{3}{25+x}=\dfrac{2}{25-x}$

4. $\dfrac{x-2}{x-5}=\dfrac{x}{x-3}$

(5–10) Solve and check. (10.2)

5. $\dfrac{x-2}{x+2}=1+\dfrac{x-3}{(x+2)^2}$

6. $\dfrac{1}{x-3}-\dfrac{1}{x+3}=\dfrac{3x}{x^2-9}$

7. $\dfrac{1}{x-3}-\dfrac{1}{x+3}=\dfrac{2x}{x^2-9}$

8. $\dfrac{3}{x-4}-\dfrac{2}{x+3}=\dfrac{x+17}{x^2-x-12}$

9. $\dfrac{2x}{5x-10}+\dfrac{3x}{5x+10}=1$

10. $\dfrac{x}{x-4}-\dfrac{2x+3}{2x-4}=\dfrac{3}{2x-8}$

11. A tourist in France bought 3 kg. of fruit for 165 francs. If 10 kg. is approximately 22 lb., what was the price in francs of 1 lb. of fruit? (10.3)

12. The ratio of sodium to chlorine in common table salt is 35 to 23. Find the amount of each of these elements in 145 gm. of salt. (10.3)

13. An ocean liner made a 3500-mi. trip and then returned. Its rate going was $\frac{2}{3}$ its rate returning. If it had traveled at a rate of 48 m.p.h. both going and returning, it would have made the round trip in the same amount of time. Find the rate going and the rate returning. (10.4)

14. A new machine can manufacture certain parts 5 times as fast as an old one. With both machines operating, 500 parts can be made in 30 min. How long would it take each machine alone to make 500 parts? (10.5)

UNIT 11

GRAPHING POINTS AND LINES

Unit 11
PRETEST

(1–4) Graph on a horizontal number line. (11.1)

1. $\{x|x < 5\}$

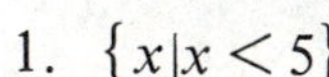

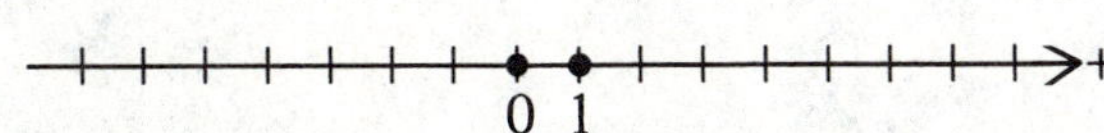

2. $\{x|x \geqslant -5\}$

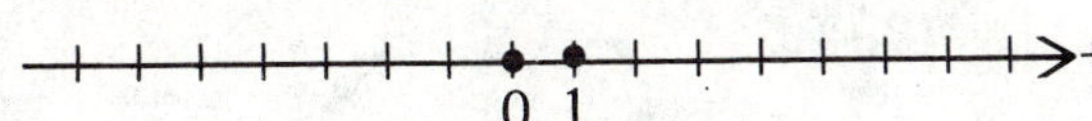

3. $\{x|-2 < x < 2\}$

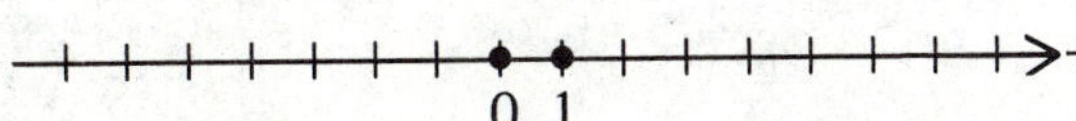

4. $\{x|-3 \leqslant x \leqslant 5\}$

(5–8) Graph on a vertical number line. (11.1)

5. $\{y|y > -4\}$

6. $\{y|y \leqslant 2\}$

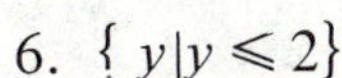

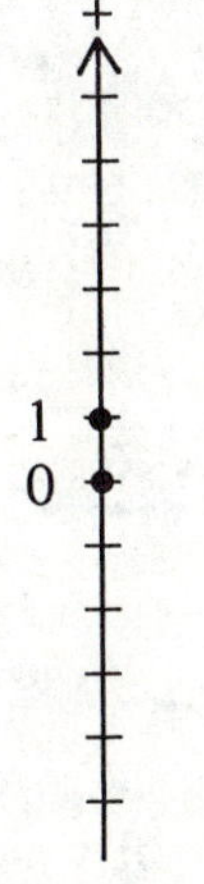

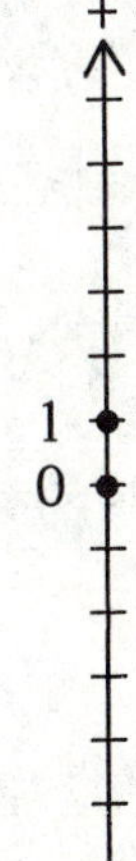

7. $\{y|-4 < y \leq 1\}$

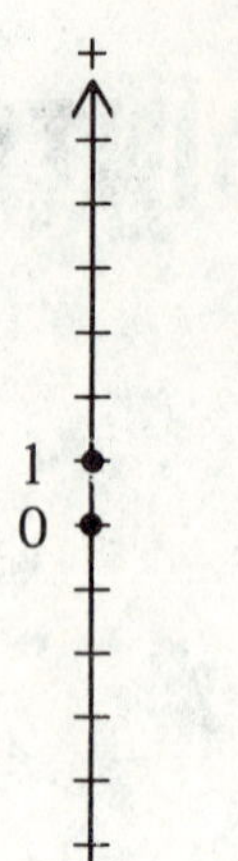

8. $\{y|0 < y < 5\}$

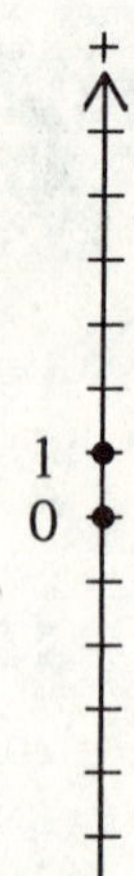

(9–13) Graph each point on the number plane at the right. (11.2)

9. A (3,4)
10. B (5,-2)
11. C (-4,-5)
12. D (-2,3)
13. E (0,4)

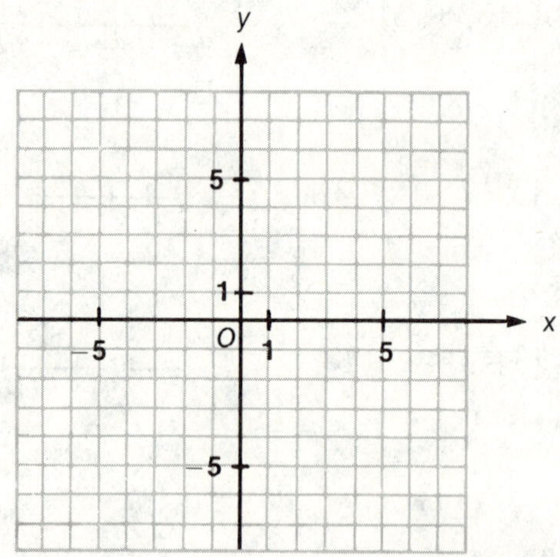

(14–18) State the coordinates of each point whose graph is shown on the number plane at the right. (11.2)

14. P (,)
15. Q (,)
16. R (,)
17. S (,)
18. T (,)

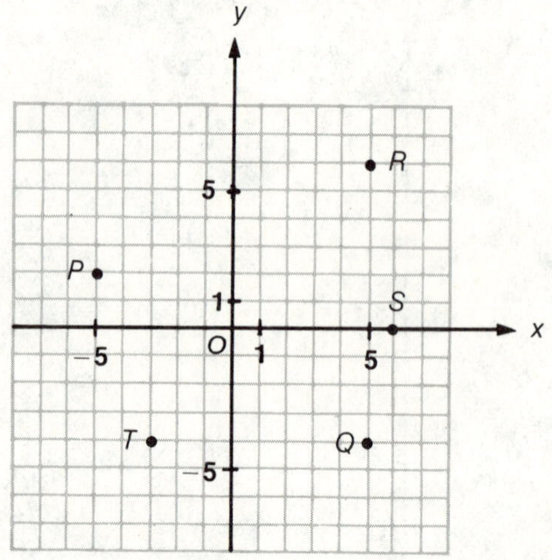

(19–24) Determine if the given ordered pair is a solution of $2x + 3y = 12$. (11.3)

19. (3,2)

20. (2,3)

21. (0,4)

22. (4,0)

23. (9,-2)

24. (6,0)

25. Find the solution of $4x + y = 17$ for which $x = 3$. (11.4)

26. Find the solution of $5x - 2y = 2$ for which $y = 4$. (11.4)

27. Graph $y = 2x - 5$ by first completing the table. (11.5)

x	y
-2	
0	
4	

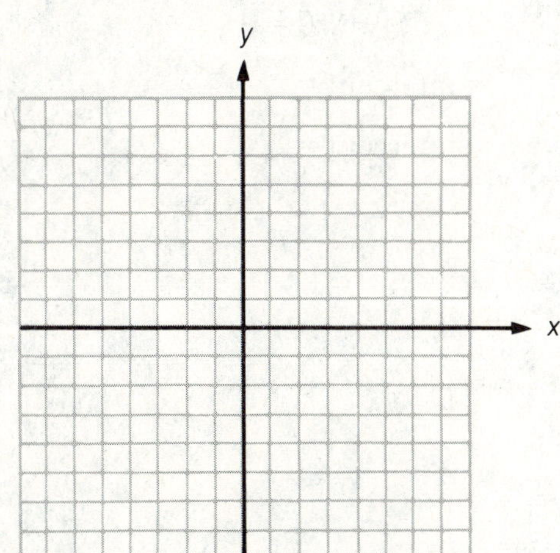

28. Graph $y = 6 - 2x$. (11.5)

x	y

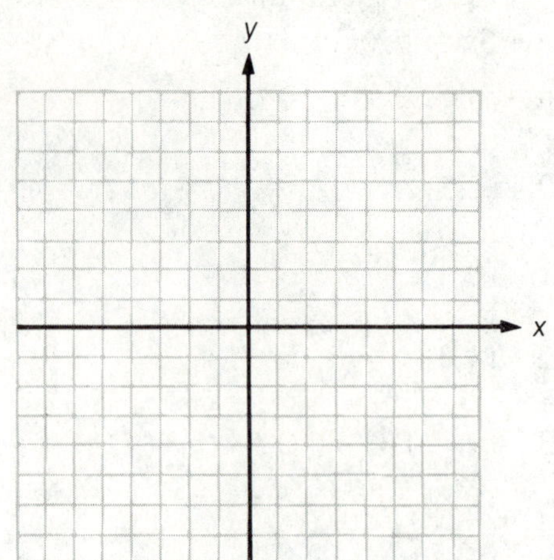

29. State the x- and y-intercepts and graph. (11.6)

$x + 3y = 9$

x-intercept =

y-intercept =

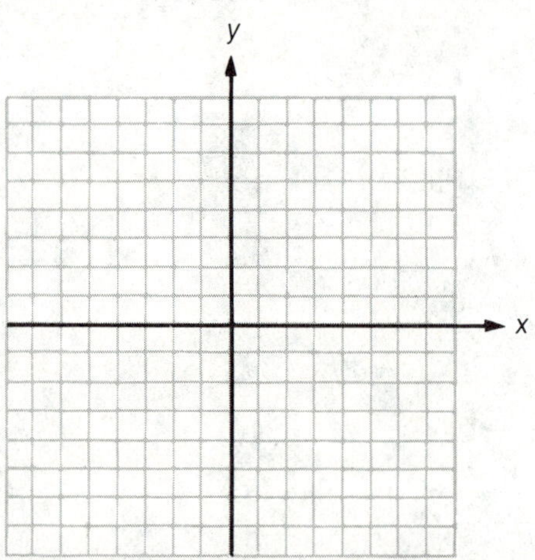

30. State the x- and y-intercepts and graph. (11.6)

$2x - 5y + 10 = 0$

x-intercept =

y-intercept =

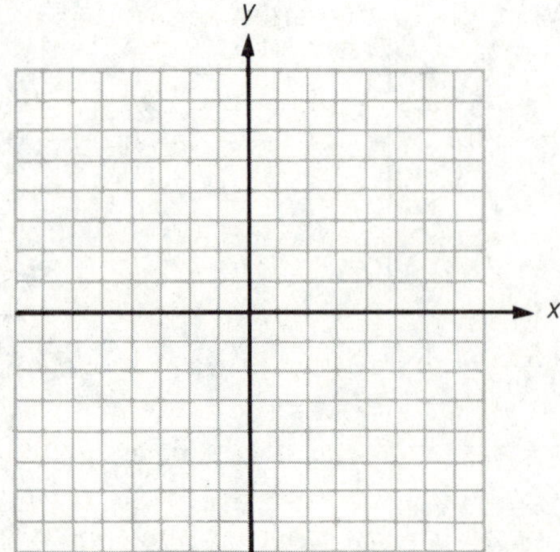

(31–32) Find the slope of the line joining the two given points. Check by graphing the line. (11.7)

31. $P(6,-3), Q(12,5)$

32. $P(-4,1), Q(-6,7)$

(33–34) State the slope and y-intercept of each line whose equation is given. Graph each line, using the slope and y-intercept. (11.8)

33. $y = 4x - 8$

slope = ______

y-intercept = ______

34. $5x + 3y - 15 = 0$

slope = ______

y-intercept = ______

11.1 GRAPHING INTERVALS ON A REAL NUMBER LINE

OBJECTIVE: To graph an interval on a real number line.

Is there a number x such that $x^2 = 2$? There is no integer whose square is 2. Moreover, there is no rational number whose square is 2.

The symbol $\sqrt{2}$ (read "positive square root of 2") is used to name a positive number whose square is 2; $(\sqrt{2})^2 = 2$.

Similarly $\sqrt{5}$ is a positive number such that $(\sqrt{5})^2 = 5$.

Numbers such as $\sqrt{2}, \sqrt{5}, -\sqrt{2}$, and $-\sqrt{5}$ are called **irrational numbers.** They occupy positions on the number line just as the rational numbers do. If the set of rational numbers is extended to include all irrational numbers, then the new set is called the set of **real numbers.** For each real number, there is one and only one point on a number line, and for each point on the line, there is one and only one real number for its coordinate.

The **set of real numbers** includes all numbers that are the coordinates of points on a number line.

An **interval** is a set of points on a number line. Important basic types of intervals are shown in the following discussion.

Type 1 Interval. $\{x|x > a\}$, the set of all points x such that x is greater than a.

Example 1. Graph $\{x|x > 2\}$ on a horizontal number line.

Solution

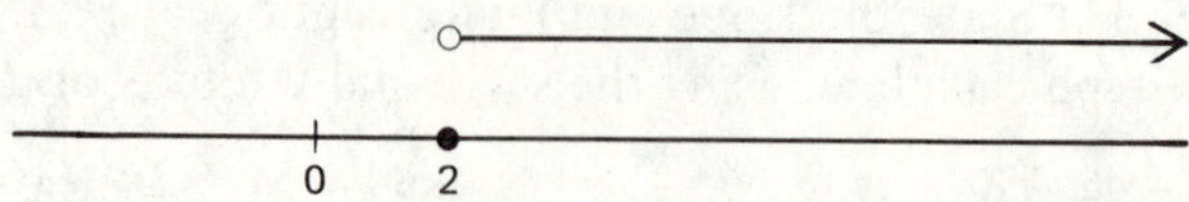

The open circle above 2 indicates that 2 is not included in the interval.

The arrowhead indicates that all points to the right of 2 are included in the interval.

Type 2 Interval. $\{x|x \geqslant a\}$, the set of all points x such that x is greater than or equal to a.

Example 2. Graph $\{x|x \geqslant -3\}$ on a horizontal number line.

Solution

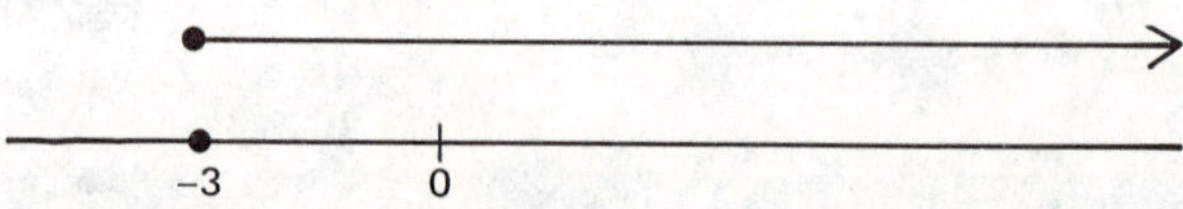

The solid circle above −3 indicates that −3 is included in the interval.

The arrowhead indicates that all points to the right of −3 are included in the interval.

Type 3 Interval. $\{x|x < a\}$, the set of all points x such that x is less than a.

Example 3. Graph $\{x|x < 1\}$ on a horizontal number line.

Solution

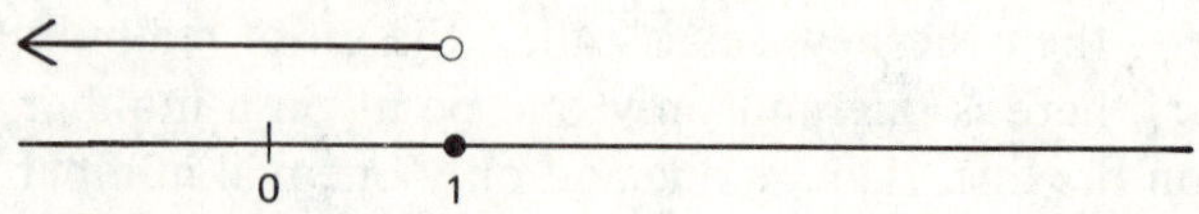

Type 4 Interval. $\{x|x \leqslant a\}$, the set of all points x such that x is less than or equal to a.

Example 4. Graph $\{x|x \leqslant -2\}$ on a horizontal number line.

Solution

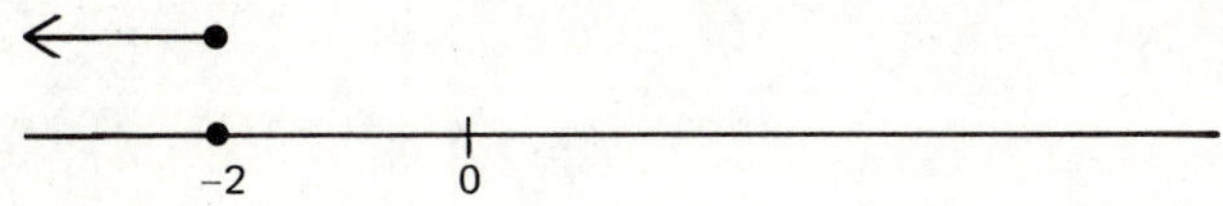

Type 5 Interval. $\{x|a < x < b\}$, the set of all points x such that x is between a and b. This interval is called an open interval.

Example 5. Graph $\{x|-1 < x < 2\}$ on a horizontal number line.

Solution

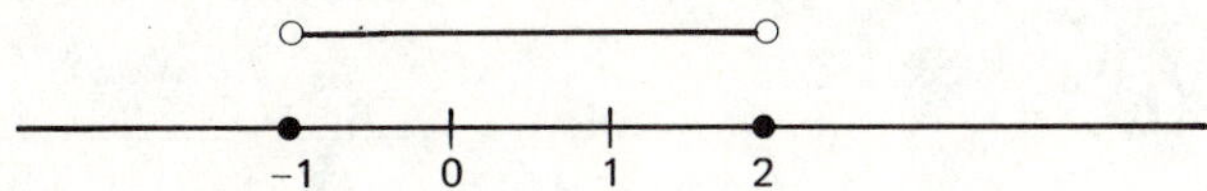

Type 6 Interval. $\{x|a \leqslant x \leqslant b\}$, the set of all points x such that x is between a and b or x is equal to a or b. This interval is called a closed interval.

Example 6. Graph $\{x|-1 \leqslant x \leqslant 2\}$ on a horizontal number line.

Solution

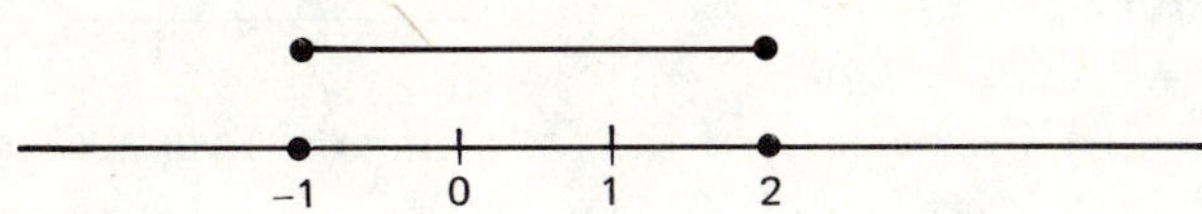

The Vertical Graph. An interval may be graphed on a vertical number line. Also, combinations or variations of the above basic types may be graphed on one number line.

Example 7. Graph $\{y|-2 < y \leqslant 3\}$ on a vertical number line. (This interval is called half open or half closed. It is open on the left and closed on the right.)

Solution

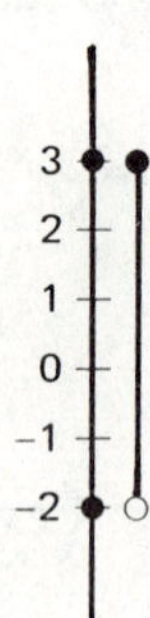

EXERCISES 11.1

(1–10) Graph each of the following on a horizontal number line.

A.

1. $\{x|x < -3\}$ ____________________

2. $\{x|x \leqslant 3\}$ ____________________

3. $\{x|x > -4\}$ ____________________

4. $\{x|x \geqslant 4\}$ ____________________

5. $\{x|-2 \leqslant x \leqslant 3\}$ ____________________

6. $\{x|0 < x < 4\}$ ____________________

7. $\{x|2 \leqslant x < 5\}$ ____________________

8. $\{x|-4 < x \leqslant 0\}$ ____________________

9. $\{x|x \leqslant -3 \text{ or } x \geqslant 3\}$ ____________________

10. $\{x|x < -2 \text{ or } x > 2\}$ ____________________

(11–20) Graph each of the following on a vertical number line.

11. $\{y|y > 2\}$

12. $\{y|y \leqslant 5\}$

13. $\{y|1 \leqslant y \leqslant 6\}$

14. $\{y|y < -3\}$

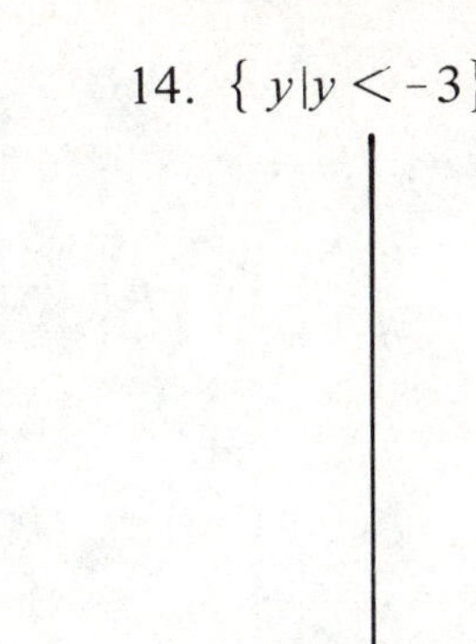

15. $\{y|y \geqslant -1\}$

16. $\{y|-3 < y < 3\}$

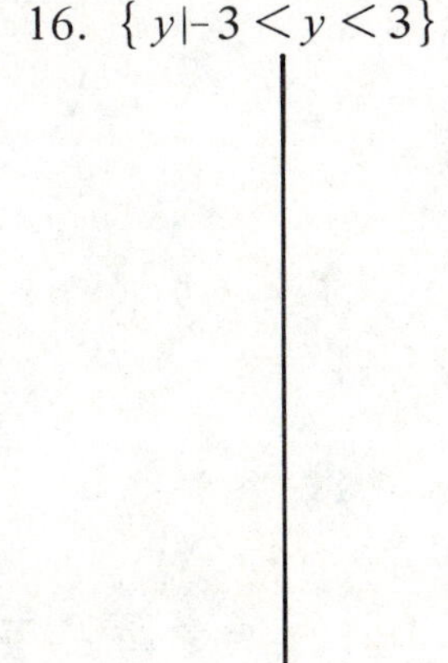

17. $\{y|y \geqslant 0\}$

18. $\{y|-2 < y \leqslant 4\}$

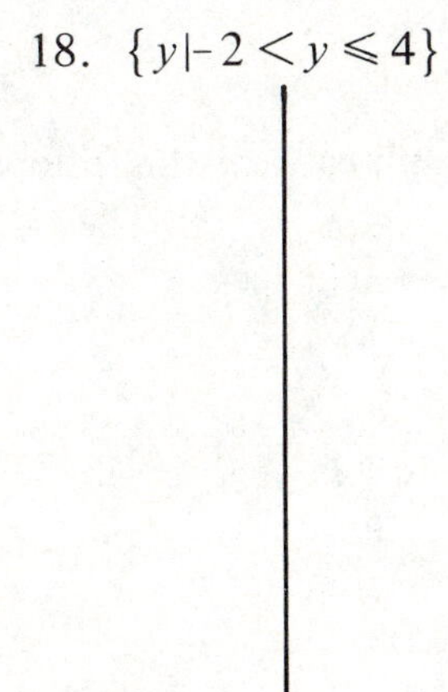

19. $\{y \mid -5 \leqslant y < -1\}$

20. $\{y \mid y \leqslant -1 \text{ or } y \geqslant 1\}$

B.

(1–10) Graph each of the following on a horizontal number line.

1. $\{x \mid x > 5\}$
2. $\{x \mid x \geqslant -5\}$
3. $\{x \mid x \leqslant 0\}$
4. $\{x \mid x < 4\}$
5. $\{x \mid -5 < x < 5\}$
6. $\{x \mid -3 \leqslant x \leqslant -1\}$
7. $\{x \mid -6 \leqslant x < 0\}$
8. $\{x \mid 3 < x \leqslant 6\}$
9. $\{x \mid x < -3 \text{ or } x \geqslant 4\}$
10. $\{x \mid x \leqslant 1 \text{ or } x > 5\}$

(11–20) Graph each of the following on a vertical number line.

11. $\{y \mid y < 0\}$

12. $\{y \mid y \geqslant -2\}$

13. $\{y \mid y \leqslant -4\}$

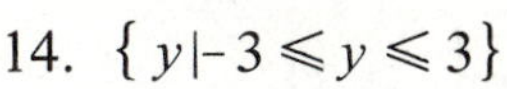

14. $\{y \mid -3 \leqslant y \leqslant 3\}$

15. $\{y \mid y \geqslant 1\}$

16. $\{y \mid -4 < y \leqslant 0\}$

17. $\{y \mid 2 \leqslant y < 6\}$

18. $\{y \mid -6 < y < -3\}$

19. $\{y \mid y < -2 \text{ or } y \geq 2\}$

20. $\{y \mid y < 0 \text{ or } y \geq 4\}$

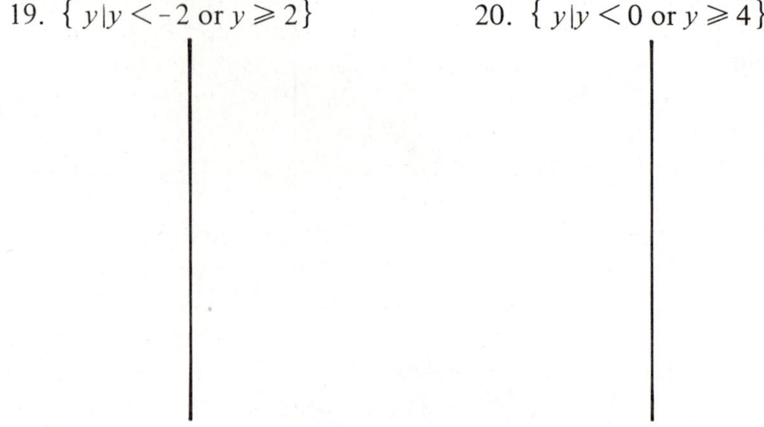

11.2 GRAPHING POINTS ON A NUMBER PLANE

OBJECTIVES: To graph an ordered pair of numbers on a number plane.

To state the coordinates of a point whose graph is shown on a number plane.

A **rectangular coordinate system** (or rectangular number plane) is formed by a vertical number line and a horizontal number line intersecting at their origins. This point of intersection is called the **origin** of the coordinate system.

To be a rectangular coordinate system, the two number lines must form right angles at their point of intersection.

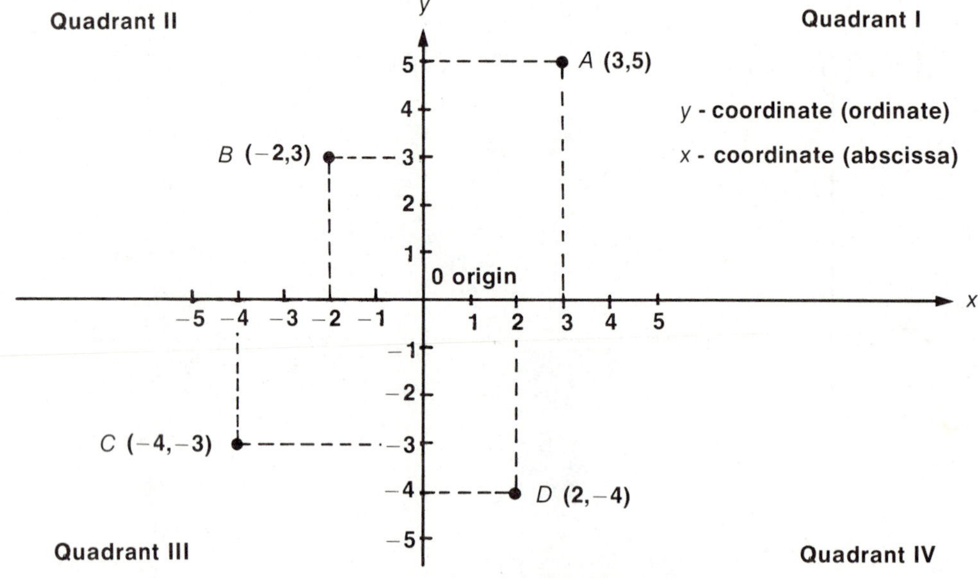

A Rectangular Coordinate System

The two number lines divide the plane into four regions called **quadrants.** The quadrants are numbered by Roman numerals in counter-clockwise order beginning with quadrant I in the upper right region.

The horizontal number line is called the *x*-**axis**, and its positive direction is conventionally taken to the right.

The vertical number line is called the *y*-**axis**, and its positive direction is conventionally taken upward.

Expressions such as (4,7) and (−3,8) are called **ordered pairs.** For the ordered pair (4,7), 4 is the first number and 7 is the second number.

An **ordered pair** is an expression having the form (a,b), where a is called the **first component** (or first member) of the ordered pair and b is called the **second component** (or second member) of the ordered pair.

The order in which the components of an ordered pair are written is important. For example, the ordered pair (3,5) is not the same as the ordered pair (5,3).

There is a **one-to-one correspondence** between the set of points on the plane and the set of ordered pairs of real numbers. The origin is assigned the ordered pair (0,0). A point on the x-axis is assigned an ordered pair of the form $(x,0)$ and a point on the y-axis is assigned an ordered pair of the form $(0,y)$. A point P is assigned the ordered pair (a,b) if and only if the vertical line through P intersects the x-axis at a and the horizontal line through P intersects the y-axis at b.

On the figure, four points are graphed, one in each quadrant; namely, A (3,5), B (−2,3), C (−4,−3), and D (2,−4).

For a point P (a,b), the numbers a and b are called the **coordinates** of P, with a called the x-coordinate, or **abscissa**, of P and b called the y-coordinate, or **ordinate**, of P.

EXERCISES 11.2

A.

(1–8) Graph each of the following points on the number plane at the right.

1. A (5, 2)
2. B (2,5)
3. C (−3,−4)
4. D (1,−5)
5. E (−4,3)
6. F (−5,0)
7. G (0,2)
8. H (0,−2)

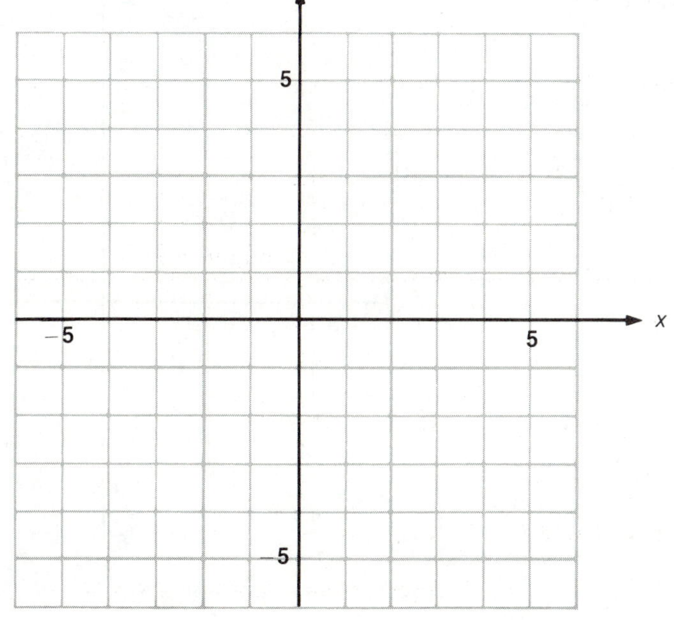

B.

(1–8) Graph each of the following points on the number plane at the right.

1. *A* (6,3)
2. *B* (-3,-6)
3. *C* (-6,-3)
4. *D* (2,-4)
5. *E* (4,-2)
6. *F* (1,0)
7. *G* (-4,0)
8. *H* (0,5)

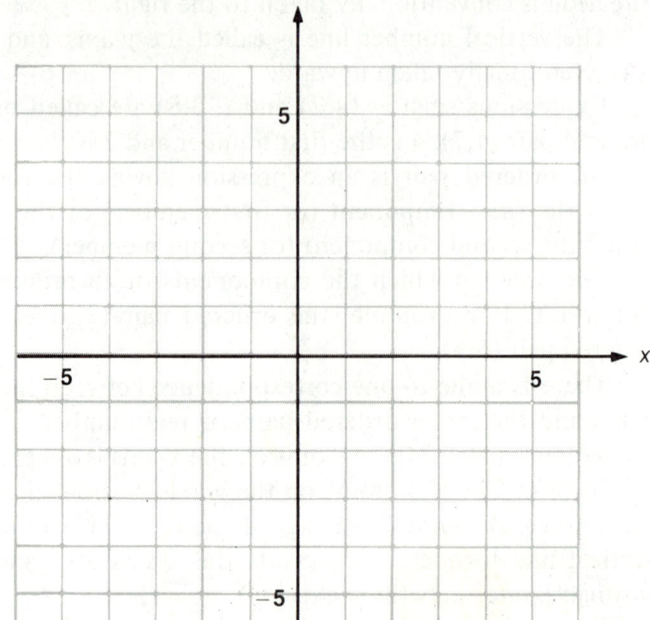

A.

(9–14) State the coordinates of each point whose graph is shown in the figure at the right.

9. *A*
10. *B*
11. *C*
12. *D*
13. *E*
14. *F*

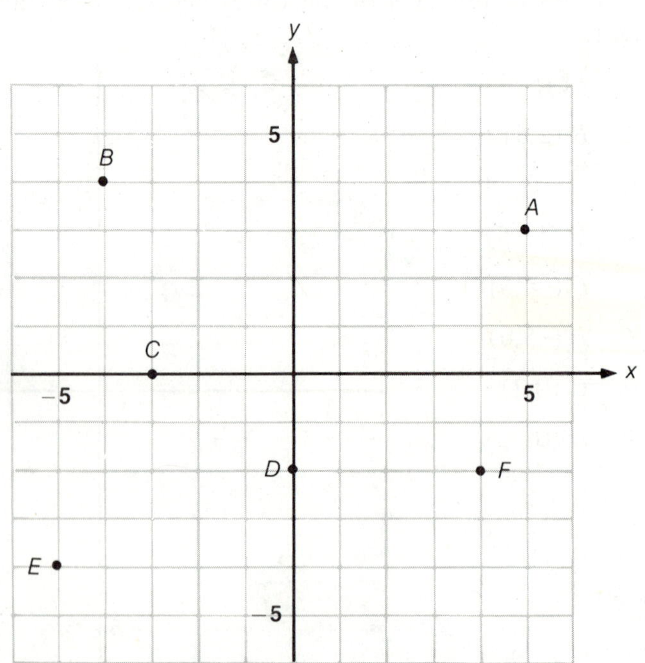

B.

(9–14) State the coordinates of each point whose graph is shown in the figure at the right.

9. *P*
10. *Q*
11. *R*
12. *S*
13. *T*
14. *U*

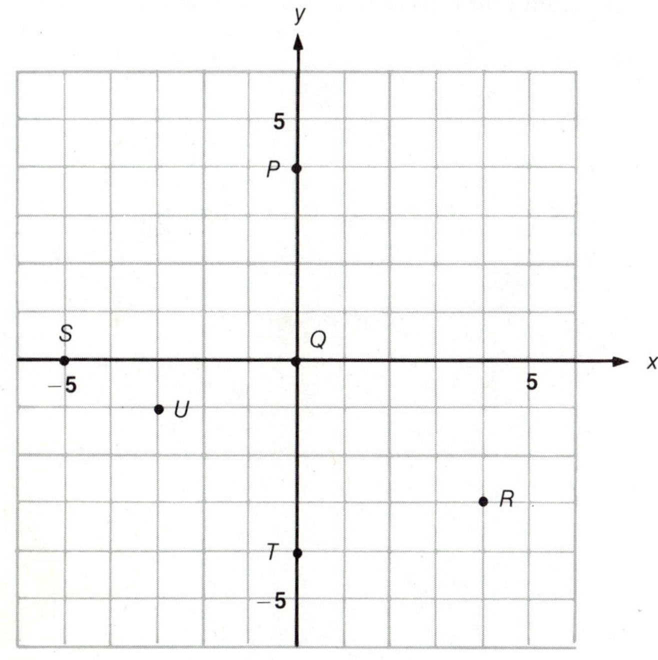

A.

15. Graph each of the following points on the number plane shown: *A* (-4,-5), *B* (0,7), *C* (4,-5), *D* (-6,2), *E* (6,2). Join *A* to *B* to *C* to *D* to *E* to *A*. Name the figure so formed.

B.

15. Graph each of the following points on the number plane shown: A (5,3), B (−5,3), C (0,−6), P (−5,−3), Q (0,6), R (5,−3). Join A to B to C to A. Then join P to Q to R to P. Name the figure so formed.

11.3 SOLUTIONS OF EQUATIONS IN TWO VARIABLES

OBJECTIVE: To determine if a given ordered pair is a solution of a given equation.

A solution of an equation in two variables, x and y, is an ordered pair (a,b) such that the equation becomes a true statement when x is replaced by a and y is replaced by b.

SAMPLE PROBLEM 1

Show that (3,2) and (−6,8) are solutions of $2x + 3y = 12$.

SOLUTION

For (3,2), $x = 3$ and $y = 2$.

$$2x + 3y = 12$$
$$2\,(3) + 3\,(2) = 12$$
$$6 + 6 = 12 \text{ (true)}$$

(3,2) is a solution.

For (−6,8), $x = -6$ and $y = 8$

$$2x + 3y = 12$$
$$2\,(-6) + 3\,(8) = 12$$
$$-12 + 24 = 12 \text{ (true)}$$

(−6,8) is a solution.

SAMPLE PROBLEM 2

Determine if the given ordered pair is a solution of $y = 4 - x^2$.

a. (−2,0) b. (4,0)

SOLUTION

a. $x = -2$ and $y = 0$.

$$y = 4 - x^2$$
$$0 = 4 - (-2)^2$$
$$0 = 4 - 4 \text{ (true)}$$

(−2,0) is a solution.

b. $x = 4$ and $y = 0$.

$$y = 4 - x^2$$
$$0 = 4 - 4^2$$
$$0 = 4 - 16$$
$$0 = -12 \text{ (false)}$$

(4,0) is not a solution.

EXERCISES 11.3

A.

(1–8) Determine if the given ordered pair is a solution of $2x + y = 6$.

1. (1,4)
2. (4,1)
3. (2,2)
4. (0,6)
5. (3,0)
6. (0,3)
7. (-4,10)
8. (5,-4)

(9–13) Determine if the given ordered pair is a solution of $3x - y = 15$.

9. (6,3)
10. (3,6)
11. (3,-6)
12. (5,0)
13. (0,15)

(14–30) Determine if the given ordered pair is a solution of the given equation.

14. $5x + 2y = 10$; (4,-5)
15. $5x + 2y = 10$; (5,2)
16. $4x - 3y = 12$; (0,3)
17. $4x - 3y = 12$; (3,4)
18. $4x - 3y = 12$; $\left(\frac{3}{2}, -2\right)$
19. $y = 5 - x$; (0,5)
20. $y = 5 - x$; (-2,3)

21. $y = 5 - x$; $(-3,8)$

22. $y = 2x - 3$; $(0,-3)$

23. $y = 2x - 3$; $\left(\frac{5}{2},2\right)$

24. $xy = 6$; $(2,3)$

25. $xy = 6$; $(-6,-1)$

26. $xy = 6$; $\left(-12,-\frac{1}{2}\right)$

27. $xy = 6$; $(12,-6)$

28. $xy = 6$; $(0,6)$

29. $y = x^2$; $(3,9)$

30. $y = x^2$; $(-3,9)$

B.

(1–10) Determine if each given ordered pair is a solution of the given equation.

1. $x + y = 10$; $(2,8)$, $(-3,-7)$, $(12,-2)$, $(5,-15)$

2. $x - y = 5$; $(7,2)$, $(-3,-8)$, $(5,0)$, $(0,5)$

3. $2x + y = 8$; $(4,2)$, $(2,4)$, $(0,8)$, $(-4,0)$

4. $x - 2y = 7$; $(5,-1)$, $(3,2)$, $(0,9)$, $(-3,-5)$

5. $3x - 2y = 12$; $(3,6)$, $(6,3)$, $(-2,-3)$, $(-2,-9)$

6. $4x + 3y = 24$; (−3,4), (0,8), (6,0), (−6,16)

7. $y = x^2 - 1$; (−5,24), (0,−1), (−1,0), (4,3)

8. $y = 9 - x^2$; (−3,0), (−2,5), (4,5), (4,−7)

9. $x^2 + y^2 = 25$; (3,−4), (4,−3), (0,−5), (5,0)

10. $x^2 - y^2 = 16$; (−5,−3), (−3,−5), (0,−4), (−4,0)

11.4 FINDING SOLUTIONS OF $Ax + By + C = 0$

OBJECTIVE: To find a solution of a linear equation in two variables, given one member of the ordered pair.

A **linear equation in two variables**, x and y, is an equation of the form $Ax + By + C = 0$, where A, B, and C are constants and A is not zero or B is not zero.

To find a solution (a,b) of $Ax + By + C = 0$, select any value for one of the variables, replace the variable by this value, and then solve the resulting equation for the other variable.

SAMPLE PROBLEM 1

Find the solution of $2x + y = 5$ for which $x = 4$.

SOLUTION

Use $x = 4$ in the equation.

$$2x + y = 5$$

$$2(4) + y = 5$$

$$y = 5 - 8$$

$$y = -3$$

The solution is (4,−3).

SAMPLE PROBLEM 2

Find the solution of $2x - 3y = 7$ for which $y = 3$.

SOLUTION

Use $y = 3$ in the equation.

$$2x - 3y = 7$$

$$2x - 3(3) = 7$$

$$2x = 7 + 9$$

$$2x = 16$$

$$x = 8$$

The solution is (8,3).

SAMPLE PROBLEM 3

Find the solution of $5x - 3y + 15 = 0$ for which $x = 0$.

SOLUTION

Replace x by 0 and solve for y.

$$5x - 3y + 15 = 0$$

$$5(0) - 3y + 15 = 0$$

$$0 - 3y + 15 = 0$$

$$-3y = -15$$

$$y = 5$$

The solution is (0,5).

SAMPLE PROBLEM 4

Find the solution of $5x - 3y + 15 = 0$ for which $y = 0$.

SOLUTION

Replace y by 0 and solve for x.

$$5x - 3y + 15 = 0$$
$$5x - 3(0) + 15 = 0$$
$$5x = -15$$
$$x = -3$$

The solution is (−3,0).

EXERCISES 11.4

Find the solution of each given equation corresponding to the given value for one of the variables.

A.

1. $x + y = 12; x = 7$ 1. ____________

2. $x + y = 12; y = -2$ 2. ____________

3. $x - y = 5; y = -3$ 3. ____________

4. $x - y = 5; x = -4$ 4. ____________

5. $2x + y = 8; x = 1$ 5. ____________

6. $3x + y = 7; x = -2$ 6. ____________

7. $3x - y = 5; x = -1$ 7. ____________

8. $2x - y = 6; y = -2$ 8. ____________

9. $2x - 5y = 10; x = 0$ 9. ____________

10. $2x - 5y = 10; y = 0$ 10. ____________

B.

1. $x - 4y = 12; y = -3$ 1. ____________

2. $x + 5y = 10; y = -2$ 2. ____________

3. $2x + y - 9 = 0; x = 4$ 3. ____________

4. $2x + y - 9 = 0; y = 3$ 4. ____________

5. $3x - y + 9 = 0; x = 5$ 5. ____________

6. $3x - y + 9 = 0; y = 5$ 6. ____________

7. $2x - 3y - 8 = 0; y = -2$ 7. ____________

8. $4x + 5y + 10 = 0; y = 0$ 8. ____________

9. $4x + 5y + 10 = 0; x = 0$ 9. ____________

10. $6y - 2x + 5 = 0; y = 1$ 10. ____________

11.5 GRAPHING THE LINEAR FUNCTION

OBJECTIVE: To graph an equation of the form $y = mx + b$.

A **linear function** is a set of ordered pairs (x,y) defined by the rule $y = mx + b$.

The **graph of a linear function** is the graph of

$$\{(x,y) | y = mx + b\}.$$

There are infinitely many ordered pairs that are solutions of $y = mx + b$, and thus there are infinitely many points on the graph of $y = mx + b$.

The graph of a linear function is a straight line.

SAMPLE PROBLEM 1

Graph $y = 2x - 6$.

SOLUTION

(1) Select any three values for x; say, $-2, 0, 3$.

(2) Find each corresponding y value by using the given equation.

For $x = -2, y = 2x - 6 = 2(-2) - 6 = -10$.

For $x = 0, y = 2x - 6 = 2(0) - 6 = -6$.

For $x = 3, y = 2x - 6 = 2(3) - 6 = 0$.

(3) Make a table for the three ordered pairs.

x	y
-2	-10
0	-6
3	0

(4) Plot the three ordered pairs on a rectangular coordinate system and draw a straight line joining the three plotted points (see Fig. 11.1).

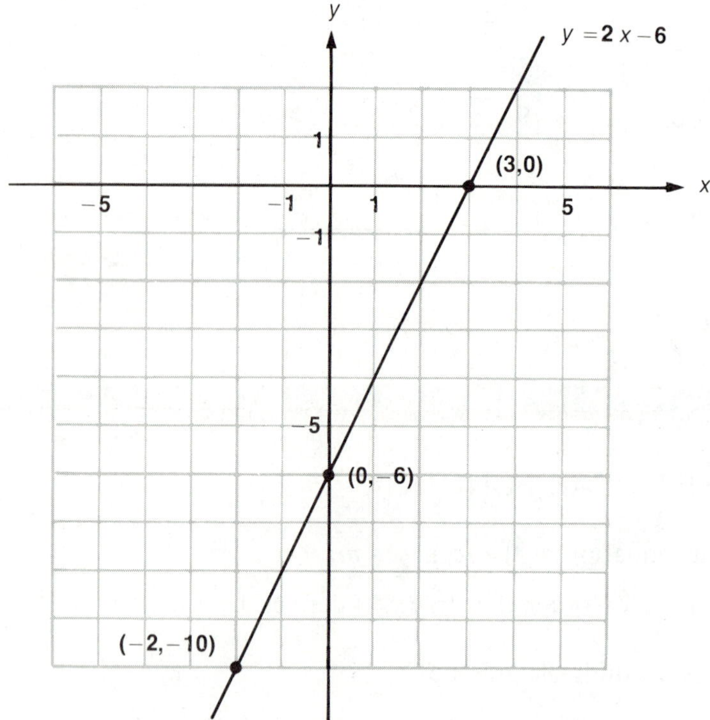

FIGURE 11.1

(Note: While only two points are needed to graph a line, the third point serves as a check point in case an error was made in the calculations.)

SAMPLE PROBLEM 2

Graph $y = 8 - 4x$.

SOLUTION

Select any three values for x, say 0, 2, 4.

Find each y value and make a table of ordered pairs.

x	y	$= 8 - 4x$
0	8	$8 - 4(0) = 8 - 0 = 8$
2	0	$8 - 4(2) = 8 - 8 = 0$
4	-8	$8 - 4(4) = 8 - 16 = -8$

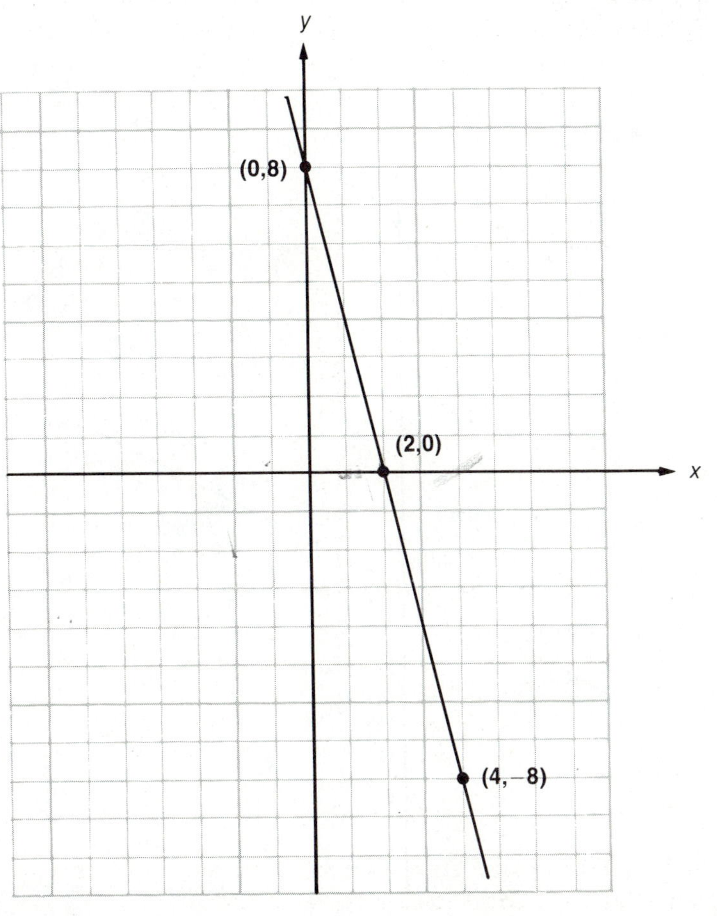

SAMPLE PROBLEM 3

Graph $y = \frac{x}{2}$.

SOLUTION

x	y	$= \frac{x}{2}$
6	3	$\frac{6}{2} = 3$
0	0	$\frac{0}{2} = 0$
-4	-2	$\frac{-4}{2} = -2$

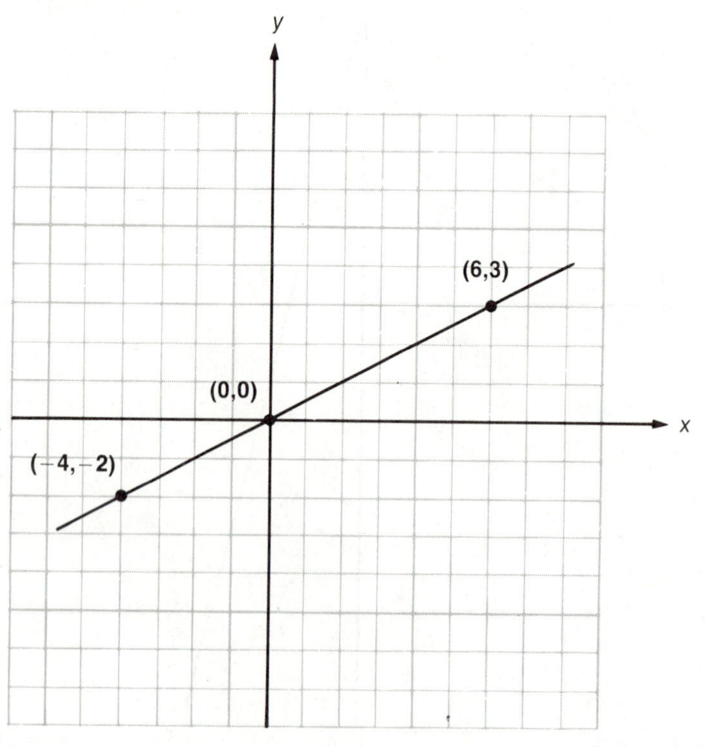

A.

EXERCISES 11.5

(1–3) Complete each table by finding the corresponding y value for each given x value. Plot the three points so found on the coordinate system at the right and draw a straight line through the plotted points.

1. $y = x + 4$

x	y
-5	
0	
3	

$y = x + 4$

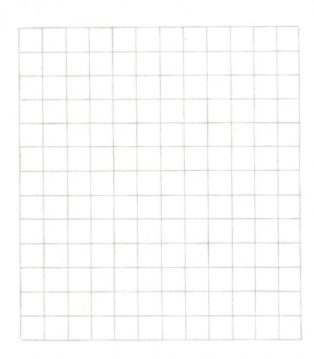

2. $y = 2x + 3$

x	y
-4	
0	
2	

$y = 2x + 3$

3. $y = 5 - x$

x	y
-2	
0	
6	

$y = 5 - x$

B.

1. $y = 3x - 12$

x	y
-1	
0	
5	

$y = 3x - 12$

2. $y = 4 - 2x$

x	y
-3	
0	
4	

$y = 4 - 2x$

3. $y = \frac{-x}{3}$

x	y
-3	
0	
6	

(4–8) Graph each of the following.

A.

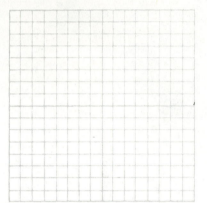

4. $y = 2x$

5. $y = -\frac{1}{2}x$

6. $y = 4 - x$

B.

4. $y = x$

5. $y = x + 5$

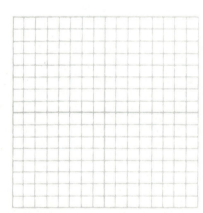

6. $y = 2x - 6$

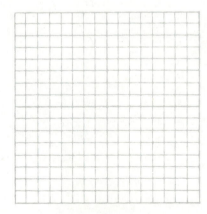

7. $y = 3x + 12$

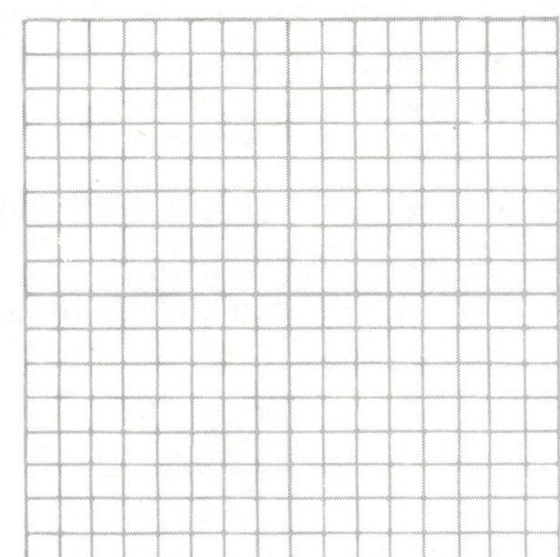

7. $y = 8 - 4x$

8. $y = 6 - 3x$

8. $y = \frac{x}{2} + 4$

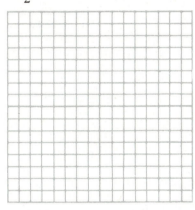

A.

9. Graph $C = \frac{5(F - 32)}{9}$, where F = degrees Fahrenheit and C = degrees centigrade. Let the horizontal axis represent F and the vertical axis C. Graph for $-40 \leqslant F \leqslant 230$, using 1 square = 10 degrees.

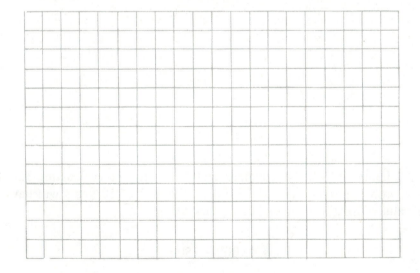

10. In competitive business, the price y for each unit of a commodity depends on the number of units x demanded by the consumers. A demand law is an equation relating the price and the number of units.

a. Graph the demand law $y = 12 - \frac{3x}{4}$ for $x \geqslant 0$ and $y \geqslant 0$.

b. Find the highest price that will be paid for this commodity. In other words, find the value of y for $x = 0$ (the y-intercept).

c. Find the greatest amount that will be demanded. In other words, find the value of x for $y = 0$ (the x-intercept).

B.

9. Using the straight-line depreciation method, a car costing \$3000 with a probable scrap value of \$500 at the end of 10 years will have a book value b dollars at the end of n years where

$$b = 3000 - 250n$$

a. Graph this equation for $0 \leqslant n \leqslant 10$, letting the horizontal axis represent n and the vertical axis b.

b. On the same set of axes, graph $a = 250n$, the amount in the depreciation fund.

c. When is the amount in the depreciation fund equal to the book value?

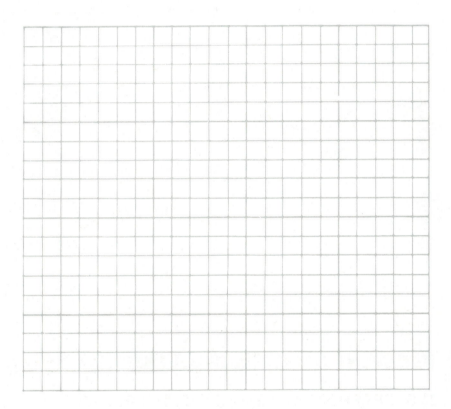

10. The cost C to mail a package weighing x pounds by parcel post is given by $C = 7x + 10$.

a. Graph this equation for $0 \leqslant x \leqslant 50$.

b. On the same set of axes, graph $C = 5x + 70$, which gives the express cost.

c. For what weight is it cheaper to mail a package by parcel post?

11.6 GRAPHING $Ax + By + C = 0$, INTERCEPTS

OBJECTIVES: To find the x- and y-intercepts of a line whose equation is given, if they exist.

To graph an equation of the form $Ax + By + C = 0$.

An equation of the form $Ax + By + C = 0$, where x and y are variables and A, B, and C are constants such that A and B are not both zero, is called a **linear equation.**

The **graph of a linear equation** is a straight line passing through the points whose coordinates satisfy the equation.

A line that crosses the x-axis has an **x-intercept.** If a is the x-intercept of a line, then the point $(a,0)$ is on the line.

A line that crosses the y-axis has a **y-intercept.** If b is the y-intercept of a line, then the point $(0,b)$ is on the line.

SAMPLE PROBLEM 1

Find the x-intercept and the y-intercept of the line whose equation is $2x - 5y = 20$.

SOLUTION

1. **To find the x-intercept, set $y = 0$.**

$$2x - 5(0) = 20$$
$$2x = 20$$
$$x = 10$$

The x-intercept is 10. The point (10,0) is on the line.

2. **To find the y-intercept, set $x = 0$.**

$$2(0) - 5y = 20$$
$$-5y = 20$$
$$y = -4$$

The y-intercept is -4. The point $(0,-4)$ is on the line.

SAMPLE PROBLEM 2

Sketch the graph of $2x - 5y = 20$ by plotting the two points corresponding to the x- and y-intercepts.

SOLUTION

Referring to sample problem 1, the points are (10,0) and $(0,-4)$.

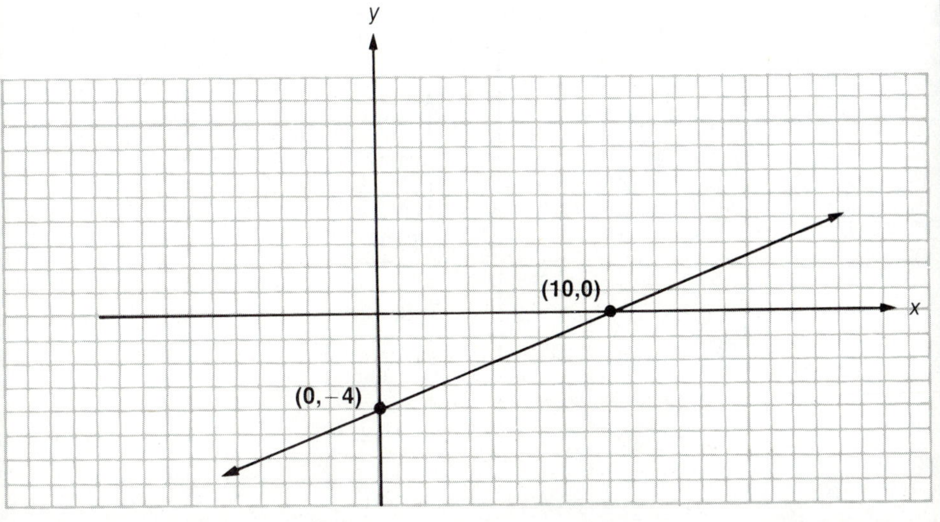

The graph of an equation having the form $x = a$ is a **vertical line** passing through the point $(a,0)$ on the x-axis. Its x-intercept is a, and it has no y-intercept for $a \neq 0$.

The graph of an equation having the form $y = b$ is a **horizontal line** passing through the point $(0,b)$ on the y-axis. Its y-intercept is b, and it has no x-intercept for $b \neq 0$.

SAMPLE PROBLEM 3

State the x- and y-intercepts, if they exist, and graph $x = 3$.

SOLUTION

There is no restriction on y, so y may have any real value. The x value must be 3.

Since $x \neq 0$, there is no y-intercept.

Since (3,0) is on the line, the x-intercept is 3.

Other points on the line are (3,4) and (3,−2).

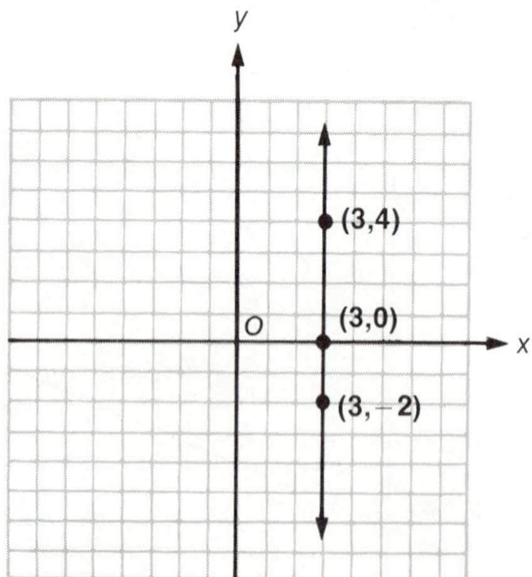

SAMPLE PROBLEM 4

State the x- and y-intercepts, if they exist, and graph $y = -2$.

SOLUTION

There is no restriction on x, so x may have any real value. The y value must be -2.

Since $y \neq 0$, there is no x-intercept.

Since $(0, -2)$ is on the line, the y-intercept is -2.

Other points on the line are $(-3, -2)$ and $(4, -2)$.

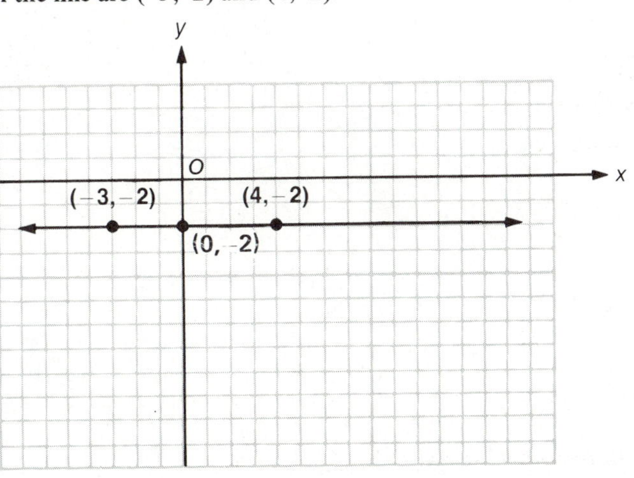

EXERCISES 11.6

Find the x- and y-intercepts, if they exist, and graph the equation.

A.

1. $x = 2$

   x-intercept =

   y-intercept =

2. $y = -3$

   x-intercept =

   y-intercept =

3. $x = 0$

x-intercept =

y-intercept =

4. $y = 0$

x-intercept =

y-intercept =

5. $2x + y = 4$

x-intercept =

y-intercept =

6. $y - 3x = 9$

x-intercept =

y-intercept =

7. $x - y = 5$

x-intercept =

y-intercept =

8. $5x + 2y - 10 = 0$

x-intercept =

y-intercept =

9. $3x + 4y + 12 = 0$

x-intercept =

y-intercept =

10. $2x - 5y - 10 = 0$

x-intercept =

y-intercept =

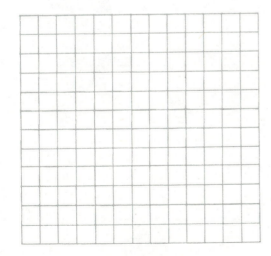

B.

1. $x = -5$

   x-intercept =

   y-intercept =

2. $y = 4$

   x-intercept =

   y-intercept =

3. $x + 2y = 6$

   x-intercept =

   y-intercept =

4. $x - 2y = 0$

   x-intercept =

   y-intercept =

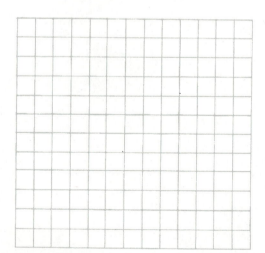

5. $3x - y = 6$

   x-intercept =

   y-intercept =

6. $5x - 2y + 10 = 0$

   x-intercept =

   y-intercept =

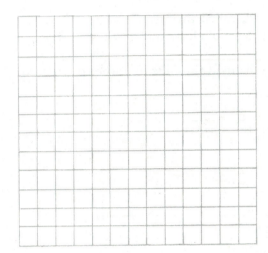

7. $5x - 6y - 30 = 0$

   x-intercept =

   y-intercept =

8. $2x + 7y + 14 = 0$

   x-intercept =

   y-intercept =

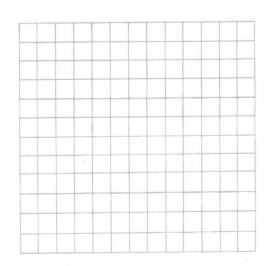

11.7 SLOPE

OBJECTIVE: To find the slope of a line, given two points on the line.
To draw a line, given its slope and a point on the line.

The steepness or slope of a road, called the grade of the road, is the ratio of the number of feet in the change of elevation for 100 ft. measured horizontally.

If a road has a 5% grade, then the road rises 5 ft. for every run of 100 ft. (the horizontal change).

$$\text{Grade} = \frac{\text{rise}}{\text{run}} = \frac{5}{100}$$

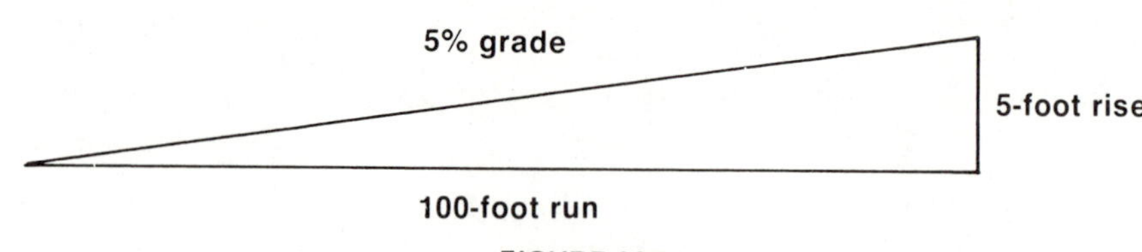

FIGURE 11.2

The slope of a line measures the steepness of the line and it is also the ratio of the rise to the run of the line.

Let A (2,4) and B (3,6) be two points on a line.

If m is the slope of the line, then

$$m = \frac{\text{rise}}{\text{run}} = \frac{6-4}{3-2} = \frac{2}{1} = 2.$$

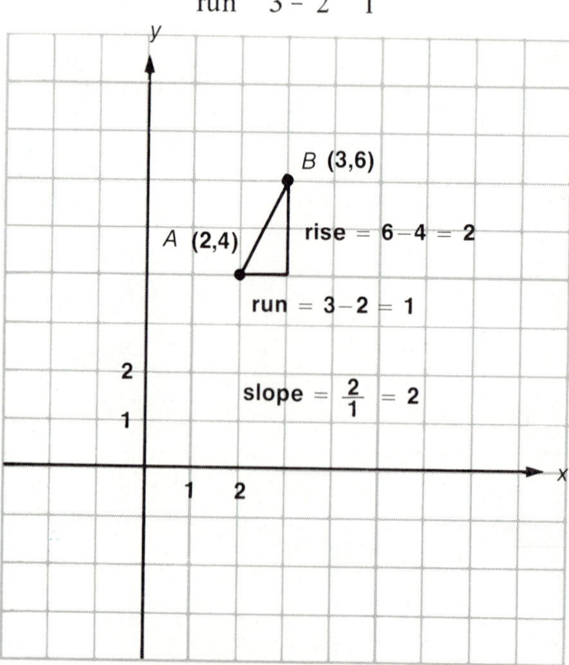

FIGURE 11.3

The slope m of a line is a constant. This means that the slope is the same no matter what two points on the line are used to calculate its value.

Definition of Slope. The slope m of the line through points A and B having different x-coordinates is defined as follows:

$$m = \frac{(y\text{-coordinate of } B) - (y\text{-coordinate of } A)}{(x\text{-coordinate of } B) - (x\text{-coordinate of } A)}$$

In sample problems 1–4 find the slope of the line determined by the two given points and sketch the line.

SAMPLE PROBLEM 1

A **(1,2) and** *B* **(3,8)**

SOLUTION

$$m = \frac{(y \text{ of } B) - (y \text{ of } A)}{(x \text{ of } B) - (x \text{ of } A)} = \frac{8-2}{3-1} = \frac{6}{2} = 3$$

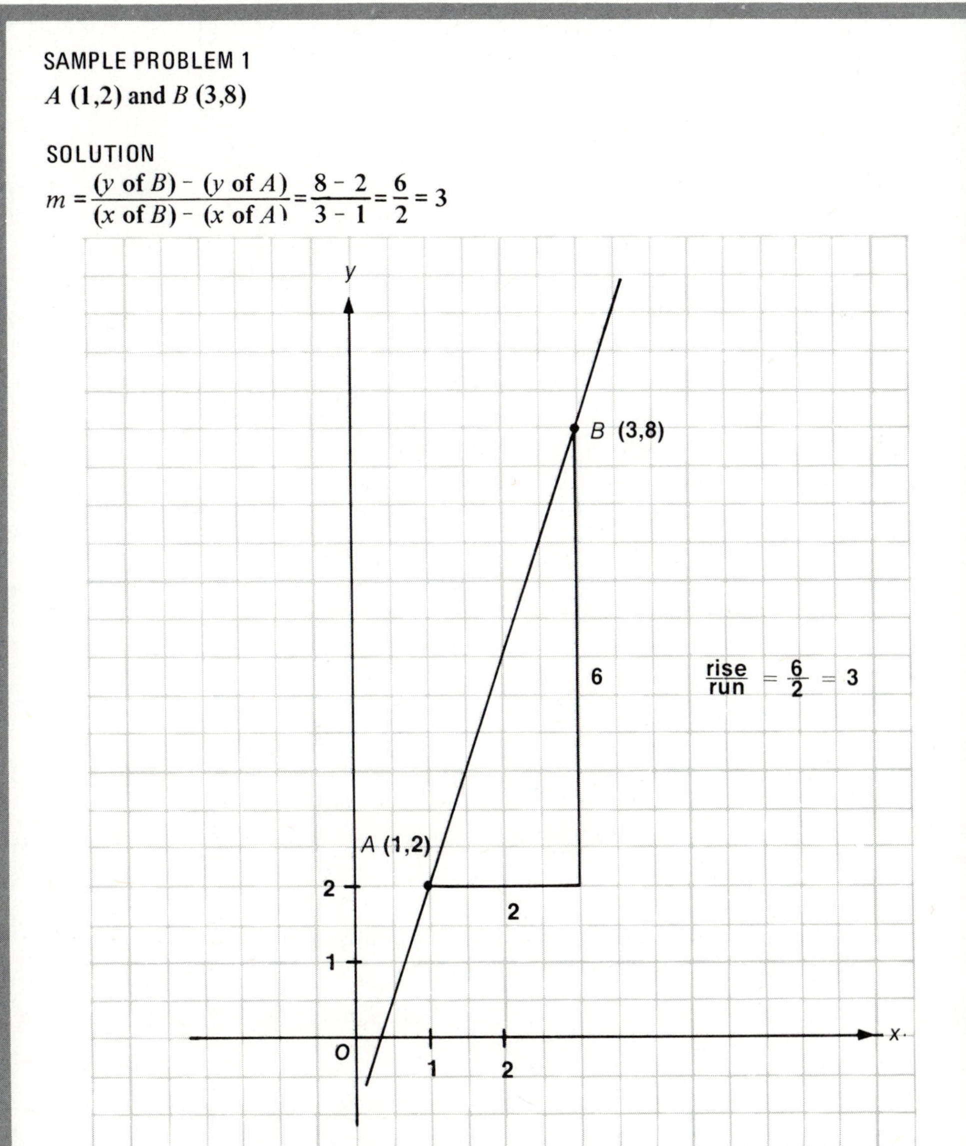

FIGURE 11.4 Line with positive slope.

ALTERNATE SOLUTION

$$m = \frac{(y \text{ of } A) - (y \text{ of } B)}{(x \text{ of } A) - (x \text{ of } B)} = \frac{2-8}{1-3} = \frac{-6}{-2} = 3$$

SAMPLE PROBLEM 2

P (3,4) and Q (4,2)

SOLUTION

$$m = \frac{(y \text{ of } Q) - (y \text{ of } P)}{(x \text{ of } Q) - (x \text{ of } P)} = \frac{2 - 4}{4 - 3} = \frac{^-2}{1} = {}^-2$$

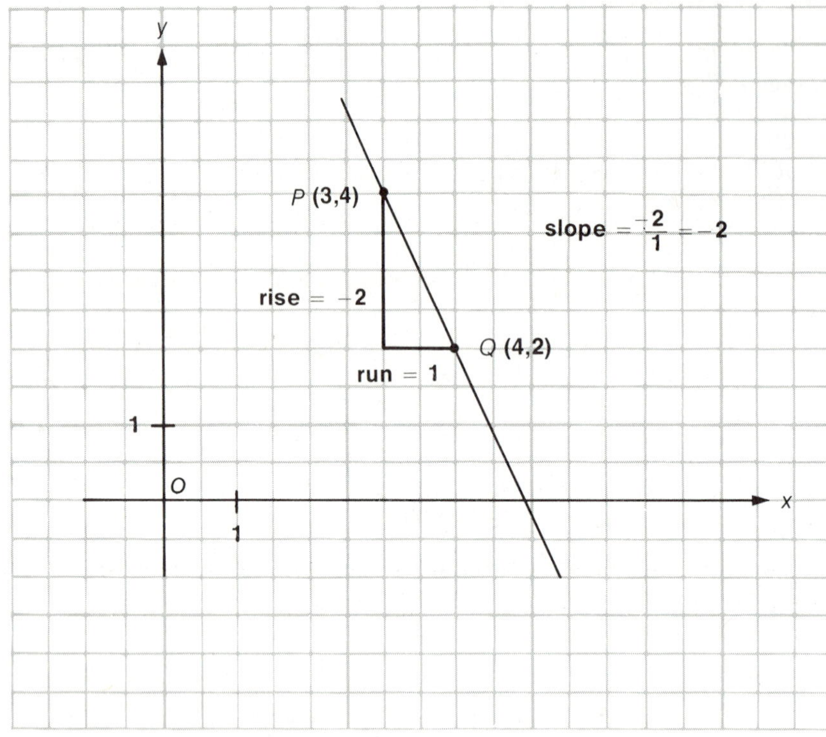

FIGURE 11.5 Line with negative slope.

Note in sample problems 1 and 2 that a line with positive slope "leans" to the right and a line with negative slope "leans" to the left.

SAMPLE PROBLEM 3

C (3,2) and D (-5,2)

SOLUTION

$$m = \frac{(y \text{ of } D) - (y \text{ of } C)}{(x \text{ of } D) - (x \text{ of } C)} = \frac{2 - 2}{-5 - 3} = \frac{0}{-8} = 0$$

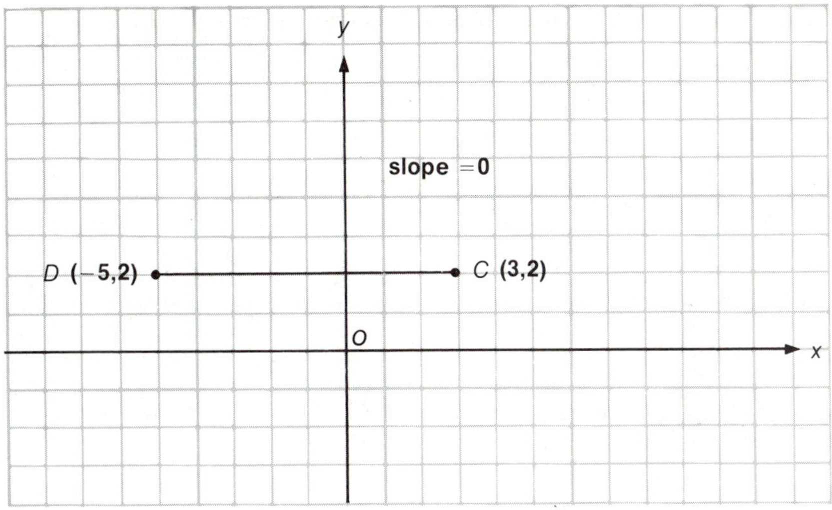

FIGURE 11.6 Horizontal line, slope = 0.

Note that this line is horizontal. In general, a horizontal line has a slope of 0.

SAMPLE PROBLEM 4

R (4,6) and S (4,2)

SOLUTION

$$m = \frac{(y \text{ of } S) - (y \text{ of } R)}{(x \text{ of } S) - (x \text{ of } R)} = \frac{2 - 6}{4 - 4} = \frac{-4}{0} \text{ (undefined)}$$

Since division by zero is undefined, this line has no slope. Note that this line is vertical. In general, the slope of a vertical line is undefined.

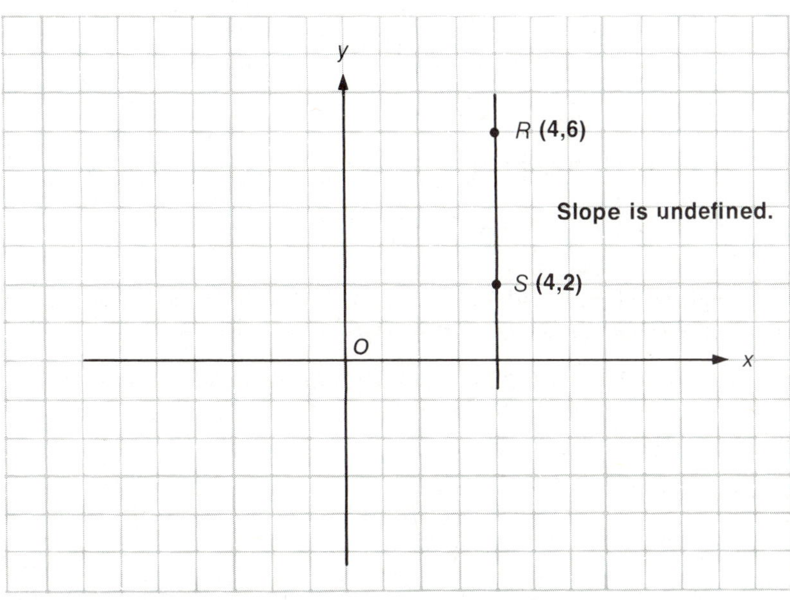

FIGURE 11.7 Vertical line, slope is undefined.

If the slope m of a line and a point on the line are known, then the line is completely determined. Sample problems 5 and 6 show how to draw a line, given its slope and a point on the line.

SAMPLE PROBLEM 5

Draw the line through $P\ (3,-2)$ having slope $m = \frac{5}{4}$

SOLUTION

(1) Graph $P\ (3,-2)$.

(2) $m = \frac{\text{rise}}{\text{run}} = \frac{5}{4}$; therefore, rise = 5 and run = 4.

Measure 5 units upward and 4 units to the right to obtain another point Q on the line.

(3) Draw the line through P and Q.

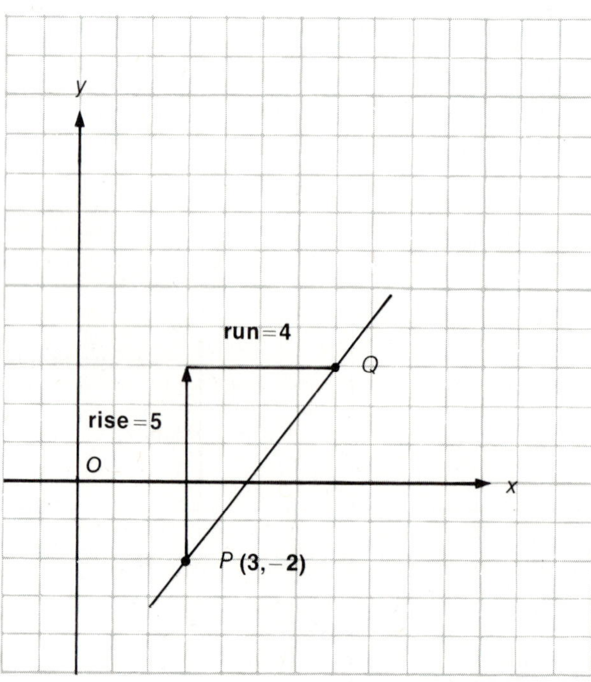

FIGURE 11.8 Graph of line through (3,–2), slope $= \frac{5}{4}$.

SAMPLE PROBLEM 6

Draw the line through P (4,7) with slope $m = -3$.

SOLUTION

(1) **Graph P (4,7).**

(2) $m = \dfrac{\text{rise}}{\text{run}} = -3 = \dfrac{-3}{1}$; **therefore, rise = −3 and run = 1.**

Measure 3 units downward and 1 unit to the right to reach point Q.

(3) **Draw the line through P and Q.**

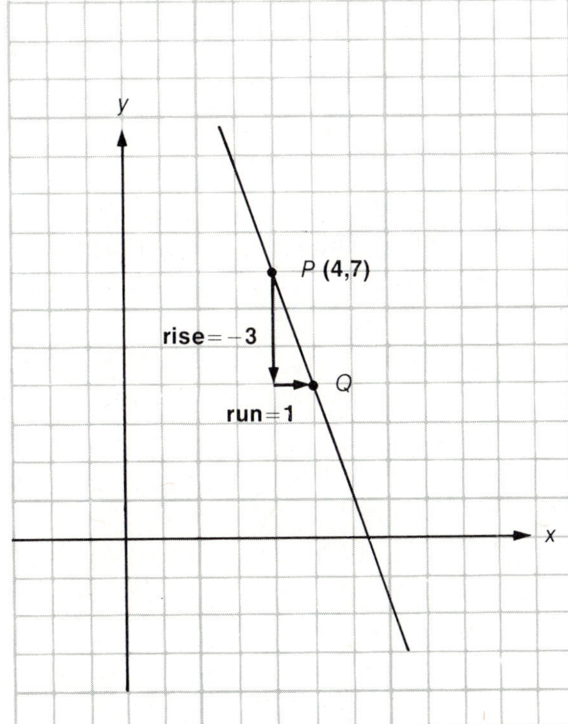

FIGURE 11.9 Graph of line through (4,7), slope = −3.

A.

EXERCISES 11.7

(1–8) For each given pair of points, find the slope m by using the definition. Check by graphing the line and finding the slope graphically.

1. A (2,7) and B (3,9)
2. A (1,5) and B (2,1)

3. A (3,−1) and B (6,0)

4. P (−2,9) and Q (0,6)

5. P (0,−5) and Q (4,0)

6. C (5,3) and D (−4,3)

7. C (0,0) and D (5,−2)

8. R (−2,4) and S (−2,9)

(9–15) Draw the line through the given point having the given slope m.

9. P (5,3); $m = 2$

10. A (4,2); $m = -2$

11. B (−3,4); $m = \frac{2}{3}$

12. C (6,−4); $m = \frac{-3}{4}$

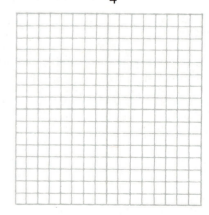

13. R (5,7); m is undefined

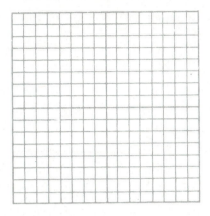

14. R (5,7); $m = 0$

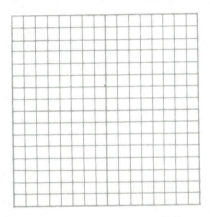

15. A highway has a $3\frac{1}{2}\%$ grade. How many feet does it rise in a $\frac{1}{2}$-mi. run? (1 mi. = 5280 ft.)

16. A highway has a $-2\frac{1}{4}\%$ grade. How many feet does it drop in a 2-mi. run?

17. A certain county specification requires that an inclined water pipe must have a slope greater than or equal to $\frac{1}{4}$. Which of the water pipes whose rises and runs are given below meets this specification?

a. Rise = 20 ft., run = 64 ft.
b. Rise = 125 ft., run = 500 ft.
c. Rise = 60 ft., run = 250 ft.

B.

(1–8) For each given pair of points, find the slope m by using the definition. Check by graphing the line and finding the slope graphically.

1. A (1,−2) and B (3,4)
2. A (4,2) and B (1,8)
3. A (4,5) and B (−2,−2)
4. P (0,−2) and Q (−6,8)
5. P (−3,0) and Q (0,−2)
6. C (−6, −1) and D (0,0)
7. R (−6,−1) and S (−6,4)
8. R (−7,−4) and S (−2,−4)

(9–15) Draw the line through the given point having the given slope m.

9. $P(-4,-5)$; $m = 4$

10. $Q(7,2)$; $m = \frac{-7}{3}$

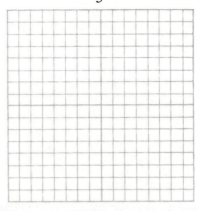

11. $D(-3,6)$; $m = \frac{1}{4}$

12. $E(2,-4)$; $m = -1$

13. $S(2,8)$; m is undefined

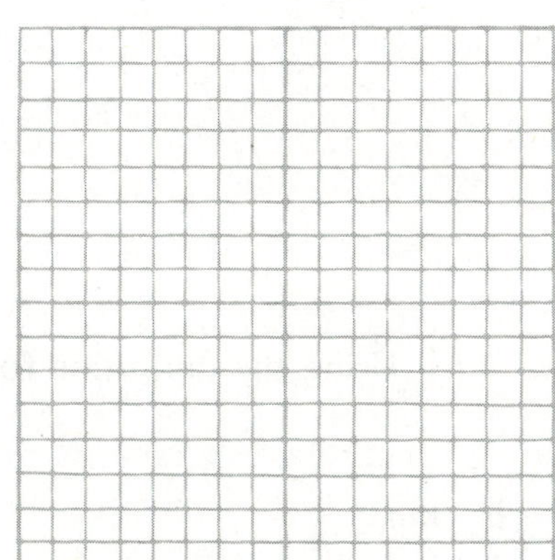

14. $S(2,8)$; $m = 0$

15. The pitch (slope) of a roof is the ratio of its vertical rise to its horizontal half-span. If the pitch of a roof is 3 to 12, find the rise for a half-span of 18 ft.

16. If the pitch of a roof is 4 to 12, find the half-span for a rise of 6 ft.

11.8 SLOPE-INTERCEPT EQUATION

OBJECTIVE: To find the slope and y-intercept of a line, given its equation.

Every equation of the form $Ax + By + C = 0$, where $B \neq 0$ can be written in the form $y = mx + b$ by solving for y.

$$Ax + By + C = 0$$

$$By = -Ax - C$$

$$y = \frac{-A}{B}x + \frac{-C}{B}$$

$$y = mx + b$$

One point on the line is $(0, b)$.

Another point on the line is $(a, ma + b)$. Using the definition for the slope of a line,

$$\text{slope} = \frac{(ma + b) - b}{a - 0} = \frac{ma}{a} = m.$$

Therefore, m, the coefficient of x, is the slope of the line.

The equation $y = mx + b$ is called the **slope-intercept form** of a linear equation.

The equation $Ax + By + C = 0$ is called the **standard form** of a linear equation.

The slope-intercept form is useful because the slope and y-intercept of a line can be read directly from this equation.

$$y = mx + b$$

↑ slope ↑ y-intercept

Also, the y-intercept and the slope can be used to graph the equation, as shown in the sample problems.

In sample problems 1–3 find the slope and y-intercept of each line, given its equation. Graph the line by using the slope and y-intercept.

SAMPLE PROBLEM 1

$y = 2x + 4$

SOLUTION

Comparing $y = mx + b$ with $y = 2x + 4$, $m = 2$ and $b = 4$. The slope is 2. The y-intercept is 4.

The graph is sketched below.

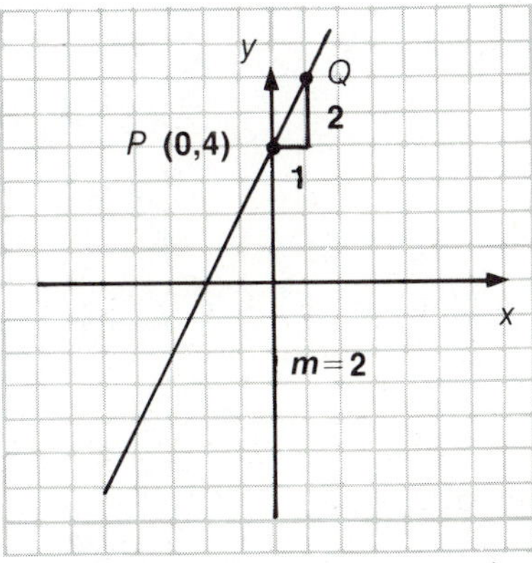

SAMPLE PROBLEM 2

$y = 4 - x$

SOLUTION

Since $4 - x = -x + 4$, the equation can be written as $y = -x + 4$ and also as $y = (-1)x + 4$.

Using $y = mx + b$, $m = -1$ and $b = 4$. The slope is -1. The y-intercept is 4.

The graph is sketched below.

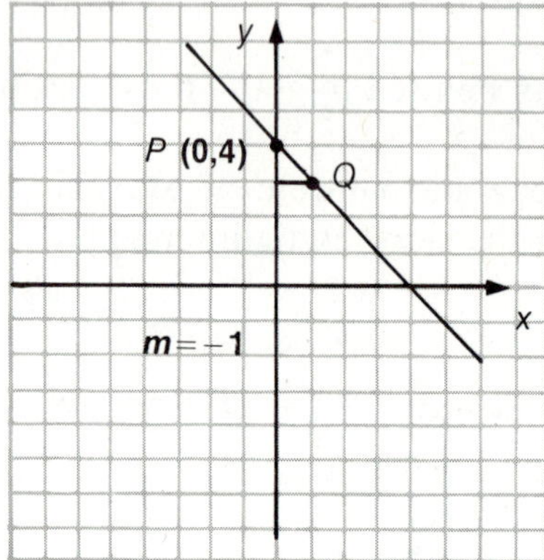

SAMPLE PROBLEM 3

$5x - 2y - 10 = 0$

SOLUTION

Solve for y.

$$5x - 10 = 2y$$
$$2y = 5x - 10$$
$$y = \frac{5}{2}x - 5$$

Using $y = mx + b$, $m = \frac{5}{2}$ and $b = -5$.

The slope is $\frac{5}{2}$ and the y-intercept is -5

The graph is shown here.

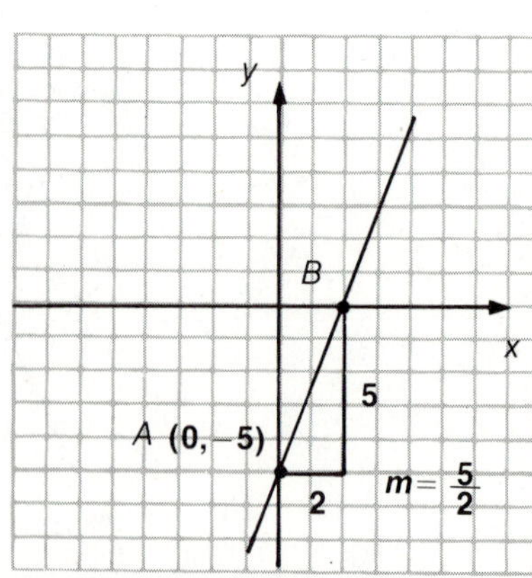

SAMPLE PROBLEM 4

State the slope and y-intercept for $y = 3x$.

SOLUTION

Using $y = mx + b$, $y = 3x + 0$; $m = 3$ and $b = 0$.

The slope is 3 and the y-intercept is 0.

SAMPLE PROBLEM 5

State the slope and y-intercept for $y = 3$.

SOLUTION

Using $y = mx + b$, $y = 0x + 3$; $m = 0$ and $b = 3$.

The slope is 0 and the y-intercept is 3.

EXERCISES 11.8

State the slope m and the y-intercept b for each of the following, if they exist. If they do not exist, write "undefined." Graph each line, using the slope and y-intercept when they exist.

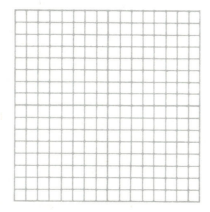

A.

1. $y = 2x + 6$

   $m =$, $b =$

2. $y = -2x + 6$

   $m =$, $b =$

3. $y = 2x$

$m =$, $b =$

4. $y = 2$

$m =$, $b =$

5. $x = 2$

$m =$, $b =$

6. $y = \frac{1}{2}x - 4$

$m =$, $b =$

7. $2x + 3y = 6$

$m =$, $b =$

8. $2x - 3y + 6 = 0$

$m =$, $b =$

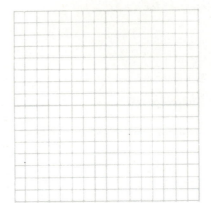

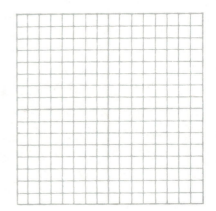

9. $2x + 3y + 6 = 0$

$m =$, $b =$

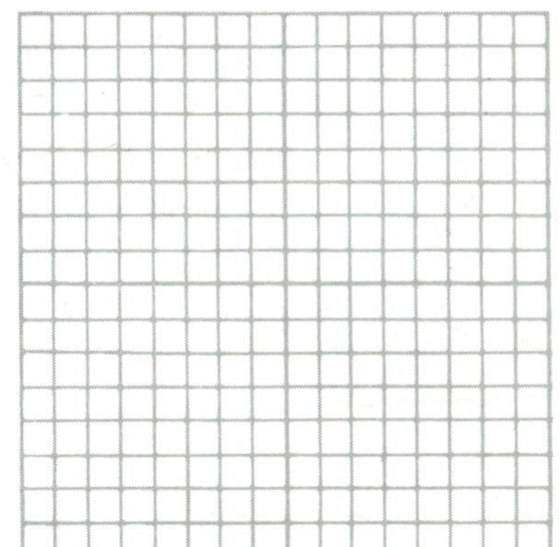

10. $3y - 2x = 12$

$m =$, $b =$

B.

1. $y = 3x - 6$

$m =$, $b =$

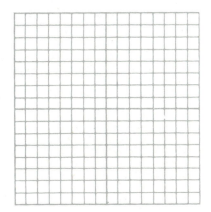

2. $y = 8 - 2x$

$m =$, $b =$

3. $y = x - 6$

$m =$, $b =$

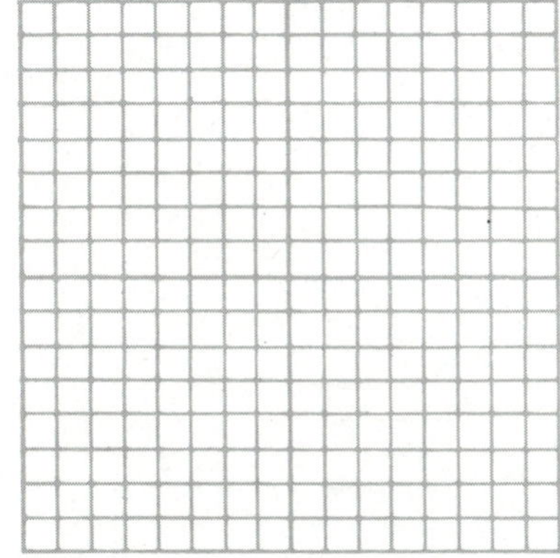

4. $y = 12 - 2x$

$m =$, $b =$

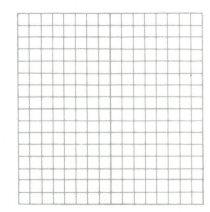

5. $2y + 5 = 0$

$m =$, $b =$

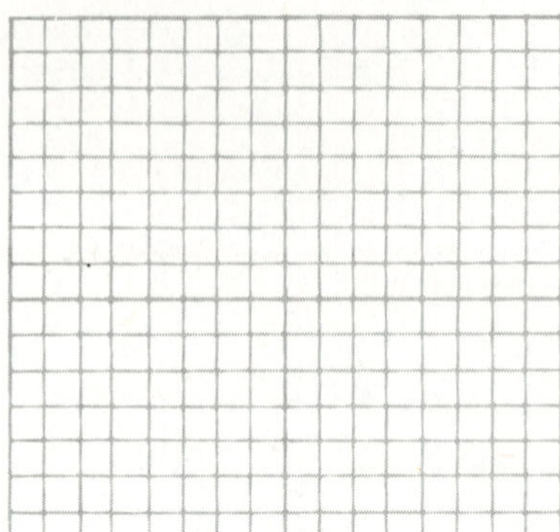

6. $x - 2y + 8 = 0$

$m =$, $b =$

7. $y - 2x + 4 = 0$

$m =$, $b =$

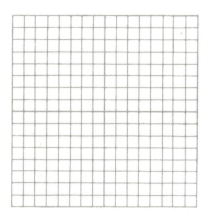

8. $2y - 2x = 5$

$m =$, $b =$

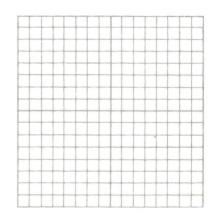

9. $x + 2y = 0$

$m =$, $b =$

10. $3y - 2x = 0$

$m =$, $b =$

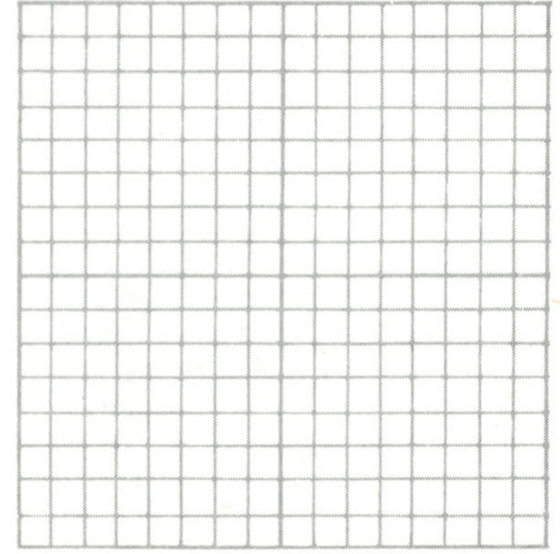

UNIT 11 POSTTEST

(1–4) Graph on a horizontal number line. (11.1)

1. $\{x|x \geqslant 4\}$ ⟶ +
2. $\{x|x < -3\}$ ⟶ +
3. $\{x|-3 \leqslant x \leqslant 3\}$ ⟶ +
4. $\{x|2 < x < 6\}$ ⟶ +

(5–8) Graph on a vertical number line. (11.1)

5. $\{y|y < 5\}$

6. $\{y|y \geqslant -4\}$

7. $\{y|-2 \leqslant y < 0\}$

8. $\{y|-4 < y < -2\}$

(9–13) Graph each of the following on the number plane at the right. (11.2)

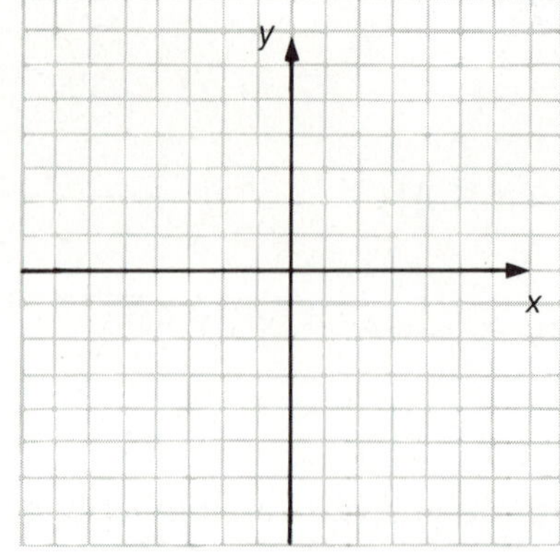

9. A (4,2)
10. B (2,−4)
11. C (−5,1)
12. D (−3,0)
13. E (−2,−5)

(14–18) State the coordinates of each point whose graph is shown on the figure at the right. (11.2)

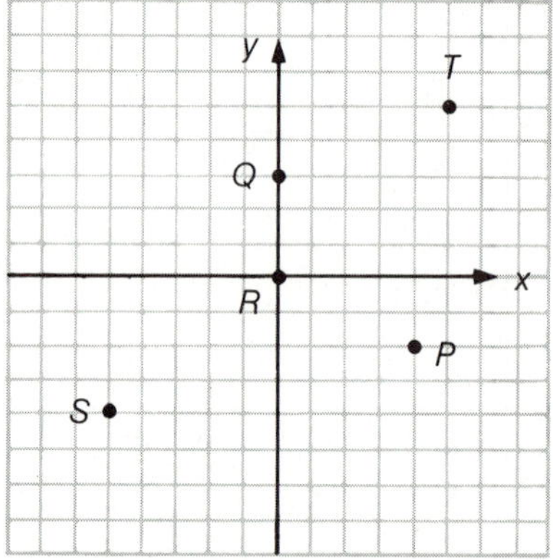

14. P (,)
15. Q (,)
16. R (,)
17. S (,)
18. T (,)

(19–24) Determine if each ordered pair is a solution of $3x - 5y = 15$. (11.3)

19. (5,0)

20. (0,5)

21. (0,3)

22. (0,−3)

23. (10,3)

24. (−5,−6)

25. Find the solution of $2x + y = 4$ for which $x = 5$. (11.4)

26. Find the solution of $4x - 3y = 10$ for which $y = 2$. (11.4)

27. Graph $y = 3x - 2$. (11.5)

28. Graph $y = -2x - 4$. (11.5)

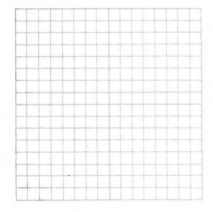

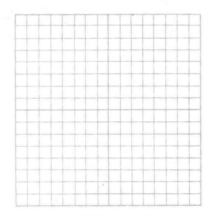

(29–30) State the x- and y-intercepts and graph. (11.6)

29. $x - 4y = 8$

x-intercept =

y-intercept =

30. $5x + y + 10 = 0$

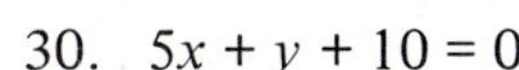

x-intercept =

y-intercept =

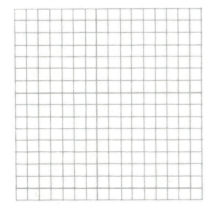

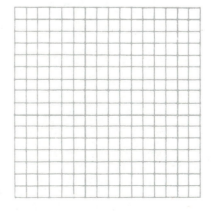

(31–32) Find the slope of the line joining the two given points. Check by graphing the line. (11.7)

31. $P(8,-5)$ and $Q(4,-3)$

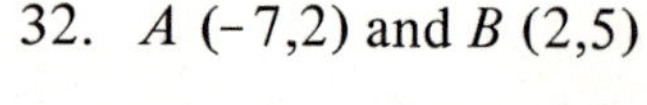

32. $A(-7,2)$ and $B(2,5)$

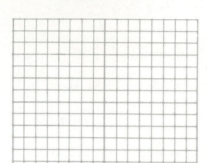

(33–34) State the slope and y-intercept for each of the following. Graph each line, using the slope and y-intercept. (11.8)

33. $y = 2x + 6$

slope =

y-intercept =

34. $2x + 4y - 16 = 0$

slope =

y-intercept =

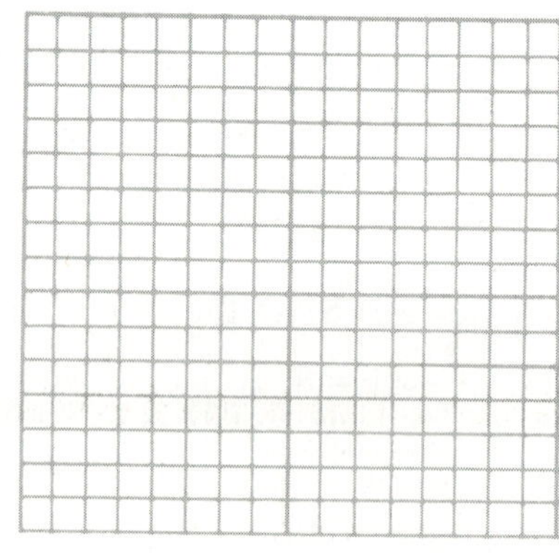

UNIT 12

LINEAR SYSTEMS, TWO VARIABLES

Unit 12 PRETEST

(1–4) Find the slopes and y-intercepts of the graphs of each pair and determine if the lines intersect, are parallel, or are coincident. (12.1)

1. $x = 4$ and $y = 5$
2. $y = 2x - 6$ and $y = 3x - 6$
3. $x = y - 6$ and $y = x - 6$
4. $x + y = 5$ and $2x + 2y = 10$

(5–6) Determine if each given ordered pair is a solution of the system. (12.2)

5. $3x - y = 5$ and $x + 4y = 6$
   a. (2,1)
   b. (3,4)
6. $x - 2y = 4$ and $2y = x - 4$
   a. (6,2)
   b. (0,−2)

7. Solve graphically and check. (12.3)

   $x + 4y = 8$ and $y = 3x + 15$

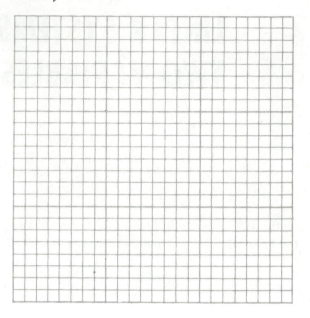

(8–9) Solve by the substitution method and check. (12.4)

8. $3y - 4x = 4$ and $y = 2x - 2$

9. $5x + 4y = 8$ and $x + 2y = 10$

10. Solve by the addition method and check. (12.5)

    $3x + 2y = 12$ and $5x - y = 7$

(11–13) Solve and identify the graph of each system as intersecting lines, parallel lines, or coincident lines. (12.6)

11. $4x - 2y = 5$ and $y = 2x + 1$

12. $x - 2y = 12$ and $2x - y = 12$

13. $2x = 3y + 2$ and $3y = 2x - 2$

14. Two dozen eggs and 5 qt. of milk cost \$3.30. If the price of eggs per dozen increased by 20% and the price of milk decreased by 10%, the same purchase would cost \$3.36. Find the original cost of 1 dozen eggs and the original cost of a quart of milk. (12.7)

15. A motorboat goes 90 mi. in 3 hr. traveling with a tide and requires 5 hr. to return traveling against the same tide. Find the rate of the tide. (12.8)

12.1 INTERSECTING, PARALLEL, AND COINCIDENT LINES

OBJECTIVE: Given two linear equations in two variables, to find the slopes and y-intercepts and to determine if the lines intersect, are parallel, or are coincident.

Two straight lines in a plane either intersect in exactly one point or are parallel and never intersect no matter how far the lines are extended.

Given two linear equations in two variables, one of three things can happen.

1. The graphs are two intersecting lines. The lines do not have the same slope.

2. The graphs are two parallel lines. The lines have the same slope (or both lines are vertical) and have different intercepts.

3. The graphs are coincident lines (the two equations have the same line for their graphs). The lines have the same slope and the same y-intercept, or both lines are vertical with the same x-intercept.

In sample problems 1–3 determine if the graphs of the two given equations are intersecting lines, parallel lines, or coincident lines. Check by graphing.

SAMPLE PROBLEM 1

$2x - y - 5 = 0$ and $x + 2y - 5 = 0$

SOLUTION

Write each equation in the $y = mx + b$ form and compare the slopes and y-intercepts.

$$2x - y - 5 = 0 \qquad y = 2x - 5 \qquad m = 2, b = -5$$

$$x + 2y - 5 = 0 \qquad y = -\frac{1}{2}x + \frac{5}{2} \qquad m = -\frac{1}{2}, b = \frac{5}{2}$$

Since the slopes are different, $2 \neq -\frac{1}{2}$, the lines intersect in exactly one point.

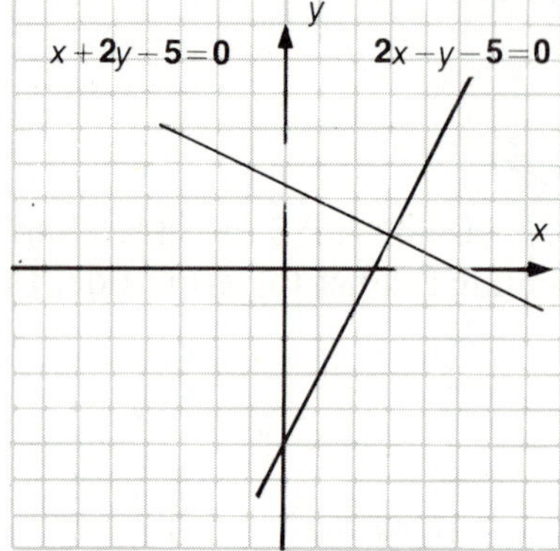

SAMPLE PROBLEM 2

$2x + 4y = 2$ and $3x + 6y = 3$

SOLUTION

$$2x + 4y = 2 \qquad y = -\frac{1}{2}x + \frac{1}{2} \qquad m = -\frac{1}{2}, b = \frac{1}{2}$$

$$3x + 6y = 3 \qquad y = -\frac{1}{2}x + \frac{1}{2} \qquad m = -\frac{1}{2}, b = \frac{1}{2}$$

Since the slopes and the y-intercepts are equal, the lines are coincident.

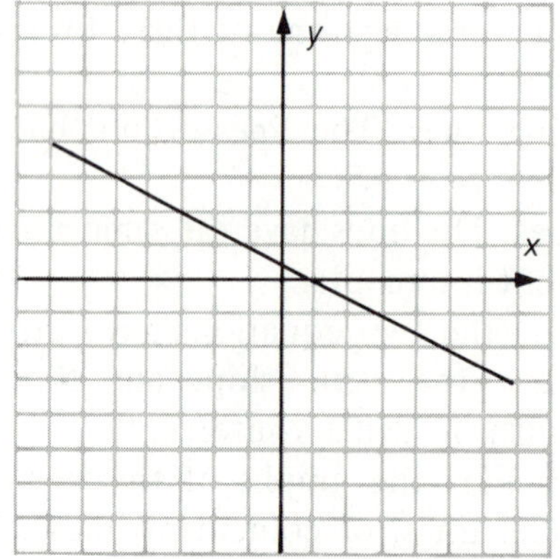

SAMPLE PROBLEM 3

$x + 2y = 5$ and $2x + 4y = 3$

SOLUTION

$$x + 2y = 5 \qquad\qquad 2x + 4y = 3$$

$$y = -\frac{1}{2}x + \frac{5}{2} \qquad\qquad y = -\frac{1}{2}x + \frac{3}{4}$$

$$m = -\frac{1}{2}, b = \frac{5}{2} \qquad\qquad m = -\frac{1}{2}, b = \frac{3}{4}$$

Since the slopes are equal and the y-intercepts are different, the lines are parallel.

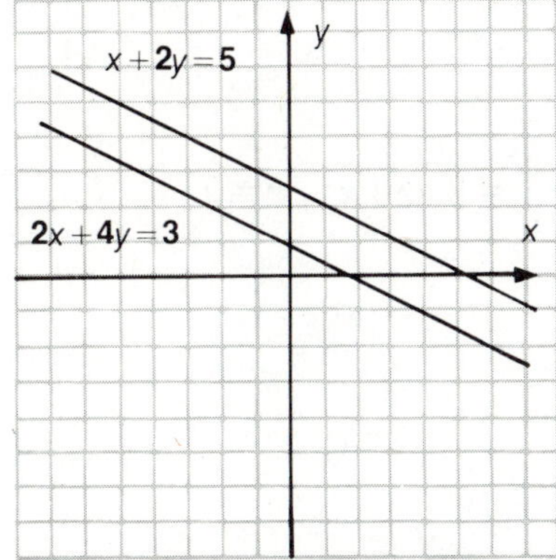

EXERCISES 12.1

Find the slopes and y-intercepts and determine if the graphs of the two given equations are intersecting lines, parallel lines, or coincident lines.

Check the result by graphing each pair of equations on the same rectangular coordinate system.

A.

1. $y = 2x - 3$ and $y = \frac{1}{2}x - 3$

2. $x + y = 5$ and $2x + y = 4$

3. $x - y = 2$ and $2x - 2y = 5$

4. $x + y = 8$ and $x - y = 2$

5. $x + 2y = 3$ and $2x + 4y = 6$

6. $4x - y = 4$ and $y = 4x + 8$

7. $4x - y = 8$ and $x + 4y = 2$

8. $3x - 2y + 6 = 0$ and $2y - 3x = 6$

B.

1. $5x + 2y = 10$ and $x - 2y = 14$

2. $5x - 2y = 10$ and $2y - 5x = 10$

3. $4x - 3y = 18$ and $2x + y = 4$

4. $2x + 3 = 0$ and $2x + y + 3 = 0$

5. $3x - 6 = 0$ and $4y + 8 = 0$

6. $2x + 10 = 0$ and $2y + 10 = 0$

7. $2y + 8 = 0$ and $2y - 8 = 0$

8. $2x = 3y + 6$ and $2x - 3y - 6 = 0$

12.2 SOLUTIONS OF LINEAR SYSTEMS

OBJECTIVE: Given a system of two linear equations in two variables, to determine whether a given ordered pair is a solution of the system.

A **solution of a system of two linear equations in two variables,** x and y, is an ordered pair (a, b) such that each equation becomes true when x is replaced by a and y by b.

The solution set of a system of two linear equations in two variables is the intersection of the solution sets of the two equations; that is, the set of all ordered pairs that are solutions of both equations.

SAMPLE PROBLEM 1

Show that $(5, -2)$ is a solution of the system $2x + 3y = 4$ and $x - 2y = 9$.

SOLUTION

Show that $(5, -2)$ is a solution of each equation.

For $x = 5$ and $y = -2$,

$$2x + 3y = 2(5) + 3(-2) = 10 - 6 = 4 \text{ (true)}$$

$$x - 2y = 5 - 2(-2) = 5 + 4 = 9 \text{ (true)}$$

Since $(5, -2)$ is a solution of each equation, it is a solution of the system.

SAMPLE PROBLEM 2

Show that $(4, 5)$ is not a solution of the system $2x - y = 3$ and $x + 2y = 7$.

SOLUTION

$$2x - y = 3$$
$$2(4) - 5 = 3$$
$$8 - 5 = 3$$
$$3 = 3 \text{ (true)}$$

$(4, 5)$ is a solution of $2x - y = 3$.

$$x + 2y = 7$$
$$4 + 2(5) = 7$$
$$4 + 10 = 7$$
$$14 = 7 \text{ (false)}$$

$(4, 5)$ is not a solution of $x + 2y = 7$.

Since $(4, 5)$ is not a solution of both equations, it is not a solution of the system.

SAMPLE PROBLEM 3

Show that (−3, 7), (0, 1), and (2, −3) are solutions of the system $4x + 2y = 2$ and $6x + 3y = 3$.

SOLUTION

For (−3, 7),

$$4x + 2y = 2 \qquad 6x + 3y = 3$$
$$4(-3) + 2(7) = 2 \qquad 6(-3) + 3(7) = 3$$
$$-12 + 14 = 2 \qquad -18 + 21 = 3$$
$$2 = 2 \text{ (true)} \qquad 3 = 3 \text{ (true)}$$

For (0, 1),

$$4x + 2y = 2 \qquad 6x + 3y = 3$$
$$4(0) + 2(1) = 2 \qquad 6(0) + 3(1) = 3$$
$$2 = 2 \text{ (true)} \qquad 3 = 3 \text{ (true)}$$

For (2, −3),

$$4x + 2y = 2 \qquad 6x + 3y = 3$$
$$4(2) + 2(-3) = 2 \qquad 6(2) + 3(-3) = 3$$
$$8 - 6 = 2 \qquad 12 - 9 = 3$$
$$2 = 2 \text{ (true)} \qquad 3 = 3 \text{ (true)}$$

EXERCISES 12.2

Determine if each given ordered pair is a solution of the given system or not.

A.

1. $x + y = 8$ and $x - y = 2$

   a. (5, 3) b. (3,5) c. (4, 2)

2. $2x - y = 9$ and $x + 2y = 2$

   a. (5, 1) b. (4,−1) c. (−2, 2)

3. $x = 4$ and $y = -1$

a. (4, 1) b. (-1,4) c. (4, -1)

4. $x = -5$ and $y = 2$

a. (-5, 0) b. (0, 2) c. (-5, 2)

5. $y = 2x$ and $3x - y = 3$

a. (-3, -6) b. (3, 6) c. (-6, -12)

B.

1. $x = 4y$ and $2x - 9y = 2$

a. (-2, -8) b. (-8, -2) c. (0, 0)

2. $x - 3y = 2$ and $x - y = 8$

a. (2, -6) b. (11, 3) c. (9, -3)

3. $2x + 3y = 12$ and $3x = 2y$

a. $(0, 4)$ b. $(-2, -3)$ c. $(3, 2)$

4. $x + 2y = 3$ and $3x + 6y = 9$

a. $(3, 0)$ b. $\left(-2, 2\frac{1}{2}\right)$ c. $\left(0, -\frac{1}{2}\right)$

5. $2x - 2y = 5$ and $5x - 5y = 4$

a. $\left(3, \frac{1}{2}\right)$ b. $(2, 2)$ c. $(0, -5)$

12.3 GRAPHICAL SOLUTION

OBJECTIVE: To solve a system of two linear equations in two variables graphically.

When the graphs of two linear equations intersect in exactly one point, then the solution set of the system of the two linear equations consists of exactly one ordered pair of real numbers, the coordinates of the point of intersection of the two lines.

To solve a system of two linear equations, graphically:

(1) Graph the two equations on the same coordinate system.

(2) Read the coordinates of the point of intersection of the two lines.

The corresponding ordered pair is the solution of the system.

SAMPLE PROBLEM

Solve the system $x + 3y = 8$ and $2x - y = 9$ graphically and check.

SOLUTION

$x + 3y = 8$ $\qquad$ $2x - y = 9$

x	y
8	0
2	2

x	y
0	-9
2	-5

Note that $-(-9) = 9$.

Note that $2(2) - (-5) = 4 + 5 = 9$.

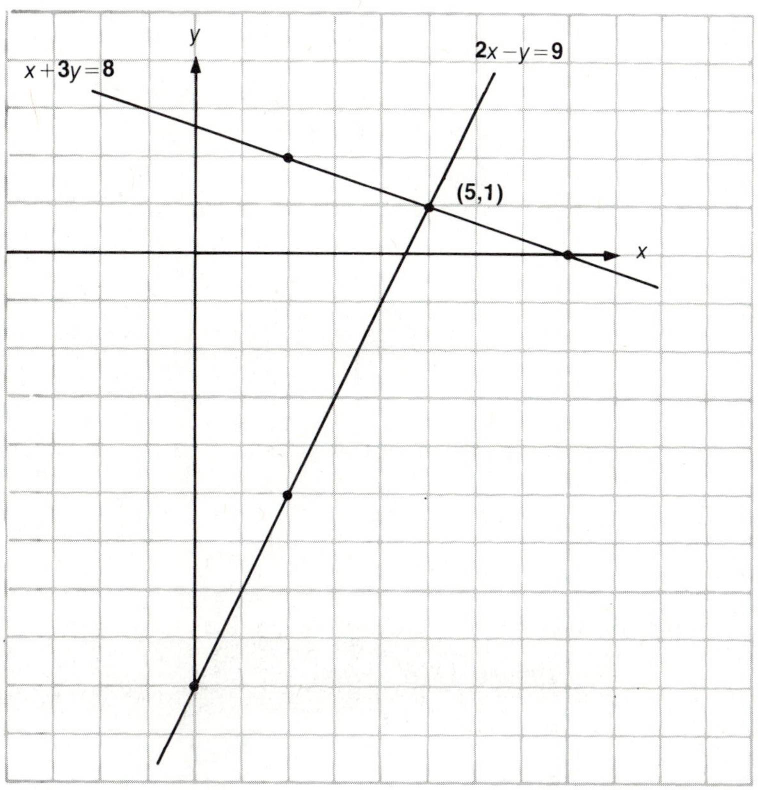

From the graph, the solution is read as (5, 1).

CHECK

For $x + 3y = 8$, $5 + 3(1) = 8$.

For $2x - y = 9$, $2(5) - 1 = 9$.

(Note: Only two points are needed for the graph of each equation. The point of intersection serves as the check point.)

EXERCISES 12.3

Solve each system graphically and check.

A.

1. $x + y = 12$ and $2x - y = 9$

Solution: ______________

Check:

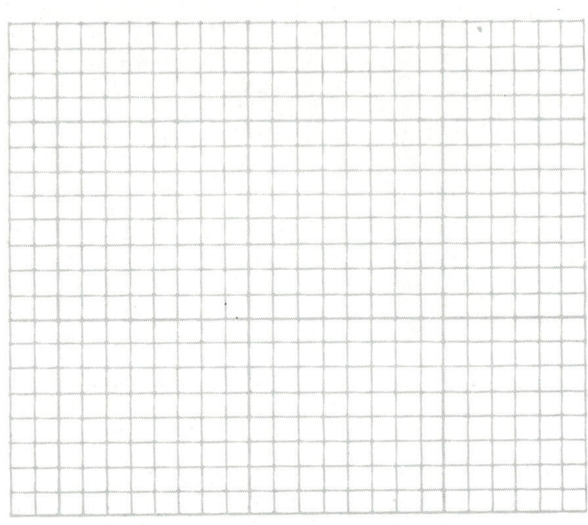

2. $2x + y = 14$ and $x - y = 4$

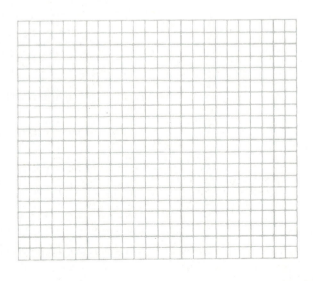

Solution: ______________

Check:

3. $x - 3y = 8$ and $x + 2y = 3$

Solution: ________________

Check:

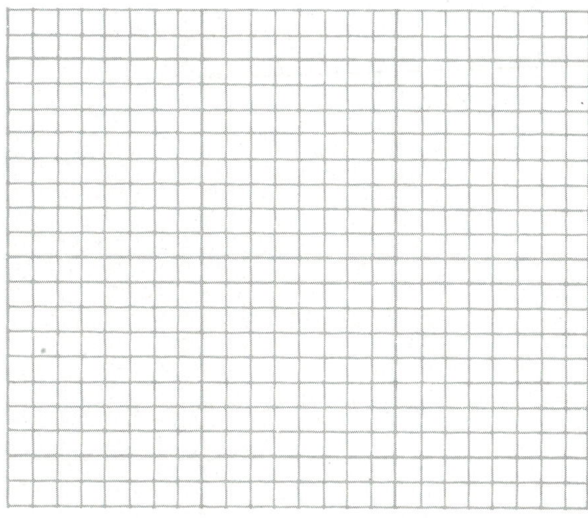

4. $3x + y = 4$ and $2x + y = 1$

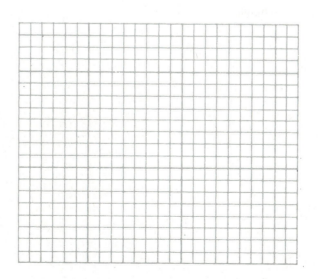

Solution: ________________

Check:

5. $3x - y + 2 = 0$ and $y = 2x - 1$

Solution: ________________

Check:

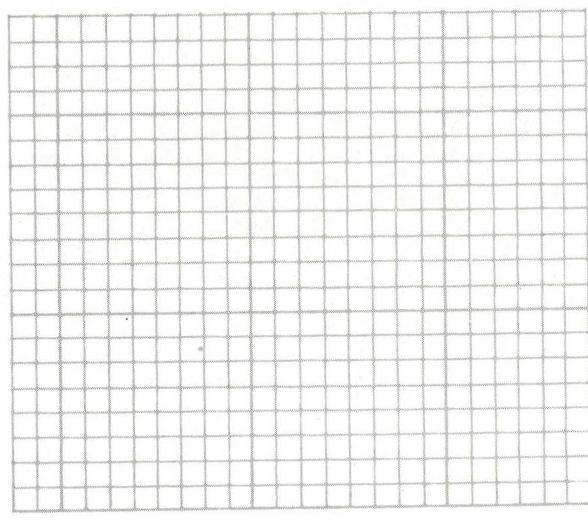

B.

1. $y = 3x + 10$ and $x - 3y = 2$

Solution: ________________

Check:

2. $3x + 2y = 6$ and $2x - y = 4$

Solution: ______________

Check:

3. $x + 2y = 15$ and $y = x - 3$

Solution: ______________

Check:

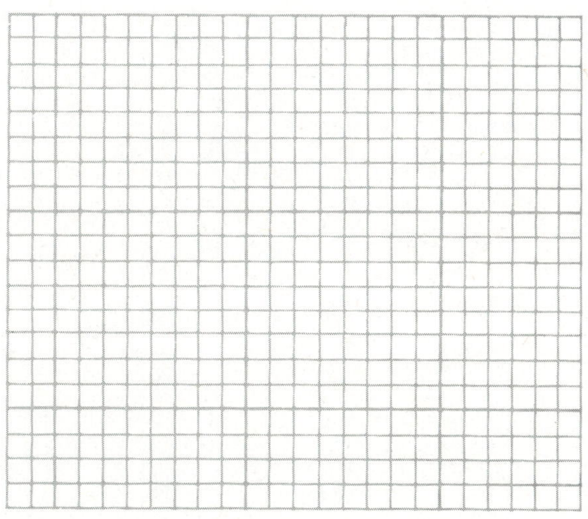

4. $y = 2x - 3$ and $x = 2y - 3$

Solution: ________________

Check:

5. $5x + 2y = 10$ and $2y - 5x = 10$

Solution: ________________

Check:

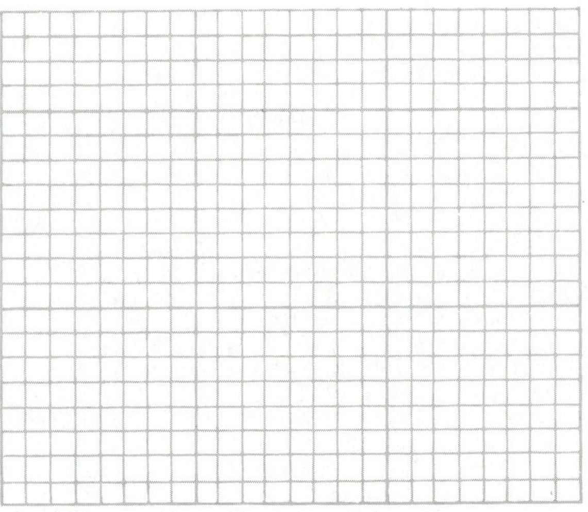

12.4 SUBSTITUTION METHOD

OBJECTIVE: To solve a linear system by the substitution method.

It is not necessary to graph a linear system to find its solution set. The solution set may also be found by algebraic methods. One such method is called the substitution method. In this method, one of the equations is solved for either x or y, and the resulting expression is substituted into the other equation.

If one equation is solved for y, then the method can be outlined as follows.

(1) Solve one of the equations for y, obtaining $y = ax + b$.

(2) In the other equation, replace y by $ax + b$ and solve the resulting equation for x.

(3) In $y = ax + b$, replace x by the value found in step (2) and find the value for y.

SAMPLE PROBLEM 1

Solve the system $2x + 3y = 2$ and $3x + y = 10$ by the substitution method.

SOLUTION

(1) Solve $3x + y = 10$ for y: $y = 10 - 3x$.

(2) Substitute in the other equation.

$$2x + 3y = 2$$
$$2x + 3(\quad) = 2$$
$$2x + 3(10 - 3x) = 2$$
$$2x + 30 - 9x = 2$$
$$-7x = -28$$
$$x = 4$$

(3) Use $y = 10 - 3x$: $y = 10 - 3(4) = -2$

The solution is $(4, -2)$.

CHECK

$2x + 3y = 2$; $2(4) + 3(-2) = 8 - 6 = 2$

$3x + y = 10$; $3(4) + (-2) = 12 - 2 = 10$

In the substitution method, the first equation used can be solved for either x or y. The choice depends on the coefficients of x and y in the two equations. If x has a coefficient of 1 in one of the equations, and y does not, then this equation is solved for x. The procedure is as follows.

(1) Solve one equation for x, obtaining $x = cy + d$.

(2) In the other equation, replace x by $cy + d$ and solve for y.

(3) In $x = cy + d$, replace y by the value found in step (2) and find the value for x.

SAMPLE PROBLEM 2

Solve the system $x - 2y = 1$ and $2x - 3y = 5$ by the substitution method.

SOLUTION

(1) Solve for x.

$$x - 2y = 1$$
$$x = 2y + 1$$

(2)

$$2x - 3y = 5$$
$$2(2y + 1) - 3y = 5$$
$$4y + 2 - 3y = 5$$
$$y = 3$$

(3) $x = 2y + 1$ and $y = 3$

$$x = 2(3) + 1$$
$$= 7$$

The solution is (7, 3).

CHECK

For $x - 2y = 1$,

$$7 - 2(3) = 7 - 6 = 1$$

For $2x - 3y = 5$,

$$2(7) - 3(3) = 14 - 9 = 5$$

A.

EXERCISES 12.4

(1–2) Solve each of the following systems by using the indicated format.

1. $2x + y = 14$ and $3x - 2y = 7$

   (1) Solve $2x + y = 14$ for y. $y =$

   (2) In $3x - 2y = 7$, replace y by the value found in step (1).

   $3x - 2(\qquad) = 7$

   Solve for x.

   (3) Replace x by the value found in step (2), using either original equation. Solve for y.

   The solution is the ordered pair (,).

   Check: $2x + y = 14$

   $3x - 2y = 7$

2. $2x - 3y = 1$ and $x - y = 3$

   (1) Solve $x - y = 3$ for x. $x =$

   (2) Replace x in $2x - 3y = 1$, using the result of step (1). $2(\qquad) - 3y = 1$

   Solve for y.

   (3) Find x.

   The solution is the ordered pair (,).

   Check:

(3–12) Solve each system by the substitution method. Check.

3. $2x + 3y = 7$ and $y = 5x - 9$

4. $5x - y = 9$ and $x = y - 3$

5. $2x + 10y = 1$ and $x = -4y$

6. $3x - 2y = 9$ and $y = 3x$

7. $5x + y = 10$ and $3x - 2y - 6 = 0$

8. $5x - 3y - 50 = 0$ and $3x + y = 2$

9. $x - 4y = 1$ and $3x + 4y = 19$

10. $2x - 3y = 21$ and $x + 6y = 3$

11. $2x - y = 2$ and $3x + 2y + 25 = 0$

12. $3x - y = 19$ and $2x - 3y = 1$

B.

Solve each system by the substitution method. Check.

1. $y - 3x = 8$ and $y = 5x$

2. $2x + y = 18$ and $3x + 2y = 29$

3. $x + 4y = 25$ and $2x - 3y = 6$

4. $4y - 3x = 17$ and $x - 2y + 7 = 0$

5. $5x + 2y = 38$ and $3x + y = 24$

6. $y - 2x = 2$ and $3x - 2y = 1$

7. $3x - y = 7$ and $4x + 3y = 18$

8. $5y - x = 10$ and $3x - 5y + 30 = 0$

9. $x + 3y = 45$ and $x - 4y = 10$

10. $2x + 5y = 8$ and $4x + y + 20 = 0$

12.5 ADDITION METHOD

OBJECTIVE: To solve a linear system by the addition method.

It can be checked that (2, 5) is the solution of the system $3x + 2y - 16 = 0$ and $x - y + 3 = 0$.

For $3x + 2y - 16 = 0$, $3(2) + 2(5) - 16 = 6 + 10 - 16 = 0$.

For $x - y + 3 = 0$, $2 - 5 + 3 = 0$

Now consider

$$a(3x + 2y - 16) + b(x - y + 3) = 0,$$

where a and b are any real numbers.

No matter what numbers a and b are, (2, 5) will always be a solution of $a(3x + 2y - 16) + b(x - y + 3) = 0$, since $3x + 2y - 16 = 0$ and $x - y + 3 = 0$ for $x = 2$ and $y = 5$.

This means that if each equation of a linear system is multiplied by a constant and the resulting equations are added, then the solution of the system is also a solution of the new equation.

The addition method for solving a linear system is based on this idea. The goal is to choose the constants a and b so that one of the variables is missing in the new equation obtained by addition.

To solve a system of two linear equations in two variables by the addition method:

(1) Multiply each term of the first equation by a constant and multiply each term of the second equation by a constant so that the resulting coefficient of y (or x) in one equation is the negative of the resulting coefficient of y (or x) in the other equation.

(2) Add the resulting equations and solve for the variable.

(3) Repeat steps (1) and (2) to find the value of the other variable, or use the substitution principle to find the other variable.

SAMPLE PROBLEM 1

Solve the system $x + 2y = 10$ and $3x - y = 9$ by the addition method.

SOLUTION

(1) **The y terms are $2y$ and $-y$. If the first equation is multiplied by 1 and the second equation by 2, the new terms will be $2y$ and $-2y$.**

$1(x + 2y = 10)$ becomes $x + 2y = 10$

$2(3x - y = 9)$ becomes $6x - 2y = 18$

(2) **Add and solve.** $7x = 28$

$x = 4$

(3) Multiply the first equation by 3 and the second by -1.

$3\ (x + 2y = 10)$ becomes $3x + 6y = 30$

$-1\ (3x - y = 9)$ becomes $-3x + y = -9$

(4) Add and solve.

$$7y = 21$$
$$y = 3$$

The solution is (4, 3).

CHECK

$x + 2y = 10; 4 + 2\ (3) = 4 + 6 = 10$

$3x - y = 9; 3\ (4) - 3 = 12 - 3 = 9$

SAMPLE PROBLEM 2

Use the addition method to solve the system $3x + 2y = 11$ and $5x - 3y = 31$.

SOLUTION

(1) $3\ (3x + 2y = 11)$ becomes $9x + 6y = 33$

$2\ (5x - 3y = 31)$ becomes $10x - 6y = 62$

(2)
$$19x = 95$$
$$x = 5$$

(3) Substitute 5 for x in $3x + 2y = 11$.

$$3\ (5) + 2y = 11$$
$$2y = 11 - 15$$
$$2y = -4$$
$$y = -2$$

The solution is (5, −2).

CHECK

For $3x + 2y = 11$, $3\ (5) + 2\ (-2) = 15 - 4 = 11$.

For $5x - 3y = 31$, $5\ (5) - 3\ (-2) = 25 + 6 = 31$.

A.

EXERCISES 12.5

(1–2) Solve each of the following systems by using the indicated format.

1. $x + 3y = 18$ and $2x - y = 8$

   Multiply each term by 1: $1\ (x + 3y = 18) \rightarrow$

   Multiply each term by 3: $3\ (2x - y = 8) \rightarrow$

   Add the two resulting equations:

   Solve for x:

   Multiply each term by 2: $2\ (x + 3y = 18) \rightarrow$

   Multiply each term by -1: $-1\ (2x - y = 8) \rightarrow$

   Add the two resulting equations:

   Solve for y:

   State the solution, (,)

   Check the solution in each equation of the system.

   For $x + 3y = 18$,

   For $2x - y = 8$,

2. $x + 6y = 3$ and $2x - 3y = 21$

   Multiply each term by 2: $2\ (x + 6y = 3) \rightarrow$

   Multiply each term by -1: $-1\ (2x - 3y = 21) \rightarrow$

   Add the resulting equations:

   Solve for y:

   Substitute the y value found in $x + 6y = 3$

   Solve for x:

   State the solution, (,)

   Check the solution in each equation of the system.

   For $x + 6y = 3$,

   For $2x - 3y = 21$,

(3–10) Solve each of the following systems by the addition method. Check each solution.

3. $x + y = 14$

   $x - y = 10$

4. $3x + 2y = 19$

   $x - 2y = 1$

5. $2x + y = 13$

   $x + 3y = 7$

6. $4x - y = 32$

   $x - 5y = 27$

7. $2x + 3y = 31$

   $3x - 2y = 1$

8. $5x - 4y = 60$

   $4x + 5y = 7$

9. $2x - y - 2 = 0$

   $3x + 2y + 25 = 0$

10. $3x + y - 2 = 0$

$5x - 3y + 13 = 0$

B.

(1–10) Solve each system by the addition method. Check each solution.

1. $x - y = 18$ and $x + y = 42$

2. $2x + 3y = 23$ and $5x - 3y = 5$

3. $4x - y = 32$ and $4x + 5y = 8$

4. $x + 2y = 12$ and $3x + y + 9 = 0$

5. $2x + 5y = 46$ and $5x + 3y = 58$

6. $3x - 4y = 6$ and $5x - 8y = 14$

7. $3x + 5y = 60$ and $3y - 5x = 36$

8. $x + 3y = 9$ and $4x - 2y = 1$

9. $5x + y - 10 = 0$ and $3x - 2y - 6 = 0$

10. $3x - y - 19 = 0$ and $2x - 3y - 29 = 0$

12.6 ALGEBRAIC METHODS, GENERAL CASE (OPTIONAL)

OBJECTIVE: To solve any type of linear system and identify the graphs of the equations as intersecting lines, parallel lines, or coincident lines.

There are three types of solution sets for a linear system of the form $Ax + By + C = 0$ and $Dx + Ey + F = 0$.

(1) The algebraic solution yields equations of the form $x = a$ and $y = b$. This solution set consists of exactly one ordered pair, (a, b), and the graphs of the two equations intersect in exactly one point, the point (a, b). This linear system is said to be **determinative.**

(2) The algebraic solution yields an equation that is false for all values of the variables. This solution set is the empty set $\emptyset$. The graphs of the two equations are parallel lines, and the linear system is said to be **inconsistent**.

(3) The algebraic solution yields an equation that is true for all values of the variables. This solution set is infinite, consisting of all solutions of either equation of the system. The graphs of the two equations are coincident lines, and the linear system is said to be **dependent**.

The substitution method is convenient to use when either one of the given equations has the form $y = mx + b$ or $x = cy + d$, or when the coefficient of x or y is 1 in one of the given equations. In this latter case, the equation may be easily written in the form $y = mx + b$ or $x = cy + d$.

The addition method is convenient to use when each of the two given equations has the form $ax + by = c$, and no coefficient of x or y is 1.

In sample problems 1–3 find the solution set of the system and state whether the graphs of the two given equations are intersecting lines, parallel lines, or coincident lines.

SAMPLE PROBLEM 1

$6x - 3y = 5$ and $y = 2x - 3$

SOLUTION

Use the substitution method: $y = 2x - 3$.

$$6x - 3y = 5$$
$$6x - 3(2x - 3) = 5$$
$$6x - 6x + 9 = 5$$
$$9 = 5 \quad \text{(false for all values of } x \text{ and } y\text{)}$$

The solution set is the empty set $\emptyset$.

The graphs are parallel lines.

SAMPLE PROBLEM 2

$2x + 4y = 6$ and $3x + 6y = 9$

SOLUTION

Use the addition method.

$$3(2x + 4y = 6) \rightarrow 6x + 12y = 18$$
$$-2(3x + 6y = 9) \rightarrow -6x - 12y = -18$$
$$0 = 0 \quad \text{(true for all values of } x \text{ and } y\text{)}$$

The solution set is $\{(x,y) | 2x + 4y = 6\}$.

The graphs are coincident lines.

SAMPLE PROBLEM 3

$5x + 2y = 20$ and $2x - y = 8$

SOLUTION

Use the addition method.

$$\begin{aligned} 1\,(5x + 2y = 20) &\rightarrow 5x + 2y = 20 \\ 2\,(2x - \ \ y = \ 8) &\rightarrow 4x - 2y = 16 \\ & \quad\ \ 9x \qquad = 36 \\ & \qquad\qquad x = 4 \end{aligned}$$

Replace x by 4 in $5x + 2y = 20$.

$$\begin{aligned} 5\,(4) + 2y &= 20 \\ 2y &= 0 \\ y &= 0 \end{aligned}$$

The solution set is $\{(4, 0)\}$.

The graphs are intersecting lines.

EXERCISES 12.6

A.

Find the solution set of each system and state whether the graphs of the two given equations are intersecting lines, parallel lines, or coincident lines.

1. $3x - y - 7 = 0$ and $x + 3y - 29 = 0$

2. $2x - 4y = 6$ and $3x = 6y + 9$

3. $6x - 3y = 1$ and $y = 2x - 5$

4. $x + 4y = 7$ and $2x + 8y + 5 = 0$

5. $3x + 4y = 12$ and $x - 2y = 4$

6. $5x - 2y = 20$ and $3x + 5y = 12$

7. $x = y - 6$ and $y = x + 6$

8. $8x + 5y = 9$ and $x - y = 6$

B.

1. $5x - 2 = 0$ and $2x - 5 = 0$

2. $2x - 5y = 10$ and $6x - 15y = 30$

3. $2x - y = 20$ and $4x + 3y = 290$

4. $3x - 5y = 100$ and $5x - 3y = 300$

5. $10x + 2y - 30 = 0$ and $5x + y - 15 = 0$

6. $4y + 7 = 0$ and $7y + 4 = 0$

7. $3x - y = 0$ and $3y = x$

8. $3x + y = 6$ and $x + 3y = 6$

12.7 MIXTURE PROBLEMS

OBJECTIVE: To solve a stated mixture problem using two variables.

Some mixture problems are easier to solve when two variables are used. Examples of these types are shown in the sample problems.

SAMPLE PROBLEM 1

Five cans of tomato sauce and two cans of mushrooms cost 97¢. Eight cans of tomato sauce and three cans of mushrooms cost $1.50. Find the cost per can for the tomato sauce and for the mushrooms.

SOLUTION

Let x = the cost per can of tomato sauce.

Let y = the cost per can of mushrooms.

Then, $5x + 2y = 97$ and $8x + 3y = 150$

$-3\ (5x + 2y = 97)$ becomes

$$-15x - 6y = -291$$

$2\ (8x + 3y = 150)$ becomes

$$16x + 6y = 300$$

$$x = 9$$

$$5x + 2y = 97$$

$$5\ (9) + 2y = 97$$

$$2y = 97 - 45 = 52$$

$$y = 26$$

The cost of one can of tomato sauce was 9¢.

The cost of one can of mushrooms was 26¢.

SAMPLE PROBLEM 2

The cost of printing 400 programs was \$7.50 while the cost of printing 1000 programs was \$10.50. The printer charged a fixed fee for typesetting and then an additional charge for each program. Find the typesetting fee.

SOLUTION

Let x = the typesetting fee.

Let y = the charge per program.

Expressing the cost in cents, $x + 400y = 750$ and $x + 1000y = 1050$.

Multiply the first equation by -1 and add the result to the second equation.

$$\begin{aligned} x + 1000y &= 1050 \\ -x - 400y &= -750 \quad (+) \\ \hline 600y &= 300 \\ y &= \frac{1}{2} \end{aligned}$$

Solve for x.

$$\begin{aligned} x + 400y &= 750 \\ x + 400\left(\frac{1}{2}\right) &= 750 \\ x + 200 &= 750 \\ x &= 550 \end{aligned}$$

The typesetting fee was \$5.50.

EXERCISES 12.7

Solve each of the following by using two variables.

A.

1. Two yards of lace and 3 yd. of seam binding cost 96¢. With no change in unit prices, 5 yd. of the same lace and 6 yd. of the same seam binding cost \$2.22. Find the cost per yard of each.

2. One quart of oil and 5 gal. of gasoline cost $2.45. With no change in unit prices, 2 qt. of the same oil and 7 gal. of the same gasoline cost $3.76. Find the cost of 1 qt. of oil and the cost of 1 gal. of gasoline.

3. When 36 lb. of copper ore A are combined with 28 lb. of copper ore B, then 39 lb. of copper are obtained. Fifty pounds of copper ore A combined with 40 lb. of copper ore B produce 55 lb. of copper. Find the percentage of copper in each ore.

4. A man received an income of $450 from $5000 invested in stocks and $3000 invested in bonds. His son received $195 from an investment of $2000 in the same stocks and $1500 invested in the same bonds. Find the percentage they received from each investment.

5. One week when John worked 52 hr. and Jill worked 40 hr., their combined income was $210. The next week, John worked 40 hr. and Jill worked 25 hr. and their combined income was $150. Find the hourly wage of each.

B.

1. A certain carpenter receives \$8 per hour and his helper receives \$5 per hour. Their combined wages for 1 week was \$550. The next week the carpenter worked half as many hours as the week before and his helper worked 10 hr. more than the week before. Their combined wages for the second week was \$400. Find the number of hours that each worked the first week.

2. A certain theater charges \$2.80 for each adult and \$1.20 for each child at all performances. One day, a total of \$696 was collected for a matinee performance. At the evening performance, when twice as many adult tickets were sold and half as many children's tickets were sold, a total of \$852 was collected. How many adults and how many children attended the matinee performance?

3. A certain amusement park charges a fixed fee for each car and an additional fee for each person in the car. The total cost for 8 cars and 30 persons was \$42.50, while the total cost for 10 cars and 40 persons was \$55. Find the fixed fee per car and the additional fee per passenger.

4. Two wine merchants enter Paris, one with 64 casks of wine and the other with 20 casks. Since they don't have enough money to pay the customs duties, the first pays 5 casks of wine and 40 francs and the second pays 2 casks of wine and receives 40 francs in change. What is the price of each cask of wine and what is the duty on it? (A problem in a book written by the French mathematician Nicolas Chuquet in 1484.)

5. The mixed price of 9 citrons and 7 fragrant wood-apples is 107; again, the mixed price of 7 citrons and 9 wood-apples is 101. O you arithmetician, tell me quickly the price of a citron and of a wood-apple here, having distinctly separated those prices well. (From the works of the Hindu Mahavira, about A.D. 850.)

12.8 UNIFORM MOTION PROBLEMS

OBJECTIVE: To solve a stated uniform motion problem using two variables.

There are some uniform motion problems whose solution is often easier when two variables are used.

Let r be the uniform rate that an object travels when there is no water current or wind.

Let s be the rate of the current or the wind.

Then $r + s$ = the rate of the object traveling with the current or wind, and $r - s$ = the rate of the object traveling against the current or wind.

SAMPLE PROBLEM 1

A boat traveling downstream (with a current) goes 36 mi. in 2 hr. Traveling upstream (against the same current), the same boat takes 3 hr. to go 36 mi. Find the rate of the current.

SOLUTION

Let x = the rate of the boat in still water.

Let y = the rate of the current.

Then $x + y$ = the rate of the boat downstream, and $x - y$ = the rate of the boat upstream.

Use the formula $rt = d$ (rate times time equals distance).

Downstream,

$$2(x + y) = 36$$

Upstream,

$$3(x - y) = 36$$

Thus,

$$x + y = 18$$
$$x - y = 12$$
$$2x = 30$$
$$x = 15 \text{ m.p.h.}$$
$$y = 3 \text{ m.p.h., rate of current}$$

SAMPLE PROBLEM 2

A plane, whose speed in still air is 180 m.p.h., travels 500 mi. with a tailwind in the same time that it travels 400 mi. against the same wind (now a headwind). Find the speed of the wind.

SOLUTION

Let x = the speed of the wind.

Let y = the time for each trip.

	r	t	d
with wind	$180 + x$	y	$(180 + x)y = 500$
against wind	$180 - x$	y	$(180 - x)y = 400$

$$180y + xy = 500$$

$$180y - xy = 400$$

Add.

$$360y = 900$$

$$y = \frac{5}{2}$$

Substitute in the first equation.

$$180\left(\frac{5}{2}\right) + x\left(\frac{5}{2}\right) = 500$$

$$180 + x = 200$$

$$x = 20 \text{ m.p.h., the speed of the wind}$$

EXERCISES 12.8

A.

Solve each of the following by using two variables.

1. In 3 hr., an airplane flies 630 mi. with a tailwind. Returning against the same wind, the airplane takes $3\frac{1}{2}$ hr. to fly 630 mi. Find the speed of the wind.

2. A boat, whose speed in still water is 20 m.p.h., can travel 12 mi. downstream in the same time that it takes to travel 8 mi. upstream. Find the rate of the water current.

3. A motorist made a 200-mi. trip, averaging 50 m.p.h. on a level road and 25 m.p.h. on a mountain road. His time spent on the mountain road was 1 hr. less than his time on the level road. How many miles of the trip was on the mountain road?

4. It takes one plane half the time to fly 852 mi. with a tailwind as it does another plane to fly 1560 mi. against the same wind. If both planes have the same speed in still air, a speed of 408 m.p.h., find the speed of the wind.

5. Flying with a wind at full speed, a traffic helicopter flies 31 mi. in 10 min. Cruising at half-speed against the same wind, the helicopter flies 18 mi. in 20 min. Find the speed of the wind in miles per hour.

B.

1. An airplane traveling with a tailwind of 18 m.p.h. flies 456 mi. in the same time it takes to fly 384 mi. against the same wind. If the speed of the airplane in still air is the same for both trips, find this speed.

2. A boat takes 30 min. to go $7\frac{1}{2}$ mi. downstream. The return trip takes 45 min. Find the rate of the boat in still water and find the rate of the water current.

3. To reach his destination, a man has to row across a river and then walk a certain distance along the shore. If he lands at dock A, he rows for 1 hr. and walks for 2 hr. for a total distance of 12 mi. If he lands at dock B, he rows for $1\frac{1}{2}$ hr. and walks for 1 hr. for a total distance of 11 mi. Find his rate of rowing and his rate of walking.

4. A man can row a boat in still water at the rate of 5 m.p.h. One day when the water current was 3 m.p.h., he left a dock, rowed downstream to a bridge, and then returned to the dock. If the entire trip took 30 min., how far from the dock was the bridge?

5. A cyclist travels 1 mi. uphill and 2 mi. downhill in 24 min. The return trip takes 30 min. Find his rates of speed uphill and downhill.

12.9 Assorted Stated Problems

Solve each of the following by using two variables.

1. An electrician makes $2.50 more per hour than his assistant does. For a 40-hr. week, their combined earnings was $420. Find the hourly wage of each.

2. A rectangular field is 15 ft. longer than it is wide. If the length and the width are each increased by 5 ft., then the new area exceeds the old area by 550 sq. ft. Find the dimensions of the original field.

3. A boat went 14 mi. downstream in 30 min. The return trip upstream took 42 min. Find the rate of the current in miles per hour.

4. A certain tank has a capacity of 550 gal. An inlet pipe can deliver water at the rate of 8 gal. per minute while an outlet pipe can drain the water at the rate of 5 gal. per minute. One day in order to fill the empty tank, the inlet pipe was opened. Later it was discovered that the outlet pipe had also been left open. The tank was full 20 min. after the outlet pipe was closed. How long did it take to fill the tank and how long were both pipes open?

5. A man has \$6000 invested with company A and \$9000 with company B. One year his total profit from both sources was \$690. The next year company A doubled his profit rate while company B decreased his profit rate by 1%. His total profit for this year was \$1020. Find the original rates of profit.

6. Of a total of 500 sheets, some were irregular and were classified as seconds. The irregular sheets were sold for \$2.50 each and the other sheets were sold for \$4.00 each. A total of \$1820 was received for the 500 sheets. How many sheets were seconds?

7. A size A card of buttons contains 6 small buttons and 3 large buttons. The size B card of buttons contains 3 small buttons and 1 large button. How many cards of size A and how many cards of size B should be purchased to obtain exactly 51 small buttons and exactly 22 large buttons?

8. A trip of 1810 mi. was made partly on a bus traveling at an average rate of 40 m.p.h. and partly on an airplane traveling at an average rate of 500 m.p.h. The time spent on the plane was 2 hr. longer than the time spent on the bus. Find the time spent on the plane and the time spent on the bus.

9. A restaurant bought 80 glasses and 60 coffee mugs for a total cost of $34. Later because of breakage, 25 glasses and 10 coffee mugs were bought for a total cost of $8. If there was no change in the prices, find the cost of 1 glass and the cost of 1 coffee mug.

10. A jet plane traveling with a tailwind made a 2640-mi. trip from California to Hawaii in 4.4 hr. The same type of jet plane having the same speed in still air made the return trip against the same wind in 4.8 hr. Find the speed of the wind.

UNIT 12 POSTTEST

(1–4) Find the slopes and y-intercepts of the graphs of each pair and determine if the lines intersect, are parallel, or are coincident. (12.1)

1. $y = 5x - 10$ and $y = 10 - 5x$
2. $x - 2y = 4$ and $x = 2y + 4$
3. $3x + 5y = 15$ and $y = 3x + 3$
4. $x - 3y = 9$ and $3x = 9y + 3$

(5–6) Determine whether each given ordered pair is a solution of the given system. (12.2)

5. $4x + y = 6$ and $3x - y = 15$
   a. $(3, -6)$
   b. $(3, 6)$
6. $2x + 5y = 2$ and $4y - 5x = 12$
   a. $(6, -2)$
   b. $(-4, 2)$

7. Solve graphically and check. (12.3)

   $2x + 5y = 10$ and $y = 8 + 2x$

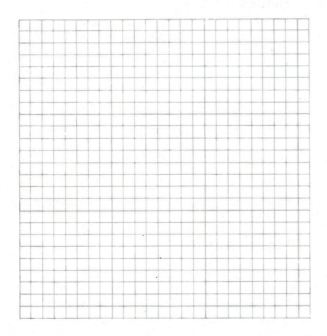

(8–9) Solve by the substitution method and check. (12.4)

8. $2x + 5y = 10$ and $y = 8 + 2x$

9. $x + 4y = 10$ and $5x + 3y + 18 = 0$

10. Solve by the addition method and check. (12.5)

    $5x - 2y = 16$ and $2x + 5y = 18$

(11–13) Solve and identify the graph of each system as intersecting lines, parallel lines, or coincident lines. (12.6)

11. $x = 3y - 2$ and $3y = x + 2$

12. $x + 2y = 8$ and $x = 2y - 8$

13. $6x - 2y = 9$ and $y - 3x = 6$

14. A salesman sold some stoves and refrigerators to a buyer from Canada for a total sales of \$5050 and to a buyer from Mexico for a total sales of \$5350. Each stove was sold at \$350 and each refrigerator was sold at \$450. The salesman lost his itemized records, but he remembered that the Mexican bought the same number of stoves as the Canadian bought refrigerators and the same number of refrigerators as the Canadian bought stoves. How many stoves and how many refrigerators should he ship to Canada? How many to Mexico? (12.7)

15. On a certain trip an airplane, having a speed in still air of 175 m.p.h., flew for 3 hr. with a tailwind. On the return trip traveling against the same wind, the plane had to land at the end of 2 hr. having gone only halfway. Find the speed of the wind. (12.8)

UNIT 13

QUADRATIC EQUATIONS

Unit 13 PRETEST

(1–2) Solve by factoring. (13.1)

1. $x^2 + 5x = 24$
2. $2x(5x - 4) = (x - 4)^2$

(3–4) Simplify. (13.2)

3. $\sqrt{25}\sqrt{36}$
4. $\dfrac{5 - \sqrt{49}}{4}$

(5–6) By using a table, approximate each of the following to the nearest hundredth. (13.2)

5. $\sqrt{46}$
6. $\dfrac{7 + \sqrt{14}}{2}$

(7–8) Solve. (13.2)

7. $4x^2 = 21$
8. $100x^2 = 3721$

(9–10) Simplify. (13.3)

9. $\sqrt{12}\sqrt{75}$
10. $\sqrt{180}$

(11–12) Simplify. (13.4)

11. $\sqrt{\dfrac{3}{5}}$
12. $\dfrac{6}{\sqrt{8}}$

(13–14) Simplify. (13.5)

13. $\sqrt{147} - \sqrt{98}$
14. $\dfrac{15 + \sqrt{54}}{3}$

(15–16) Solve by completing the square. (13.6)

15. $x^2 + 44 = 14x$

16. $5x^2 + 4x - 2 = 0$

(17–18) Solve by using the quadratic formula. (13.7)

17. $3x^2 + 4x - 5 = 0$

18. $2x^2 = 10x - 11$

19. John and Bill decide that each will mow half of a lawn that is in the shape of a 45-meter by 28-meter rectangle. How wide a strip must John mow around the outer edge in order to leave half of the lawn for Bill to finish? (13.8)

20. At a water treatment plant, it is found that a 2-in. pipe takes 10 min. longer to fill a tank than a $2\frac{1}{2}$-in. pipe. After another $2\frac{1}{2}$-in. pipe is installed, the 3 pipes together fill the tank in 21 min. How long would it take one $2\frac{1}{2}$-in. pipe alone to fill the tank? (13.8)

13.1 SOLUTION BY FACTORING

OBJECTIVE: To solve a quadratic equation by writing the equation in standard form, factoring the quadratic polynomial, and applying the zero product property.

The following are examples of quadratic equations in the variable x.

$$3x^2 - 5x + 2 = 0 \qquad x^2 = 8 - 2x$$

$$4x^2 - 9x = 0 \qquad x^2 = 9$$

A **quadratic equation in one variable** is a polynomial equation in which the highest power of the variable is the second.

The **standard form of a quadratic equation** in the variable x is

$$ax^2 + bx + c = 0,$$

where a, b, and c are constants and a is positive.

In sample problems 1–5 write each equation in standard form and state the values of a, b, and c.

SAMPLE PROBLEM 1

$5x^2 + 2x - 6 = 0$

SOLUTION

The equation is in standard form: $a = 5, b = 2, c = -6$.

SAMPLE PROBLEM 2

$x^2 = 4x + 12$

SOLUTION

$x^2 - 4x - 12 = 0; a = 1, b = -4, c = -12$.

SAMPLE PROBLEM 3

$2x\,(x + 3) = 3\,(2x - 1)$

SOLUTION

Remove parentheses: $2x^2 + 6x = 6x - 3$.

Write in standard form: $2x^2 + 3 = 0$.

$a = 2, b = 0, c = 3$

SAMPLE PROBLEM 4

$(x + 5)(x + 2) = 10$

SOLUTION

$x^2 + 7x + 10 = 10$

$x^2 + 7x = 0; a = 1, b = 7, c = 0$

SAMPLE PROBLEM 5

$x(x - 7) = 3(x + 2)(x - 2)$

SOLUTION

$$x^2 - 7x = 3x^2 - 12$$

$$-2x^2 - 7x + 12 = 0$$

$$2x^2 + 7x - 12 = 0 \qquad \textbf{(standard form)}$$

$a = 2, b = 7, c = -12$

The factoring methods studied in Unit 8 can be used to solve some types of quadratic equations. Also needed is a property called the zero product property of real numbers. This property states that if a product of two factors is zero, then one factor or the other factor must equal zero.

Zero product property: **If $AB = 0$, then $A = 0$ or $B = 0$.**

To solve a quadratic equation by factoring:

1. Write the equation in standard form.
2. Factor the quadratic polynomial.
3. Set each linear factor equal to zero.
4. Solve the resulting equations.
5. Check each solution in the original equation.
6. State the solution set.

In general, a quadratic equation has two solutions, also called **roots** of the equation. If two linear factors of the quadratic polynomial are identical, then only one solution is obtained and it is called a **double root.**

In sample problems 6–10 solve by factoring.

SAMPLE PROBLEM 6

$x^2 - 9x + 20 = 0$

SOLUTION

(1) $x^2 - 9x + 20 = 0$

(2) $(x - 5)(x - 4) = 0$

(3) $x - 5 = 0$ or $x - 4 = 0$

(4) $x = 5$ $\quad x = 4$

(5) $5^2 - 9(5) + 20 = 0$ $\quad 4^2 - 9(4) + 20 = 0$

$25 - 45 + 20 = 0$ $\quad 16 - 36 + 20 = 0$

$0 = 0$ $\quad 0 = 0$

(6) The solution set is $\{5, 4\}$.

SAMPLE PROBLEM 7

$(x + 3)(x + 5) = 3$

SOLUTION

(1) $x^2 + 8x + 15 = 3$

$x^2 + 8x + 12 = 0$

(2) $(x + 6)(x + 2) = 0$

(3) $x + 6 = 0$ or $x + 2 = 0$

(4) $x = -6$ $\quad x = -2$

(5) $(-6 + 3)(-6 + 5) = 3$ $\quad (-2 + 3)(-2 + 5) = 3$

$(-3)(-1) = 3$ $\quad (1)(3) = 3$

$3 = 3$ $\quad 3 = 3$

(6) The solution set is $\{-6, -2\}$.

SAMPLE PROBLEM 8

$3x^2 = 15x$

SOLUTION

(1) $3x^2 - 15x = 0$

(2) $3x(x - 5) = 0$

(3) $3x = 0$ or $x - 5 = 0$

(4) $x = 0$ $\qquad x = 5$

(5) $3(0^2) = 15(0)$ $\qquad 3(5^2) = 15(5)$

$0 = 0$ $\qquad 75 = 75$

(6) The solution set is $\{0, 5\}$.

SAMPLE PROBLEM 9

$2x(2x - 5) = 5(5 - 2x)$

SOLUTION

(1) $4x^2 - 10x = 25 - 10x$

$4x^2 - 25 = 0$

(2) $(2x + 5)(2x - 5) = 0$

(3) $2x + 5 = 0$ or $2x - 5 = 0$

(4) $2x = -5$ $\qquad 2x = 5$

$x = \frac{-5}{2}$ $\qquad x = \frac{5}{2}$

(5)
$$2\left(\frac{-5}{2}\right)\left(2 \cdot \frac{-5}{2} - 5\right) = 5\left(5 - 2 \cdot \frac{-5}{2}\right)$$
$$-5(-5 - 5) = 5(5 + 5)$$
$$-5(-10) = 5(10)$$
$$50 = 50$$
$$2\left(\frac{5}{2}\right)\left(2 \cdot \frac{5}{2} - 5\right) = 5\left(5 - 2 \cdot \frac{5}{2}\right)$$
$$5(5 - 5) = 5(5 - 5)$$
$$0 = 0$$

(6) The solution set is $\left\{\frac{-5}{2}, \frac{5}{2}\right\}$.

SAMPLE PROBLEM 10

$4(3x - 1) = 9x^2$

SOLUTION

(1) $12x - 4 = 9x^2$

$9x^2 - 12x + 4 = 0$

(2) $(3x - 2)(3x - 2) = 0$

(3) $3x - 2 = 0$ or $3x - 2 = 0$

(4) $x = \frac{2}{3}$ $\quad x = \frac{2}{3}$

(5) $4\left(3 \cdot \frac{2}{3} - 1\right) = 9\left(\frac{2}{3}\right)^2$

$4(2 - 1) = 9\left(\frac{4}{9}\right)$

$4 = 4$

(6) The solution set is $\left\{\frac{2}{3}\right\}$, where $\frac{2}{3}$ is a double root.

EXERCISES 13.1

Solve by factoring and check.

A.

1. $x^2 - 2x - 8 = 0$

2. $x^2 + 5x = 14$

3. $x^2 - 25 = 0$

4. $36x^2 = 49$

5. $4x(x - 6) = 0$

6. $3 - 5x = 2x^2$

7. $x^2 + 16 = 8x$

8. $(x + 5)(x - 4) = 10$

9. $2x(2x - 25) = 50(2 - x)$

10. $6(x + 2)(x + 3) = 0$

11. $12x = 4x^2$

12. $2y = 3 - 5y^2$

13. $100t - 25t^2 = 0$

14. $(y + 2)(y + 3) = 2$

15. $6(2r + 1) = r(5r - 1)$

16. $(u - 5)^2 = (2u + 5)^2$

17. $3(t + 1) = 8t(t + 1)$

18. $10y(y + 1) = 3(1 - y)$

19. $(r - 8)^2 = 64$

20. $(4x + 7)^2 = 7(8x + 7)$

B.

1. $x^2 + 6x + 5 = 0$

2. $x^2 - 3x = 18$

3. $2x^2 - 18 = 0$

4. $64 = 81x^2$

5. $5x(x + 7) = 0$

6. $2x + 8 = 3x^2$

7. $(x + 5)(x + 10) = x + 1$

8. $x(x - 7) = (2x + 1)(x - 4)$

9. $(x - 4)(x + 2) = 40$

10. $(x + 3)^2 = 2(3x + 5)$

11. $9x = 5 - 2x^2$

12. $81y = 9y^2$

13. $t(t + 3) = t - 1$

14. $(y - 4)(y + 5) = 22$

15. $(r + 4)^2 = (3r - 4)^2$

16. $6u^2 = 7u + 3$

17. $(t + 7)(t - 7) = 3(t - 3)$

18. $5(y - 1) = 6y(y - 2)$

19. $(2x + 9)^2 = 18x + 81$

20. $(5x - 6)^2 = 36$

13.2 SQUARE ROOT RADICALS

OBJECTIVES: To simplify a square root radical that represents an integer.

To use a table to approximate expressions containing a square root radical.

To solve an equation of the form $ax^2 = b$ by the factoring method and by using a table.

By solving $x^2 = 25$ by the factoring method, it can be seen that 5 and -5 are the roots of this equation. Since $5^2 = 25$ and since $(-5)^2 = 25$, 5 and -5 are called the **square roots** of 25. Similarly, 4 and -4 are the **square roots** of 16 since $4^2 = 16$ and $(-4)^2 = 16$.

Definition: **r is a square root of s if and only if $r^2 = s$.**

Each positive real number has two square roots, two real numbers that have the same absolute value but opposite signs. The one and only one square root of 0 is 0. Negative real numbers do not have real numbers for square roots since the product of two positive real numbers is positive and the product of two negative real numbers is positive.

The radical sign, $\sqrt{}$, is used to indicate the positive real square root of a positive real number. Thus, $\sqrt{25} = +5$, the positive square root of 25. The negative square root of a positive real number is indicated by writing a minus sign before the radical. Thus, $-\sqrt{25} = -5$, the negative square root of 25.

The number under the radical sign is called the **radicand.** For example, 25 is the radicand of $\sqrt{25}$.

Not all rational numbers have rational square roots. For example, there is no rational number for which $x^2 = 2$. (A rational number is a number that can be expressed as a ratio of two integers.) The square roots of 2 are denoted by $\sqrt{2}$ for the positive root and by $-\sqrt{2}$ for the negative root. Stated another way, $(\sqrt{2})^2 = 2$ and $(-\sqrt{2})^2 = 2$. Numbers such as $\sqrt{2}$ and $-\sqrt{2}$ are called **irrational numbers.**

Definition: If s is any positive real number, then $\sqrt{s}$ indicates the positive square root of s, and $(\sqrt{s})^2 = s$.

The negative square root of s is indicated by $-\sqrt{s}$ and $(-\sqrt{s})^2 = s$.

Every rational number can be expressed as a terminating or a repeating decimal. For example, $\frac{5}{4} = 1.25$, a terminating decimal, and $\frac{601}{165} = 3.6424242...$, where the digits 42 repeat indefinitely. The **set of irrational numbers** consists of the set of nonterminating, nonrepeating decimals. For example, $\sqrt{2} = 1.4142...$ and $\sqrt{3} = 1.7320...$, where no digit or set of digits repeats. The **set of real numbers** is the union of the set of rational numbers and the set of irrational numbers.

For most practical purposes, an irrational number is approximated by a rational number expressed in decimal form. For example, using the Table of Squares and Square Roots on the inside cover of this book, $\sqrt{10}$ may be approximated by the rational number 3.162; that is, $\sqrt{10} \approx 3.162$, where $\approx$ is read "equals approximately."

In sample problems 1–8 simplify, if possible.

SAMPLE PROBLEMS	SOLUTIONS
1. $\sqrt{36}$	1. $\sqrt{36} = 6$
2. $\sqrt{(20)^2}$	2. $\sqrt{(20)^2} = 20$
3. $\sqrt{(-6)^2}$	3. $\sqrt{(-6)^2} = \sqrt{36} = 6$
4. $-\sqrt{9}$	4. $-\sqrt{9} = -3$
5. $\sqrt{-9}$	5. $\sqrt{-9}$ is not a real number.
6. $(\sqrt{3})^2$	6. $(\sqrt{3})^2 = 3$
7. $\sqrt{25}\sqrt{36}$	7. $\sqrt{25}\sqrt{36} = 5\,(6) = 30$
8. $\sqrt{0}\sqrt{2}$	8. $\sqrt{0}\sqrt{2} = 0 \cdot \sqrt{2} = 0$

In sample problems 9–10 use the Table of Squares and Square Roots on the inside cover and approximate each of the following to the nearest hundredth.

SAMPLE PROBLEM 9

$\sqrt{38}$

SOLUTION

$\sqrt{38} \approx 6.164$

$\sqrt{38} = 6.16$ **to the nearest hundredth**

SAMPLE PROBLEM 10

$\frac{6 - \sqrt{8}}{2}$

SOLUTION

$$\frac{6 - \sqrt{8}}{2} \approx \frac{6 - 2.828}{2}$$

$$\approx \frac{3.172}{2}$$

$$\approx 1.586$$

$\frac{6 - \sqrt{8}}{2} = 1.59$ **to the nearest hundredth**

In sample problems 11–13 solve by factoring.

SAMPLE PROBLEM 11

$x^2 = 2$

SOLUTION

$$x^2 - 2 = 0$$

$$x^2 - (\sqrt{2})^2 = 0$$

$$(x - \sqrt{2})(x + \sqrt{2}) = 0$$

$$x - \sqrt{2} = 0 \quad \text{or} \quad x + \sqrt{2} = 0$$

$$x = \sqrt{2} \qquad x = -\sqrt{2}$$

The solution set is $\{\sqrt{2}, -\sqrt{2}\}$.

SAMPLE PROBLEM 12

$4x^2 = 5$

SOLUTION

$$4x^2 - 5 = 0$$

$$(2x)^2 - (\sqrt{5})^2 = 0$$

$$(2x - \sqrt{5})(2x + \sqrt{5}) = 0$$

$$2x - \sqrt{5} = 0 \quad \text{or} \quad 2x + \sqrt{5} = 0$$

$$2x = \sqrt{5} \qquad 2x = -\sqrt{5}$$

$$x = \frac{\sqrt{5}}{2} \qquad x = \frac{-\sqrt{5}}{2}$$

The solution set is $\left\{\frac{\sqrt{5}}{2}, \frac{-\sqrt{5}}{2}\right\}$.

SAMPLE PROBLEM 13

$0.01x^2 = 42.25$

SOLUTION

$x^2 = 4225$ Multiply each side by 100 to remove the decimal.

$x^2 - 4225 = 0$

$x^2 - (65)^2 = 0$ Using the table, locate 4225 in the squares column, beside the number 65.

$(x - 65)(x + 65) = 0$

$$x - 65 = 0 \quad \text{or } x + 65 = 0$$

$$x = 65 \qquad x = -65$$

The solution set is $\{65, -65\}$.

A.

EXERCISES 13.2

(1–17) Simplify, if possible. (See sample problems 1–8.)

1. $\sqrt{64}$
2. $\sqrt{81}$
3. $\sqrt{(11)^2}$
4. $\sqrt{(-10)^2}$
5. $(\sqrt{8})^2$
6. $(-\sqrt{7})^2$
7. $\sqrt{-4}$
8. $-\sqrt{16}$
9. $\sqrt{-16}$

10. $\sqrt{5}\sqrt{0}$

11. $\sqrt{25}\sqrt{4}$

12. $\dfrac{\sqrt{100}}{\sqrt{4}}$

13. $5 + \sqrt{9}$

14. $2 - \sqrt{36}$

15. $\dfrac{3 + \sqrt{49}}{2}$

16. $\dfrac{6 - \sqrt{121}}{3}$

17. $2\sqrt{81} + \dfrac{\sqrt{36}}{3}$

(18–23) By using a table, approximate each of the following to the nearest hundredth. (See sample problems 9, 10.)

18. $\sqrt{20}$

19. $\sqrt{95}$

20. $2\sqrt{5}$

21. $\dfrac{\sqrt{10}}{2}$

22. $\dfrac{4 - \sqrt{26}}{2}$

23. $\dfrac{6 + 3\sqrt{2}}{3}$

(24–29) Solve by factoring. (See sample problems 11–13.)

24. $x^2 = 5$

25. $9x^2 = 40$

26. $0.5x^2 = 4.5$

27. $x^2 = 4900$

28. $0.01x^2 = 0.5625$

29. $100x^2 = 69$

B.

(1–17) Simplify, if possible. (See sample problems 1–8.)

1. $\sqrt{9}$

2. $-\sqrt{4}$

3. $\sqrt{(13)^2}$

4. $\sqrt{(-15)^2}$

5. $(\sqrt{15})^2$

6. $\sqrt{-36}$

7. $-\sqrt{36}$

8. $(-\sqrt{36})^2$

9. $\sqrt{64}\sqrt{4}$

10. $\dfrac{\sqrt{81}}{\sqrt{9}}$

11. $\sqrt{0}\sqrt{6}$

12. $\sqrt{5}\sqrt{0}\sqrt{7}$

13. $7 - \sqrt{16}$

14. $4 + \sqrt{25}$

15. $\frac{4 - \sqrt{36}}{8}$

16. $\frac{7 + \sqrt{64}}{3}$

17. $3\sqrt{16} + \frac{\sqrt{64}}{2}$

(18–23) By using a table, approximate each of the following to the nearest hundredth. (See sample problems 9, 10.)

18. $\sqrt{56}$

19. $\sqrt{87}$

20. $3\sqrt{7}$

21. $\frac{\sqrt{15}}{3}$

22. $\frac{2 + \sqrt{45}}{2}$

23. $\frac{15 - 5\sqrt{3}}{10}$

(24–29) Solve by factoring. (See sample problems 11–13.)

24. $x^2 = 7$

25. $25x^2 = 73$

26. $0.01x^2 = 16$

27. $0.06x^2 = 24$

28. $100x^2 = 529$

29. $10{,}000x^2 = 6724$

13.3 PRODUCTS OF RADICALS

OBJECTIVE: To simplify products of radicals and to simplify radicals having the form $\sqrt{r^2 s^2}$ or the form $\sqrt{r^2 s}$.

Since $\sqrt{4}\sqrt{9} = 2 \cdot 3 = 6$ and $\sqrt{4 \cdot 9} = \sqrt{36} = 6$, it follows that $\sqrt{4}\sqrt{9} = \sqrt{4 \cdot 9}$.

Similarly, $\sqrt{16}\sqrt{25} = 4 \cdot 5 = 20$, $\sqrt{16 \cdot 25} = \sqrt{400} = 20$, and $\sqrt{16}\sqrt{25} = \sqrt{16 \cdot 25}$.

In general, if two square root radicals have positive radicands, then the product of the radicals is equal to the positive square root of the product of the radicands.

Product of radicals: $\sqrt{r}\sqrt{s} = \sqrt{rs}$, **where** $r \geqslant 0$ **and** $s \geqslant 0$.

This property and its symmetric form $\sqrt{rs} = \sqrt{r}\sqrt{s}$ can be used to simplify certain radicals.

In sample problems 1–5 simplify.

SAMPLE PROBLEM 1

$\sqrt{5}\sqrt{20}$

SOLUTION

$\sqrt{5}\sqrt{20} = \sqrt{5 \cdot 20} = \sqrt{100} = 10$

SAMPLE PROBLEM 2

$\sqrt{3}\sqrt{12}$

SOLUTION

$\sqrt{3}\sqrt{12} = \sqrt{3 \cdot 12} = \sqrt{36} = 6$

SAMPLE PROBLEM 3

$\sqrt{3025}$

SOLUTION

Factor the radicand: $3025 = 5\,(605) = 5^2\,(121) = 5^2\,(11^2)$.

$\sqrt{3025} = \sqrt{5^2\,(11^2)} = \sqrt{5^2}\,\sqrt{11^2} = 5 \cdot 11 = 55$

SAMPLE PROBLEM 4

$\sqrt{784}$

SOLUTION

$784 = 4\,(196) = 4^2\,(49) = 4^2 \cdot 7^2$

$\sqrt{784} = \sqrt{4^2}\,\sqrt{7^2} = 4 \cdot 7 = 28$

SAMPLE PROBLEM 5

$\sqrt{20}$

SOLUTION

$20 = 4 \cdot 5 = 2^2 \cdot 5$

$\sqrt{20} = \sqrt{2^2 \cdot 5} = \sqrt{2^2}\sqrt{5} = 2\sqrt{5}$

Note in the last sample problem that the original radicand is not a product of perfect square factors. However, the radicand can be written as a product of a perfect square and a number that has no perfect square factors. By applying the product property to this form, the radical can then be simplified. Note the following useful theorem:

Theorem: **When r and s are positive,** $\sqrt{r^2 s} = r\sqrt{s}$.

Definition: If r and s are positive and if s has no perfect square factors, then $r\sqrt{s}$ is the simplified form of $\sqrt{r^2 s}$.

SAMPLE PROBLEM 6

Simplify $\sqrt{108}$.

SOLUTION

Factor 108 in the form $r^2 s$.

$108 = 4\,(27) = 4\,(9)\,(3) = 36\,(3) = 6^2\,(3)$

$\sqrt{108} = \sqrt{6^2\,(3)} = \sqrt{6^2}\sqrt{3} = 6\sqrt{3}$

SAMPLE PROBLEM 7

Simplify $\sqrt{10}\sqrt{80}$.

SOLUTION

$\sqrt{10}\sqrt{80} = \sqrt{800}$

$800 = 100\,(8) = 100\,(4)\,(2) = 400\,(2) = (20)^2\,(2)$

$\sqrt{800} = \sqrt{(20)^2\,(2)} = \sqrt{(20)^2}\sqrt{2} = 20\sqrt{2}$

SAMPLE PROBLEM 8

Simplify $\sqrt{6}\sqrt{75}$.

SOLUTION

$$\begin{aligned}\sqrt{6}\sqrt{75} &= \sqrt{6\,(75)}\\ &= \sqrt{2\,(3)\,(3)\,(25)}\\ &= \sqrt{25}\sqrt{9}\sqrt{2}\\ &= 5\,(3)\sqrt{2}\\ &= 15\sqrt{2}\end{aligned}$$

EXERCISES 13.3

Simplify.

A.

1. $\sqrt{36}\sqrt{5}$
2. $\sqrt{9}\sqrt{6}$
3. $\sqrt{7}\sqrt{25}$
4. $\sqrt{3}\sqrt{49}$
5. $\sqrt{100}\sqrt{40}$
6. $\sqrt{7}\sqrt{28}$
7. $\sqrt{48}\sqrt{3}$
8. $\sqrt{2}\sqrt{50}$
9. $\sqrt{196}$

10. $\sqrt{1296}$

11. $\sqrt{7225}$

12. $\sqrt{12}$

13. $\sqrt{45}$

14. $\sqrt{600}$

15. $\sqrt{98}$

16. $\sqrt{8}$

17. $\sqrt{160}$

18. $\sqrt{30}\sqrt{6}$

19. $\sqrt{18}\sqrt{50}$

20. $\sqrt{2}\sqrt{6}\sqrt{15}$

B.

1. $\sqrt{49}\sqrt{17}$

2. $\sqrt{64}\sqrt{3}$

3. $\sqrt{10}\sqrt{81}$

4. $\sqrt{2}\sqrt{16}$

5. $\sqrt{4}\sqrt{20}$

6. $\sqrt{2}\sqrt{18}$

7. $\sqrt{6}\sqrt{24}$

8. $\sqrt{5}\sqrt{45}$

9. $\sqrt{324}$

10. $\sqrt{1936}$

11. $\sqrt{9801}$

12. $\sqrt{28}$

13. $\sqrt{54}$

14. $\sqrt{75}$

15. $\sqrt{180}$

16. $\sqrt{27}$

17. $\sqrt{192}$

18. $\sqrt{7}\sqrt{14}$

19. $\sqrt{640}\sqrt{50}$

20. $\sqrt{3}\sqrt{15}\sqrt{35}$

13.4 QUOTIENTS OF RADICALS

OBJECTIVE: To simplify an expression of the form $\frac{\sqrt{r}}{\sqrt{s}}$ or the form $\sqrt{\frac{r}{s}}$.

Since $\frac{\sqrt{100}}{\sqrt{4}} = \frac{10}{2} = 5$ and since $\sqrt{\frac{100}{4}} = \sqrt{25} = 5$, it follows that $\frac{\sqrt{100}}{\sqrt{4}} = \sqrt{\frac{100}{4}}$ and $\sqrt{\frac{100}{4}} = \frac{\sqrt{100}}{\sqrt{4}}$.

Similarly, $\frac{\sqrt{49}}{\sqrt{64}} = \frac{7}{8}$ and $\sqrt{\frac{49}{64}} = \frac{7}{8}$. Thus $\frac{\sqrt{49}}{\sqrt{64}} = \sqrt{\frac{49}{64}}$ and $\sqrt{\frac{49}{64}} = \frac{\sqrt{49}}{\sqrt{64}}$.

This property is true in general for radicals having positive radicands.

Quotient of radicals: For r and s positive, $\frac{\sqrt{r}}{\sqrt{s}} = \sqrt{\frac{r}{s}}$ and $\sqrt{\frac{r}{s}} = \frac{\sqrt{r}}{\sqrt{s}}$.

An expression is not in simplified form if it contains a radical in a denominator or if a fraction occurs in a radicand.

To simplify an expression having the form $\sqrt{\frac{r}{s}}$, first replace $\frac{r}{s}$ by an equivalent fraction having the smallest perfect square for the denominator. Then apply the quotient of radicals property and simplify the denominator.

SAMPLE PROBLEM 1

Simplify $\sqrt{\frac{1}{5}}$

SOLUTION

$\sqrt{\frac{1}{5}} = \sqrt{\frac{1(5)}{5(5)}} = \sqrt{\frac{5}{25}}$ **Multiply numerator and denominator by 5 to obtain the perfect square 25 in the denominator.**

$= \frac{\sqrt{5}}{\sqrt{25}}$ **Apply the quotient of radicals property.**

$= \frac{\sqrt{5}}{5}$ **Simplify the denominator.**

SAMPLE PROBLEM 2

Simplify $\sqrt{\frac{3}{8}}$

SOLUTION

While multiplying the denominator by 8 would produce the perfect square 64, a smaller perfect square 16 can be obtained by multiplying the denominator by 2.

Finding the smallest perfect square denominator simplifies the work that follows.

$$\sqrt{\frac{3}{8}} = \sqrt{\frac{3\,(2)}{8\,(2)}} = \sqrt{\frac{6}{16}} = \frac{\sqrt{6}}{\sqrt{16}} = \frac{\sqrt{6}}{4}$$

SAMPLE PROBLEM 3

Simplify $\frac{1}{\sqrt{20}}$

SOLUTION

$$\frac{1}{\sqrt{20}} = \sqrt{\frac{1}{20}} = \sqrt{\frac{5}{100}} = \frac{\sqrt{5}}{\sqrt{100}} = \frac{\sqrt{5}}{10}$$

SAMPLE PROBLEM 4

Simplify $\frac{2}{\sqrt{14}}$

SOLUTION

$$\frac{2}{\sqrt{14}} = \frac{2\sqrt{1}}{\sqrt{14}} = 2\sqrt{\frac{1}{14}} = 2\sqrt{\frac{14}{(14)\,(14)}} = \frac{2\sqrt{14}}{14}$$

Now $\frac{2\sqrt{14}}{14} = \frac{2\sqrt{14}}{2\,(7)} = \frac{\sqrt{14}}{7}$, the final simplified form.

Note that a resulting fraction should always be simplified, whenever possible.

SAMPLE PROBLEM 5

Simplify $\frac{\sqrt{3}}{\sqrt{15}}$

SOLUTION

$$\frac{\sqrt{3}}{\sqrt{15}} = \sqrt{\frac{3}{15}} = \sqrt{\frac{1}{5}} \quad \text{(simplify a fraction whenever possible)}$$

$$= \sqrt{\frac{5}{25}} = \frac{\sqrt{5}}{\sqrt{25}} = \frac{\sqrt{5}}{5}$$

Simplify.

EXERCISES 13.4

A.

1. $\sqrt{\dfrac{1}{2}}$

2. $\sqrt{\dfrac{2}{7}}$

3. $\sqrt{\dfrac{1}{27}}$

4. $\sqrt{\dfrac{4}{75}}$

5. $\dfrac{1}{\sqrt{6}}$

6. $\dfrac{2}{\sqrt{28}}$

7. $\dfrac{1}{\sqrt{24}}$

8. $\dfrac{\sqrt{2}}{\sqrt{28}}$

9. $\dfrac{\sqrt{162}}{\sqrt{6}}$

10. $\sqrt{\dfrac{81}{100}}$

11. $\sqrt{0.56}$

12. $\sqrt{0.012}$

13. $\dfrac{\sqrt{5}}{\sqrt{10}}$

14. $\dfrac{3\sqrt{7}}{\sqrt{6}}$

15. $\sqrt{\dfrac{75}{4}}$

B.

1. $\sqrt{\dfrac{1}{3}}$

2. $\sqrt{\dfrac{5}{6}}$

3. $\sqrt{\dfrac{1}{50}}$

4. $\sqrt{\dfrac{9}{32}}$

5. $\dfrac{2}{\sqrt{10}}$

6. $\dfrac{7}{\sqrt{14}}$

7. $\dfrac{1}{\sqrt{56}}$

8. $\dfrac{\sqrt{6}}{\sqrt{108}}$

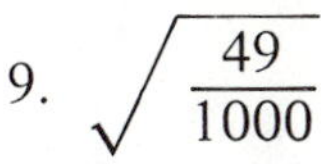

9. $\sqrt{\dfrac{49}{1000}}$

10. $\sqrt{0.042}$

11. $\dfrac{\sqrt{5}}{\sqrt{8}}$

12. $\sqrt{2\dfrac{1}{4}}$

13. $\dfrac{\sqrt{12}}{\sqrt{18}}$

14. $\dfrac{5\sqrt{2}}{\sqrt{20}}$

15. $\sqrt{\dfrac{32}{9}}$

13.5 RADICALS: SUMS AND DIFFERENCES

OBJECTIVE: To simplify an expression containing a sum or difference of one or more radicals.

Square root radicals having the same radicand are called **like radicals.** As examples, $3\sqrt{5}$ and $4\sqrt{5}$ are like radicals, and $\sqrt{7}$ and $2\sqrt{7}$ are like radicals.

On the other hand, $4\sqrt{3}$ and $4\sqrt{5}$ are unlike radicals, since the radicands are different.

Like radicals can be combined by using the distributive axiom for real numbers. Some unlike radicals may become like radicals after simplification and then these can be combined.

SAMPLE PROBLEM 1

Simplify $\sqrt{12} + \sqrt{75}$.

SOLUTION

Simplify each radical.

$\sqrt{12} = \sqrt{4}\sqrt{3} = 2\sqrt{3}$ and $\sqrt{75} = \sqrt{25}\sqrt{3} = 5\sqrt{3}$

$$\begin{aligned}\sqrt{12} + \sqrt{75} &= 2\sqrt{3} + 5\sqrt{3} \\ &= (2 + 5)\sqrt{3} \quad \text{(using the distributive axiom)} \\ &= 7\sqrt{3}\end{aligned}$$

SAMPLE PROBLEM 2

Simplify $\sqrt{36} + \sqrt{24}$.

SOLUTION

$\sqrt{36} = 6$ and $\sqrt{24} = \sqrt{4}\sqrt{6} = 2\sqrt{6}$

$\sqrt{36} + \sqrt{24} = 6 + 2\sqrt{6}$

No further simplification is possible.

SAMPLE PROBLEM 3

Simplify $\sqrt{45} - \sqrt{20}$.

SOLUTION

$$\begin{aligned}\sqrt{45} - \sqrt{20} &= \sqrt{9}\sqrt{5} - \sqrt{4}\sqrt{5} \\ &= 3\sqrt{5} - 2\sqrt{5} \\ &= (3 - 2)\sqrt{5} \\ &= 1\sqrt{5} \\ &= \sqrt{5}\end{aligned}$$

SAMPLE PROBLEM 4

Simplify $\sqrt{63} + \sqrt{147}$.

SOLUTION

$$\begin{aligned}\sqrt{63} + \sqrt{147} &= \sqrt{9}\sqrt{7} + \sqrt{49}\sqrt{3} \\ &= 3\sqrt{7} + 7\sqrt{3}\end{aligned}$$

No further simplification is possible.

The forms $\frac{r + \sqrt{s}}{t}$ and $\frac{r - \sqrt{s}}{t}$ occur in the general solution of the quadratic equation. Therefore, it is useful to learn how to simplify these forms.

SAMPLE PROBLEM 5

Simplify $\dfrac{6+\sqrt{45}}{3}$

SOLUTION

First, simplify the radical: $\sqrt{45}=\sqrt{9}\sqrt{5}=3\sqrt{5}.$

Then, simplify the resulting fraction, if possible, by finding a common factor of the denominator and *each* term of the numerator.

$$\frac{6+\sqrt{45}}{3}=\frac{6+3\sqrt{5}}{3}$$

$$=\frac{3\,(2+\sqrt{5})}{3}$$

$$=\frac{2+\sqrt{5}}{1}$$

$$=2+\sqrt{5}$$

SAMPLE PROBLEM 6

Simplify $\dfrac{28-\sqrt{800}}{12}$

SOLUTION

$$\frac{28-\sqrt{800}}{12}=\frac{28-\sqrt{400}\sqrt{2}}{12}$$

$$=\frac{28-20\sqrt{2}}{12}$$

$$=\frac{4\,(7-5\sqrt{2})}{4\,(3)}$$

$$=\frac{7-5\sqrt{2}}{3}$$

EXERCISES 13.5

Simplify.

A.

1. $\sqrt{45}+\sqrt{80}$

2. $\sqrt{98}-\sqrt{32}$

3. $\sqrt{90} - \sqrt{40}$

4. $\sqrt{72} + \sqrt{75}$

5. $\sqrt{49} + \sqrt{28}$

6. $\sqrt{6} - \sqrt{36}$

7. $\frac{15 + \sqrt{50}}{10}$

8. $\frac{10 - \sqrt{12}}{5}$

9. $\frac{4 - \sqrt{56}}{4}$

10. $\frac{7 + \sqrt{81}}{24}$

11. $\frac{12 - \sqrt{8}}{6}$

12. $\frac{6 - \sqrt{20}}{4}$

B.

1. $\sqrt{75} + \sqrt{12}$

2. $\sqrt{54} - \sqrt{24}$

3. $\sqrt{40} - \sqrt{50}$

4. $\sqrt{12}+\sqrt{32}$

5. $\sqrt{108}-\sqrt{81}$

6. $5+\sqrt{20}$

7. $\dfrac{2+\sqrt{20}}{4}$

8. $\dfrac{6-\sqrt{54}}{15}$

9. $\dfrac{14+\sqrt{60}}{2}$

10. $\dfrac{3-\sqrt{121}}{16}$

11. $\dfrac{2-\sqrt{12}}{4}$

12. $\dfrac{28-\sqrt{300}}{14}$

13.6 COMPLETING THE SQUARE

OBJECTIVE: To solve a quadratic equation by completing the square.

By the factoring method, it can be seen that if $x^2 = 2$, then $x = \sqrt{2}$ or $x = -\sqrt{2}$. Similarly, if $x^2 = 5$, then $x = \sqrt{5}$ or $x = -\sqrt{5}$. This leads to the following general statement.

Theorem: For r any positive real number,

$$\text{if } X^2 = r, \text{ then } X = \sqrt{r} \text{ or } X = -\sqrt{r}.$$

SAMPLE PROBLEM 1

Solve $(x - 5)^2 = 3$.

SOLUTION

Use the theorem $x - 5 = \sqrt{3}$ **or** $x - 5 = -\sqrt{3}$.

Add 5: $x = 5 + \sqrt{3}$ **or** $x = 5 - \sqrt{3}$.

SAMPLE PROBLEM 2

Solve $(3x + 2)^2 = 7$.

SOLUTION

$$3x + 2 = \sqrt{7} \quad \text{or} \quad 3x + 2 = -\sqrt{7}$$

$$3x = -2 + \sqrt{7} \qquad 3x = -2 - \sqrt{7}$$

$$x = \frac{-2 + \sqrt{7}}{3} \qquad x = \frac{-2 - \sqrt{7}}{3}$$

To solve a general quadratic equation by this method, it is necessary to rewrite the given equation in the form $(rx + s)^2 = t$. This involves a process called "completing the square."

In Units 7 and 8, the perfect square trinomial formula was studied. This formula is used in solving a quadratic equation by completing the square.

Perfect square trinomial formula: $X^2 + 2AX + A^2 = (X + A)^2$

Completing the Square Solution Method

To solve $X^2 + 2AX + C = 0$:

(1) Rewrite the equation as $X^2 + 2AX = -C$.

(2) Add A^2 (the square of one-half the coefficient of X) to each side to form the equation $X^2 + 2AX + A^2 = A^2 - C$.

(3) Rewrite the left side as the square of a binomial and simplify the right side.

(4) Solve the resulting equation by using the theorem illustrated in sample problems 1 and 2.

SAMPLE PROBLEM 3

Solve by completing the square: $x^2 + 6x + 7 = 0$.

SOLUTION

(1) $x^2 + 6x = -7$ (variables on the left side, constant on the right side)

(2) $x^2 + 6x + 3^2 = -7 + 3^2$ (adding 3^2 to each side; 3 is $\frac{1}{2}$ of 6, the coefficient of x in $6x$)

(3) $(x + 3)^2 = 2$ (using the perfect square trinomial formula on the left side and simplifying the right side)

(4) $x + 3 = +\sqrt{2}$ or $x + 3 = -\sqrt{2}$

$x = -3 + \sqrt{2}$ $\quad$ $x = -3 - \sqrt{2}$

SAMPLE PROBLEM 4

Solve by completing the square: $x^2 - 12x - 14 = 0$.

SOLUTION

(1) $x^2 - 12x = 14$

(2) $x^2 - 12x + (-6)^2 = 14 + 36$

(3) $(x - 6)^2 = 50$

(4) $x - 6 = +\sqrt{50}$ or $x - 6 = -\sqrt{50}$

$x = 6 + \sqrt{50}$ $\quad$ $x = 6 - \sqrt{50}$

$x = 6 + 5\sqrt{2}$ $\quad$ $x = 6 - 5\sqrt{2}$

Note that the radical is left in simplified form.

SAMPLE PROBLEM 5

Solve by completing the square: $3x^2 - 10x + 5 = 0$.

SOLUTION

In this equation the second degree term, $3x^2$, is not a perfect square. By multiplying each term of the equation by 3, the second degree term becomes $9x^2$, a perfect square.

$$3(3x^2) - 3(10x) + 3(5) = 0$$

$$9x^2 - 30x + 15 = 0$$

$$9x^2 - 30x = -15$$

$X^2 + 2AX + A^2$ (Compare with the general form).

Using $X = 3x$, then $-30x = -10(3x)$, $2A = -10$, and $A = -5$.

$$9x^2 - 30x + (-5)^2 = -15 + 25$$

$$(3x - 5)^2 = 10$$

$$3x - 5 = +\sqrt{10} \quad \text{or} \quad 3x - 5 = -\sqrt{10}$$

$$3x = 5 + \sqrt{10} \qquad 3x = 5 - \sqrt{10}$$

$$x = \frac{5 + \sqrt{10}}{3} \qquad x = \frac{5 - \sqrt{10}}{3}$$

EXERCISES 13.6

Solve by completing the square, when necessary.

A.

1. $(x - 4)^2 = 5$
2. $(x + 2)^2 = 3$
3. $(x - 8)^2 = 12$
4. $(2x + 3)^2 = 10$
5. $(5x - 1)^2 = 20$
6. $x^2 + 2x - 1 = 0$

7. $x^2 - 6x + 3 = 0$

8. $x^2 - 8x - 34 = 0$

9. $4x^2 + 28x + 47 = 0$

10. $25x^2 - 10x - 26 = 0$

11. $3x^2 + 2x - 2 = 0$

12. $3x^2 + 2x - 1 = 0$

13. $2x^2 + 10x - 1 = 0$

14. $3x^2 - 54x + 108 = 0$

B.

1. $(x + 7)^2 = 6$

2. $(x - 5)^2 = 7$

3. $(x + 9)^2 = 18$

4. $(3x - 4)^2 = 14$

5. $(6x + 1)^2 = 32$

6. $x^2 - 4x - 1 = 0$

7. $x^2 + 10x + 15 = 0$

8. $x^2 + 12x + 8 = 0$

9. $9x^2 - 48x + 59 = 0$

10. $16x^2 + 24x - 3 = 0$

11. $5x^2 + 6x - 1 = 0$

12. $5x^2 + 6x + 1 = 0$

13. $6x^2 - 10x + 3 = 0$

14. $2x^2 - 28x + 50 = 0$

13.7 QUADRATIC FORMULA

OBJECTIVE: To solve a quadratic equation by using the quadratic formula.

When the general quadratic equation in standard form is solved by completing the square, a formula is obtained that can be used to solve any special quadratic equation. Given

$$ax^2 + bx + c = 0, \text{ where } a \neq 0,$$

multiply each term by a so that the second degree term is a perfect square:

$$a^2x^2 + abx + ac = 0$$

$$(ax)^2 + b(ax) = -ac$$

Now, since the term to be added to complete the square is one-half of b, it is more convenient to multiply throughout by 4 to avoid working with fractions and to keep the second degree term a perfect square:

$$4a^2x^2 + 4abx = -4ac$$
$$(2ax)^2 + 2(b)(2ax) = -4ac$$
$$(2ax)^2 + 2(b)(2ax) + b^2 = -4ac + b^2$$
$$(2ax + b)^2 = b^2 - 4ac$$

$$2ax + b = +\sqrt{b^2 - 4ac} \quad \text{or} \quad 2ax + b = -\sqrt{b^2 - 4ac}$$
$$2ax = -b + \sqrt{b^2 - 4ac} \qquad 2ax = -b - \sqrt{b^2 - 4ac}$$
$$x = \frac{-b + \sqrt{b^2 - 4ac}}{2a} \qquad x = \frac{-b - \sqrt{b^2 - 4ac}}{2a}$$

This last result is called the quadratic formula.

Quadratic formula: If $ax^2 + bx + c = 0$, where $a \neq 0$, then

$$x = \frac{-b + \sqrt{b^2 - 4ac}}{2a} \quad \text{or} \quad x = \frac{-b - \sqrt{b^2 - 4ac}}{2a}$$

The quadratic formula is often condensed and written as follows:

$$x = \frac{-b \pm \sqrt{b^2 - 4ac}}{2a}$$

To solve a quadratic equation by using the quadratic formula:

(1) Write the equation in standard form.
(2) Determine the coefficients a, b, and c.
(3) In the formula replace a, b, and c by their numerical values.
(4) Simplify the resulting expressions.

In sample problems 1–3 solve by using the quadratic formula.

SAMPLE PROBLEM 1

$3x^2 + 6x + 2 = 0$

SOLUTION

(1) $3x^2 + 6x + 2 = 0$

(2) $a = 3, b = 6, c = 2$

(3) $x = \frac{-b \pm \sqrt{b^2 - 4ac}}{2a}$

$$x = \frac{-6 \pm \sqrt{36 - 4(3)(2)}}{2(3)}$$
$$= \frac{-6 \pm \sqrt{36 - 24}}{6}$$
$$= \frac{-6 \pm \sqrt{12}}{6}$$
$$= \frac{-6 \pm 2\sqrt{3}}{6}$$
$$= \frac{-3 \pm \sqrt{3}}{3}$$

SAMPLE PROBLEM 2

$x^2 + 23 = 10x$

SOLUTION

(1) $x^2 - 10x + 23 = 0$

(2) $a = 1, b = -10, c = 23$

(3) $x = \dfrac{-b \pm \sqrt{b^2 - 4ac}}{2a}$

$$x = \frac{-(-10) \pm \sqrt{100 - 4(1)(23)}}{2(1)}$$

$$= \frac{10 \pm \sqrt{100 - 92}}{2}$$

Note 1

$$= \frac{10 \pm \sqrt{8}}{2}$$

$$= \frac{10 \pm 2\sqrt{2}}{2}$$

Note 2

$$= \frac{2(5 \pm \sqrt{2})}{2}$$

$$= 5 \pm \sqrt{2}$$

Note 3

Note 1. It is important to note that when b is negative, then $-b$ in the formula is positive; in this case, $-(-10) = +10$.

Note 2. The radical must be simplified. This is especially important because the fraction may often be simplified.

Note 3. In simplifying the fraction, a factor common to the denominator and *each* term of the numerator must be found.

SAMPLE PROBLEM 3

$5x^2 = 3x + 2$

SOLUTION

(1) $5x^2 - 3x - 2 = 0$

(2) $a = 5, b = -3, c = -2$

(3) $x = \dfrac{-b \pm \sqrt{b^2 - 4ac}}{2a}$

$$x = \frac{-(-3) \pm \sqrt{9 - 4(5)(-2)}}{2(5)}$$

$$= \frac{+3 \pm \sqrt{9 - (-40)}}{10} \qquad \text{Note 1}$$

$$= \frac{3 \pm \sqrt{9 + 40}}{10}$$

$$= \frac{3 \pm \sqrt{49}}{10}$$

$$x = \frac{3 + 7}{10} \quad \text{or} \quad x = \frac{3 - 7}{10} \qquad \text{Note 2}$$

$$x = 1 \qquad\qquad x = \frac{-2}{5}$$

Note 1. When the constant term, c, is negative, it is important to be very careful when evaluating $b^2 - 4ac$, the quantity under the radical sign. This is a common source of errors.

Note 2. When the solutions are rational, they are always left in their simplified rational form.

EXERCISES 13.7

Solve by using the quadratic formula.

A.

1. $3x^2 + 5x + 1 = 0$
2. $x^2 = 3x + 1$
3. $x^2 = 2(3x - 2)$
4. $3x^2 = 2 - 2x$

5. $5x^2 + 9x = 18$

6. $x^2 - 80x + 1200 = 0$

7. $x^2 + x = 1$

8. $x^2 + 4x = 2$

9. $x(x - 2) = 1$

10. $16x^2 - 40x + 25 = 0$

11. $4x(x - 1) = 35$

12. $2x^2 = x + 1$

13. $x^2 + 2bx + c = 0$ (solve for x)

14. $2x^2 - 5xy + 2y^2 = 0$ (solve for x)

B.

1. $3x^2 + 9x + 5 = 0$

2. $9x^2 = 12x + 1$

3. $2x^2 = 4 - 5x$

4. $7x = 2x^2 + 3$

5. $10x^2 = 11x + 18$

6. $x^2 + 170x - 6000 = 0$

7. $x^2 = x + 1$

8. $x^2 - 10x = 3$

9. $x^2 = 12(x + 1)$

10. $36x^2 + 84x + 49 = 0$

11. $3x(3x - 2) = 1$

12. $x^2 = 20(x + 240)$

13. $x^2 - 2rx - s = 0$ (solve for x)

14. $y^2 + 2xy - 15x^2 = 0$ (solve for y)

13.8 APPLICATIONS

OBJECTIVE: To solve a verbal problem by using a quadratic equation.

Quadratic equations can be used to solve a wide variety of practical problems. The procedure for forming an equation from the words of a problem is basically the same as that used for linear equations in Unit 6.

There are many problems that require the use of the **theorem of Pythagoras.**

Theorem of Pythagoras. In a right triangle, the sum of the squares of the legs is equal to the square of the hypotenuse. In symbols, referring to Figure 13.1, this theorem is expressed as $c^2 = a^2 + b^2$.

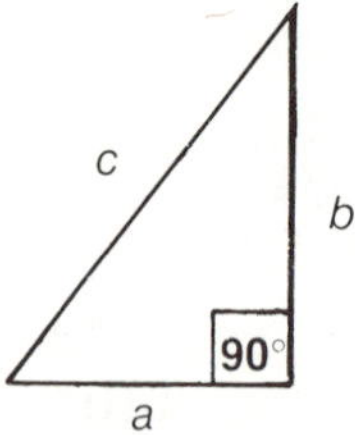

FIGURE 13.1

The **hypotenuse** is the longest side of a right triangle, the side opposite the 90-degree (right) angle. The **legs** of a right triangle are the sides that form the right angle.

For sample problems 1 and 2 refer to the right triangle in Figure 13.1.

SAMPLE PROBLEM 1

Find c if $a = 15$ in. and $b = 20$ in.

SOLUTION

Use the theorem of Pythagoras.

$$c^2 = a^2 + b^2$$

$$c^2 = 15^2 + 20^2 = 225 + 400$$

$$c^2 = 625$$

$$c = \sqrt{625} = 25 \text{ in.}$$

Only the positive root is accepted since geometric measurements are not negative.

SAMPLE PROBLEM 2

Find b if $c = 26$ meters and $a = 10$ meters.

SOLUTION

$$a^2 + b^2 = c^2$$
$$10^2 + b^2 = 26^2$$
$$b^2 = 26^2 - 10^2 = (26 - 10)(26 + 10)$$
$$b^2 = 16(36)$$
$$b = \sqrt{16}\sqrt{36} = 4(6) = 24 \text{ meters}$$

SAMPLE PROBLEM 3

A ladder 25 ft. long leans against a building with the top of the ladder touching a point on the building 24 ft. above the ground. How far from the building is the foot of the ladder?

SOLUTION

Make a sketch such as that in Figure 13.2 and use the theorem of Pythagoras.

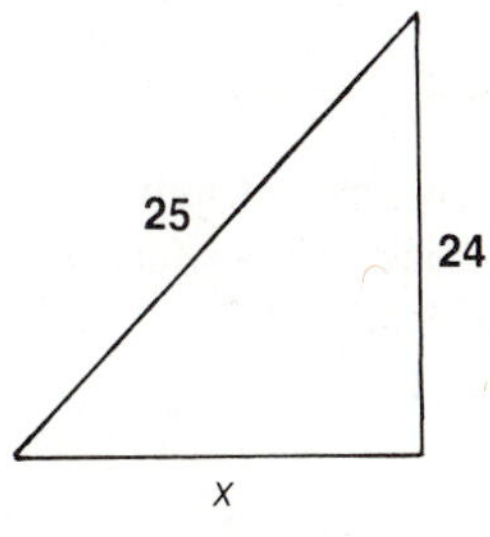

FIGURE 13.2

$$x^2 + 24^2 = 25^2$$
$$x^2 = 25^2 - 24^2 = (25 - 24)(25 + 24)$$
$$x^2 = 49$$
$$x = 7 \text{ ft.}$$

SAMPLE PROBLEM 4

A baseball diamond is a square with each side 90 ft. in length. Find the distance from home plate to second base.

SOLUTION

First make a sketch such as that in Figure 13.3.

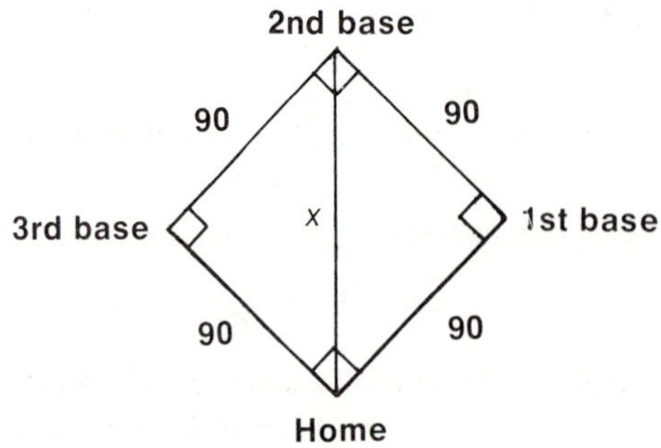

FIGURE 13.3

The distance is the length of the hypotenuse of a right triangle whose legs are each 90 ft.

$$x^2 = 90^2 + 90^2$$

$$x^2 = 16{,}200$$

$$x = \sqrt{8100\,(2)} = 90\sqrt{2}\text{ ft.}$$

$$x \approx 90\,(1.414) \approx 127.3\text{ ft.}$$

When the answer to a practical problem is an irrational number, then this number is usually approximated as a decimal.

SAMPLE PROBLEM 5

A man wants to pour a concrete strip of uniform width along one 45-ft. side of his house and along the two 30-ft. sides. How wide a strip should he pour if the total area of the concrete strip is to be 1625 sq. ft.?

SOLUTION

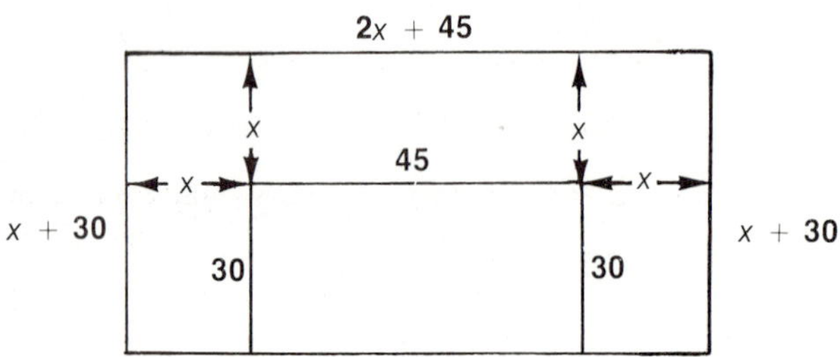

FIGURE 13.4

Let x = width of strip in feet.

Area of outer rectangle − area of house = area of strip.

$$(2x + 45)(x + 30) - 30(45) = 1625$$

$$2x^2 + 105x - 1625 = 0$$

$$(2x - 25)(x + 65) = 0$$

$$2x = 25 \quad \text{or} \quad x + 65 = 0$$

$$x = 12\tfrac{1}{2} \text{ ft.} \qquad x = -65$$

The answer is $12\frac{1}{2}$ ft.; −65 must be rejected, since the answer must be positive.

SAMPLE PROBLEM 6

On a reforestation project, it took one boy scout troop 2 hr. longer to plant 100 trees than another troop. When both troops worked together they planted 100 trees in 5 hr. How long did it take each troop working alone to plant 100 trees?

SOLUTION

Let x = time of the faster troop and $x + 2$ = time of the slower troop.

Then, $\frac{1}{x}$ = the rate of the faster troop and $\frac{1}{x+2}$ = the rate of the slower troop.

Use work = rate × time and work of A + work of B = total work.

$$\frac{5}{x} + \frac{5}{x+2} = 1$$

$$5(x + 2) + 5x = x(x + 2)$$

$$10x + 10 = x^2 + 2x$$

$$x^2 - 8x = 10$$

$$x^2 - 8x + 16 = 26$$

$$(x - 4)^2 = 26$$

$$x - 4 = \sqrt{26} \quad \text{or} \quad x - 4 = -\sqrt{26}$$

$$x = 4 + \sqrt{26} \qquad x = 4 - \sqrt{26}$$

$$x \approx 4 + 5.099$$

(rejected, since it is a negative result)

$$x \approx 9.1 \text{ hr.}$$

$$x + 2 \approx 11.1 \text{ hr.}$$

EXERCISES 13.8

A.

1. A surveyor lays off an 80-ft. by 150-ft. rectangular plot of land. To check his work, he measures the diagonal of the plot. What should this diagonal measure if the plot is truly a rectangle?

2. A boat, headed for a point directly across a river from its starting point, lands on the opposite shore 1000 ft. downstream from its original destination. If the boat traveled 1450 ft. how wide was the river?

3. An artist wants to put a frame of uniform width around a 3-ft. by 4-ft. rectangular painting. He wants the area of the frame to equal the area of the painting. How wide should he make the frame?

4. A rectangular piece of sheet metal is twice as long as it is wide. A 10-cm. square is cut from each corner. The ends and sides are then turned up to form an open box having a volume of 17,500 cu. cm. Find the length and width of the piece of sheet metal.

5. On a soil conservation project, a crew takes 15 hr. longer digging holes by hand than a machine takes to do the same amount of work. When the crew and the machine work together, the same amount of work can be completed in 4 hr. How long would it have taken the crew working alone to do the job?

6. A new machine can produce a certain number of articles in 4 hr. less time than an old machine. After the new machine had been working for 3 hr., it broke down, and the old machine completed the job in 5 more hr. How long would it have taken the new machine alone to do the whole job?

7. A man rows a boat 24 mi. downstream and then immediately returns to his starting point. If the entire trip took $4\frac{1}{2}$ hr. and if his speed in still water was 12 m.p.h., find the rate of the current.

8. A projectile is fired vertically upward with an initial speed of 800 ft. per second. The equation

$$s = 800t - 16t^2$$

gives the distance s in feet above the ground that the projectile reaches t seconds after it is fired. When is the projectile 9600 ft. above the gound? Explain the two answers.

9. A certain manufacturing company uses the profit equation

$$P = 80x - 2x^2$$

that relates the profit P in dollars that can be made by selling x units of an article that the company manufactures. Find how many units should be sold to obtain a profit of:

(a) \$750 (b) \$800

10. When 3 moles of pure ethyl alcohol are mixed with 1 mole of acetic acid at room temperature, then the number, x, of moles of ester that are formed is given by

$$\frac{x^2}{(3 - x)(1 - x)} = 4,$$

where 4 is the equilibrium constant for this reaction. Solve for x. (By chemical considerations, no more than 1 mole of ester can be formed.)

B.

1. The foot of an 18-ft. pole is 16 ft. from the foot of a 30-ft. pole. Find the distance between the tops of the poles.

2. A 30-ft. guy wire is to be attached to the top of a 24-ft. pole and to a point on the ground. How far from the foot of the pole should the guy wire be fastened?

3. Find the width x of the 18-cm. by 24-cm. L-beam shown in the figure below if the area of the L-beam is 41 sq. cm.

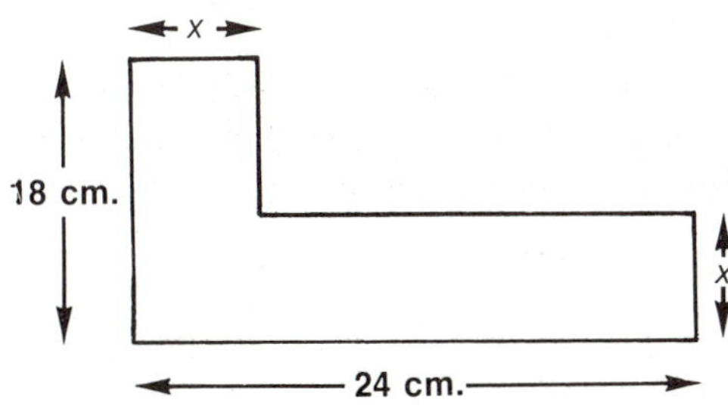

4. It takes 450 small square tiles to cover a certain floor. Using larger square tiles, whose sides are 2 in. larger than those of the smaller tiles, only 50 tiles are needed. Find the length of the sides of the smaller tiles.

5. It takes 48 min. longer to cut a certain amount of lumber by using a hand saw than it does by using a power saw. When both the hand saw and the power saw are used, the same amount of lumber can be cut in 7 min. How long does it take to cut this lumber using the hand saw alone?

6. In order to finish a job as soon as possible, a contractor rents a spray outfit that can paint a building in 9 hr. less time than the spray outfit he owns. Using both outfits, he has the building painted in 20 hr. How long would it have taken if he had used only his own spray outfit?

7. An airplane flew 1140 mi. with a tailwind of 20 m.p.h. and then flew an additional 870 mi. with a tailwind of 30 m.p.h. If the total time of the flight was $3\frac{1}{2}$ hr., find the speed of the plane in still air.

8. A bomb is released 2064 ft. above the earth by an airplane and travels 640 ft. per second vertically downward. The equation $s = 640t + 16t^2$ gives the distance s in feet below the point where the bomb was dropped t seconds after it was dropped. How long does it take the bomb to reach the earth?

9. A tourist received a certain number of Swiss francs in exchange for \$21. If he had exchanged them a month earlier when the exchange rate of a franc was 5¢ less, he would have received 10 more francs for his \$21. What exchange value per franc did he receive?

10. In a certain electronic circuit having an output of 3025 watts of power, a resistance of 4 ohms, and a line voltage of 220 volts, the current x in amps is given by $3025 = 220x - 4x^2$. Find x.

UNIT 13 POSTTEST

(1–2) Solve by factoring. (13.1)

1. $16x^2 = (2 - 3x)(2 + 3x)$
2. $4 - x = 2x(x + 3)$

(3–4) Simplify. (13.2)

3. $\frac{2\sqrt{81}}{3}$
4. $\frac{(\sqrt{6})^2 + \sqrt{100}}{5}$

(5–6) By using a table, approximate each of the following to the nearest hundredth. (13.2)

5. $5\sqrt{68}$
6. $\frac{3 - \sqrt{15}}{4}$

(7–8) Solve by factoring. (13.2)

7. $9x^2 = 35$
8. $0.01x^2 = 0.2916$

(9–10) Simplify. (13.3)

9. $\sqrt{18}\sqrt{98}$
10. $\sqrt{192}$

(11–12) Simplify. (13.4)

11. $\frac{5}{\sqrt{150}}$
12. $\sqrt{\frac{7}{12}}$

(13–14) Simplify. (13.5)

13. $\sqrt{150} + \sqrt{96}$
14. $\frac{12 - \sqrt{40}}{4}$

(15–16) Solve by completing the square. (13.6)

15. $x^2 + 8x + 4 = 0$
16. $3x^2 - 4x - 8 = 0$

(17–18) Solve by using the quadratic formula. (13.7)

17. $9x^2 = 2(6x + 1)$
18. $2x^2 + 2x - 1 = 0$

19. A glass window in a building is in the shape of a rectangle surmounted by a triangle. The height of the rectangle is 3 ft. longer than the width of the rectangle. The base and altitude of the triangle are each equal to the width of the rectangle. Find the dimensions of the rectangle so that the window will admit 72 sq. ft. of light. (13.8)

20. A motorboat takes 45 min. longer to travel 36 mi. upstream than it takes to travel the same distance downstream. If the boat can travel 20 m.p.h. in still water, find the rate of the current. (13.8)

ANSWERS

UNIT 1 PRETEST

1. $4 + x$
2. $3y$
3. $9 - x$
4. $\frac{7}{y}$
5. 6^2 or 36
6. n^2
7. 5^3 or 125
8. y^3
9. $4y - 3$
10. $9(x + 6)$
11. $10 - (n + 3)$
12. $(y + 4)^2$
13. $\frac{(t + 1)^3}{2}$
14. 15
15. 500
16. 2
17. 9
18. 2
19. 4
20. 14
21. 2
22. 36
23. 13
24. 63
25. 39

Exercises 1.1 A

1. $7y$
2. $s + 11$
3. $3 + x$
4. $5 - x$
5. $x - 5$
6. $\frac{n}{4}$
7. $\frac{x}{8}$
8. $\frac{4}{n}$
9. $q + 6$
10. $6 - q$
11. $6 - q$
12. $\frac{6}{q}$
13. y
14. $9y$
15. $3 \cdot 25$, or $3(25)$, or $(3)(25)$
16. $8 - r$
17. $\frac{7}{z}$
18. $x + 9$
19. $9x$
20. $t - 2$
21. $t - z$
22. xy
23. $2 \cdot 3$, or $2(3)$, or $(2)(3)$
24. $2x$
25. $\frac{n}{2}$
26. $\frac{2}{n}$
27. $y - x$
28. $x - y$
29. $y + z$
30. $\frac{r}{s}$
31. $2y$
32. $\frac{y}{2}$
33. $\frac{t}{3}$
34. $\frac{n}{4}$
35. $\frac{x}{5}$

Exercises 1.1 B

1. $n + 8$
2. $8n$
3. $t - 12$
4. $12 - t$
5. $\frac{x}{7}$
6. $\frac{7}{x}$
7. ab
8. $a + b$
9. $\frac{a}{b}$

10. $b - a$
11. n
12. $x - 4$
13. $\frac{t}{4}$
14. $\frac{4}{t}$
15. $9 + y$
16. $9 - y$
17. $2t$
18. $3r$
19. $\frac{z}{2}$
20. $\frac{n}{3}$

Exercises 1.2 A

1. y^2
2. n^2
3. 4^3
4. n^3
5. 6^3
6. t^3
7. t^2
8. x^3
9. 10^2 or 100
10. 9^2 or 81
11. 9^3 or 729
12. y^3
13. y^4
14. 10^4 or 10,000
15. 8^5
16. x^5
17. 4^5
18. 9^7
19. n^6
20. y^8

Exercises 1.2 B

1. b^2
2. z^3
3. 7^3, or 343
4. 18^2, or 324
5. 5^3, or 125
6. r^4
7. s^3
8. t^5
9. 10^{23}
10. x^{12}
11. 5^6
12. z^3
13. r^9
14. x^6
15. y^n

Exercises 1.3 A

1. 1
2. 26
3. 30
4. 144
5. 10
6. 9
7. 21
8. 7
9. 52
10. 18
11. 150
12. 7
13. 4
14. 6
15. 2
16. 6
17. 6
18. 38
19. 75
20. 2
21. 33
22. 33
23. 40
24. 9
25. 9
26. 27
27. 17
28. 6
29. 6
30. 0

Exercises 1.3 B

1. 30
2. 48
3. 100
4. 1000
5. 24
6. 4
7. 10
8. 0

9. 3
10. 15
11. 75
12. 49
13. 16
14. 6
15. 5

Exercises 1.4 A

1. $9(x+3)$
2. $\frac{x-3}{2}$
3. $\frac{y+2}{5}$
4. $4(y-6)$
5. $3\left(\frac{y}{8}\right)$ or $\frac{3y}{8}$
6. $4+5x$
7. $5(x+4)$
8. $6y-1$
9. $6(y-1)$
10. $s-(t+6)$
11. $x+(8-y)$
12. $x+\frac{3}{x}$
13. $6-2r$
14. $5+9z$
15. $x(y-3)$
16. $\frac{4+z}{9}$
17. $t(t+8)$
18. $(7z)^2$
19. $(x+y+z)^3$
20. n^2+3^2 or n^2+9
21. $\frac{5}{n+3}$
22. $q+7q$ or $8q$
23. $\frac{4t}{9}$
24. $\frac{t^3}{9}$
25. $\frac{n-6}{2}$

Exercises 1.4 B

1. x^3-3x
2. $5z+7$
3. $\frac{8y}{3}$
4. $9(9-s)$
5. $x+(x-6)$
6. $y-(x+6)$
7. $12-(6-x)$
8. $2(t+3)$
9. $n-(4+n)$
10. $2(xy)$, or $2xy$
11. $\frac{xy}{2}$
12. $5y^2$
13. $6x^3$
14. x^2-1
15. $1+y^3$
16. x^2-y^2
17. $(x-y)^2$
18. x^2+5^2 or x^2+25
19. y^3+2^3 or y^3+8
20. $x-(y-z)$
21. $x+(y-z)$
22. $x(x-3)$
23. $x+3(x-3)$
24. $\frac{5(y+3)}{2}$
25. $\frac{x+y}{2(x-y)}$

Exercises 1.5 A

1. 16
2. 2
3. 36
4. 360
5. 14
6. 12
7. 56
8. 1
9. 13
10. 12
11. 231
12. 56
13. 11
14. 120
15. 3

Exercises 1.5 B

1. 3
2. 48
3. 81
4. 21
5. 4096
6. 144
7. 3
8. 12
9. 11
10. 3
11. 6
12. 2
13. 18
14. 36
15. 8

Exercises 1.6 A

1. 7
2. 35
3. 3
4. 88
5. 77
6. 39
7. 196
8. 120
9. 2
10. 45
11. 17
12. 65
13. 10
14. 54
15. 7
16. 590
17. 40
18. 16
19. 0
20. 46

Exercises 1.6 B

1. 4
2. 52
3. 20
4. 11
5. 23
6. 6
7. 0
8. 32
9. 20
10. 63
11. 31
12. 20
13. 5
14. 18

Exercises 1.7 A

1. 2
2. 286
3. 20
4. 240
5. 27
6. 11
7. 24
8. 21
9. 21
10. 37

Exercises 1.7 B

1. 45
2. 12
3. 165
4. 280
5. 98
6. 28
7. 100
8. 143
9. 1
10. 1380
11. 5
12. 1

UNIT 1 POSTTEST

1. $8x$
2. $n + 6$
3. $y - 4$
4. $\frac{x}{3}$
5. 7^2, or 49
6. t^2
7. 8^3, or 512
8. n^3
9. $7(n + 8)$
10. $3x - 5$
11. $(y - 8)^2$
12. $y - (x + 2)^3$
13. $\frac{2x}{x + 2}$
14. 25
15. 10
16. 29
17. 4
18. 45
19. 9
20. 3
21. 20
22. 16
23. 2
24. 279
25. 175

UNIT 2 PRETEST

1. {5, 6, 7, 8, 9}
2. {11, 12, 13, 14, 15, . . .}
3. {1, 2, 3, . . . , 12, 13, 14}
4. {0, 1, 2, 3, 4, 5, 6}
5. {6, 12, 18, 24, 30, . . .}
6. {1, 2, 3, 4, 6, 8, 12, 24}
7. {1, 2, 5, 10, 25, 50}
8. {30, 60, 90, 120, 150, . . .}
9. {52, 54, 56, 58, 60, . . .}
10. {51, 53, 55, 57, 59}
11. {2, 7}
12. {6, 10, 15, 30}

Exercises 2.1 A

1. {0, 1, 2, 3, 4}
2. {7, 8, 9}
3. {7, 8, 9, 10, 11, . . .}
4. {1, 2, 3, 4}
5. {0, 1, 2, 3, 4}
6. {6, 7, 8, 9}
7. {6, 7, 8, 9, 10, . . .}
8. {1, 3, 5, 9, 15, 45}
9. {1, 3, 5, 9}
10. {0, 2, 4, 6, 8}
11. {2, 4, 6, 8, 10, . . .}

Exercises 2.1 B

1. {1, 2, 3, 4, 5, 6,}
2. {0, 1, 2, 3, 4, 5, 6}
3. {5, 6, 7, 8, 9}
4. {5, 6, 7, 8, 9, . . .}
5. {1, 2, 3, 4, 5, 6}
6. {1, 2, 3, 4, 5, 6, 10, 12, 15, 20, 30, 60}
7. {60, 120, 180, 240, 300, . . .}
8. {0, 3, 6, 9}
9. {3, 6, 9, 12, 15, . . .}
10. {1, 2, 4, 5, 7, 8}
11. *N*, or {1, 2, 3, 4, 5, . . .}

Exercises 2.2 A

1. {1, 2, 4, 5, 10, 20}
2. {20, 40, 60, 80, 100, . . .}
3. {1, 3, 7, 21}
4. {8, 16, 24, 32, 40, . . .}
5. {5, 7}
6. ∅, or { }
7. {30, 60, 90, 120, . . .}
8. {1, 2, 3, 5, 6, 10, 15, 30}
9. {1, 5, 7, 35}
10. *N*, or {1, 2, 3, 4, 5, . . .}

Exercises 2.2 B

1. {45, 90, 135, 180, 225, . . .}
2. {1, 3, 5, 9, 15, 45}
3. {30, 60, 90, 120, 150, . . .}
4. {1, 2, 3, 5, 6, 10, 15, 30}
5. {2, 4, 7, 14}
6. ∅ or { }
7. {1, 2, 3, 6, 7, 14, 21, 42}
8. {10, 20, 30, 40, 50, . . .}
9. N, or {1, 2, 3, 4, 5, . . .}
10. {5}

Exercises 2.3 A

1. {12, 14, 16, 18}
2. {21, 23, 25, 27, 29}
3. {23, 29}
4. {21, 22, 24, 25, 26, 27, 28}
5. {2}
6. {3, 5, 7, 11, 13, . . .}
7. {2, 3, 5}
8. {2, 4, 6, 10, 12, 20, 30, 60}
9. {1}
10. {9, 15}
11. ∅, or { }
12. {14, 28, 42, 56, 70, . . .}
13. {7, 21, 35, 49, 63, . . .}
14. ∅, or { }
15. {3, 5, 7}

Exercises 2.3 B

1. {31, 37, 41, 43, 47}
2. {36, 38, 39, 40, 42, 44}
3. {22, 44, 66, 88, 110, . . .}
4. {11, 33, 55, 77, 99, . . .}
5. {2, 7, 11}
6. {14, 22, 77, 154}
7. {1, 5, 7, 35}
8. {2, 10, 14, 70}
9. ∅, or { }
10. {2}
11. {1, 2, 3, 6, 9, 18, 27, 54}
12. {2, 3}
13. {6, 9, 18, 27, 54}
14. {2, 6, 18, 54}
15. {1, 3, 9, 27}

UNIT 2 POSTTEST

1. {14, 15, 16, 17, 18, . . .}
2. {0, 1, 2, 3, 4, 5, 6, 7}
3. {6, 12, 18, 24, 30, . . .}
4. {0, 3, 6, 9}
5. {1, 2, 3, 6, 7, 14, 21, 42}
6. {1}
7. {9, 18, 27, 36, 45, . . .}
8. ∅, or { }
9. {101, 103, 105, 107, 109}
10. {2, 4, 6, 8, 12, 16, 24, 48}
11. {41, 43, 47}
12. {4, 10, 20}

UNIT 3 PRETEST

1. symmetric
2. false
3. reflexive
4. transitive
5. transitive
6. st
7. $x + 4$
8. $5y$
9. $n + t$
10. $6(x + y)$
11. $xy + 8$
12. $(96 + 4) + 38 = 138$
13. $86(25 \cdot 4) = 8600$
14. $87 + (75 + 25) = 187$
15. $(125 \cdot 8)79 = 79{,}000$
16. $5ab$
17. $r + s + 7$
18. $2xy + 6$
19. $ab + xy + 8$
20. $7x + 14y$
21. $3xy + 3y$
22. $2(a + 4b)$
23. $8x$
24. $25 \cdot 40 + 25 \cdot 8 = 1200$
25. $49(45 + 55) = 4900$

Exercises 3.1 A

1. $5 = 5$
2. $3 + 7 = 3 + 7$
3. $x + 2 = x + 2$
4. $x + y = x + y$
5. $2(x - 5) = 2(x - 5)$
6. $8 = 5 + 3$
7. $7 - 3 = 4$
8. $5x - 1 = 6$
9. $2x = 10$
10. $x - 5 = 0$
11. $3(10 - 6) = 12$
12. $2(5^2) = 50$
13. $(x + 3)^2 = 25$
14. $x(x + 2) = 15$
15. $4x = 20$
16. reflexive
17. false
18. symmetric
19. false
20. transitive
21. false
22. transitive

Exercises 3.1 B

1. $2(x + 4) = 2(x + 4)$
2. $7xy = 7xy$
3. $(2x - y)^2 = (2x - y)^2$
4. $10 - (x - 3) = 10 - (x - 3)$
5. $10 - 3(x + 2) = 10 - 3(x + 2)$
6. $x + 7 = 0$
7. $\frac{7}{7} = 1$
8. $1 = \frac{3}{3}$
9. $2x + 2y = 2(x + y)$
10. $(x + y)^2 = x^2 + 2xy + y^2$
11. $5 - x = 7$
12. $2x - 2 = 8$
13. $x = 3$
14. $x = 2$
15. $x = 9$
16. symmetric
17. false
18. reflexive
19. transitive
20. false
21. transitive
22. symmetric

Exercises 3.2 A

1. cd
2. $c + d$
3. $7x$
4. $2\,(y + 5)$
5. $y + 5$
6. $(x + y) + 3$, or $x + y + 3$
7. $(a - b) + 5$, or $a - b + 5$
8. $x + (y + 6)$
9. $y + (z - 7)$
10. $xy + 2$

Exercises 3.2 B

1. $cd + 4$
2. $8\,(xy)$
3. $(c + d) + 4$ or $c + d + 4$
4. $4\,(c - d)$
5. $5\,(a + b)$
6. $3\,(x + y)$
7. $(3x)\,y$ or $3xy$
8. $st + 7$
9. $rs + t$
10. $b\,(a + 4)$

Exercises 3.3 A

1. $(4 + 6) + 7 = 17$
2. $3 + (2 + 6) = 11$
3. $3 + (9 + 5) = 17$
4. $(25 + 75) + 87 = 187$
5. $(3 \cdot 7)\,4 = 84$
6. $(2 \cdot 50)\,8 = 800$
7. $21\,(4 \cdot 5) = 420$
8. $57\,(25 \cdot 4) = 5700$
9. $45 + (55 + 87) = 187$
10. $96\,(8 \cdot 125) = 96{,}000$

Exercises 3.3 B

1. $(15 + 10) + 5 = 30$
2. $(8 + 7) + 4 = 19$
3. $68 + (35 + 65) = 168$
4. $59 + (49 + 51) = 159$
5. $(3 \cdot 4)\,5 = 60$
6. $(4 \cdot 25)\,9 = 900$
7. $47\,(2 \cdot 50) = 4700$
8. $92\,(4 \cdot 25) = 9200$
9. $(8 \cdot 125)\,25 = 25{,}000$
10. $(43 + 57) + 39 = 139$

Exercises 3.4 A

1. $x + y + 2$
2. $3cd$
3. $5rs$
4. $2LW$
5. $l + m + n$
6. $r + t + 2$
7. $n\,(k + 7)$
8. $x + y + 3$
9. $4xy$
10. $a + b + c + 1$
11. $2 \cdot 3\,x^2$, or $6x^2$
12. $15pq$
13. $2\,(x + y)$
14. $xy + z + 5$
15. $5\,(x + y + 2)$
16. $5mn + 8$
17. $cd + nt + 10$

Exercises 3.4 B

1. $2xy$
2. $b + c + d$
3. $x + y + 5$
4. xyz
5. $n(m + 1)$
6. $7kn$
7. $5mn$
8. $s + t + 14$
9. $8xy$
10. $x + y + 3$
11. $15xy$
12. $2xyz$
13. $6(a + b)$
14. $ab + cd + 8$
15. $rst + 9$
16. $3(x + y + 4)$
17. $ab + xy + 13$

Exercises 3.5 A

1. $4x + 28$
2. $xy + 2xz$
3. $abr + abs$
4. $5x^2 + 5x$
5. $8x + 8y + 8z$
6. $2x^2 + 12x$
7. $6xy + 3y^2$
8. $x^2 + xy + x$
9. $6x^2 + 12xy + 30x$
10. $2x^2y + xy^2 + 5xy$
11. $9(x + y)$
12. $y(x + 1)$
13. $2(3rs + 28)$
14. $7x$
15. $7(r + 2s + t)$
16. $5(x + 1)$
17. $10x$
18. $12(2y + 1)$
19. $2x(x + 3)$
20. $9y$
21. $300 + 35 = 335$
22. $72(75 + 25) = 7200$
23. $500 + 200 = 700$
24. $(97 + 3)\,64 = 6400$
25. $10{,}000 + 500 = 10{,}500$

Exercises 3.5 B

1. $6x + 24y$
2. $5x^2 + 45x$
3. $3x^2 + 6xy$
4. $2ab + 2a$
5. $9kr + 18ks + 9k$
6. $4y^2 + 40y$
7. $28x^2 + 35xy$
8. $8xy + 32y^2$
9. $6x^2y + 4xy^2 + 8xy$
10. $4x^2 + 4xy + 4z$
11. $x(y + 2z)$
12. $4(x + 2)$
13. $x(x + 1)$
14. $6y$
15. $8(x + y + 1)$
16. $3(6x + 1)$
17. $6x$
18. $10(3y + 5)$
19. $5y(2y + 1)$
20. $8x$
21. $100 + 36 = 136$
22. $96(28 + 72) = 9600$
23. $100(85) = 8500$
24. $6300 + 140 = 6440$
25. $59(100) = 5900$

UNIT 3 POSTTEST

1. transitive
2. symmetric
3. false
4. false
5. reflexive
6. $n + 5$
7. $8x$
8. $y + z$
9. xy
10. $rs + 6$
11. $9(x + 3)$
12. $(25 \cdot 4)\,56 = 5600$
13. $(69 + 31) + 97 = 197$
14. $76 + (89 + 11) = 176$
15. $95\,(125 \cdot 8) = 95{,}000$
16. $x + y + 4$
17. $9cx$
18. $cd + kn + 3$
19. $7ax + 12$
20. $4abc + 4b$
21. $5x^2 + 20x$
22. $9y$
23. $3(2x + 5y)$
24. $100\,(79) = 7900$
25. $4800 + 400 = 5200$

UNIT 4 PRETEST

1. an altitude of 750 ft. below sea level
   an altitude at sea level
2. $A\,(-4)$, $B\,(0)$, $C\,(+3)$
3. $D\,(0)$, $E\,(40)$, $F\,(-20)$
4. 20
5. 19
6. 25
7. 36
8. $x + 3 < 9$
9. $3x > 0$
10. $+7 > +5$, +7 is greater than +5
11. $-7 < +5$, −7 is less than +5
12. $-7 < -5$, −7 is less than −5
13. $-5 > -7$, −5 is greater than −7
14. 90
15. 90
16. −85
17. −50
18. −12
19. −29
20. +20
21. 0
22. −45
23. −8
24. 25
25. −25
26. 40
27. 50
28. 0
29. 0
30. 120
31. −300
32. −72
33. 0
34. −6
35. −20
36. 9
37. −6
38. −1
39. 0
40. −1
41. 16
42. 66
43. 50
44. −210
45. 1

Exercises 4.1 A

1. a withdrawal of $50
2. a loss of 8 yd.

3. an altitude 200 ft. below sea level
4. a velocity of 45 m.p.h. in the south direction
5. no gain or loss in weight
6. a decrease of 75¢ in the price of a stock
7. the year 1250 B.C.
8. a force of 20 lb. upward
9. no rotation
10. a current of 8 amps for a discharging battery

Exercises 4.1 B

1. 30° longitude east of Greenwich
2. a temperature 20° below zero
3. zero latitude (on the equator)
4. a loss of 5 yd.
5. no increase or decrease
6. a loss of $100
7. a deceleration of 15 m.p.h.
8. the distance of an image 24 cm. in back of a mirror
9. a loss of 32 points
10. a score of 17 points below average

Exercises 4.2 A

1. A (− 11), B (− 3), C (2), D (6)
2. A (− 6), B (− 8), C (0), D (− 3)
3. A (5), B (0), C (11), D (7)
4. A (8), B (3), C (− 2), D (− 8)
5. A (− 6), B (14), C (− 18), D (6)

(6–10)

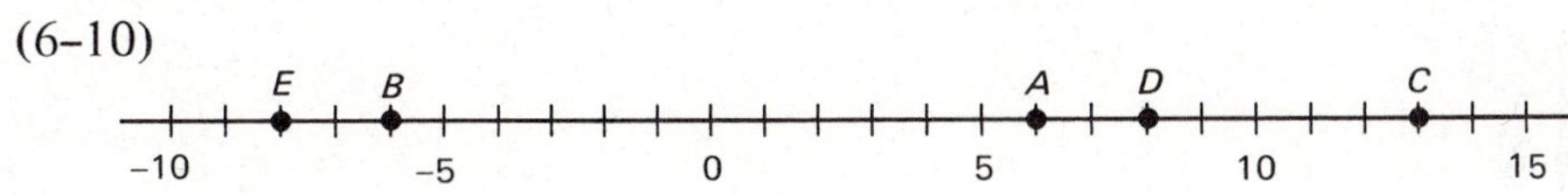

(11–15)

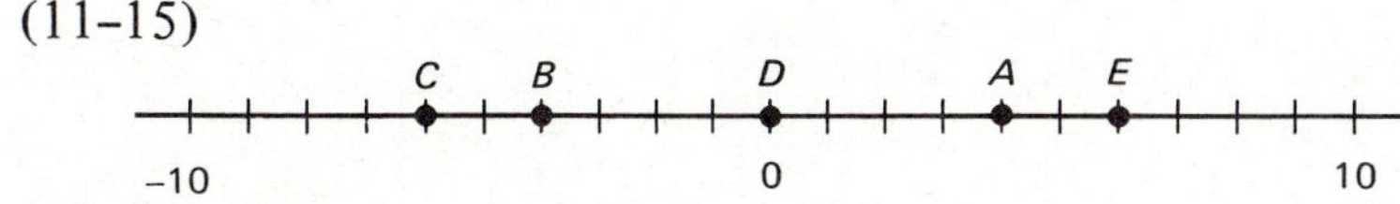

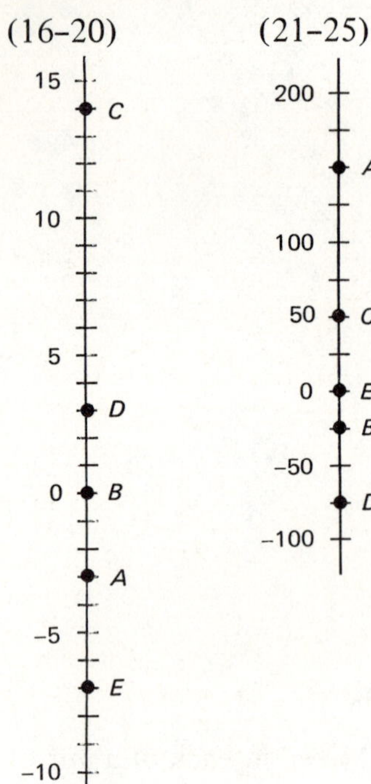

Exercises 4.2 B

1. A (6), B (−8), C (16), D (−16)
2. A (9), B (−17), C (−22), D (2)
3. A (3), B (−9), C (−3), D (8)
4. A (20), B (−40), C (35), D (−15)
5. A (−40), B (20), C (−90), D (90)

(6–10)

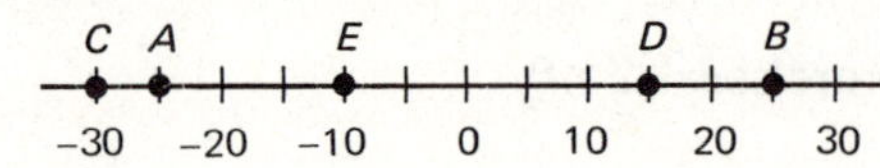

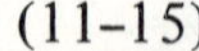

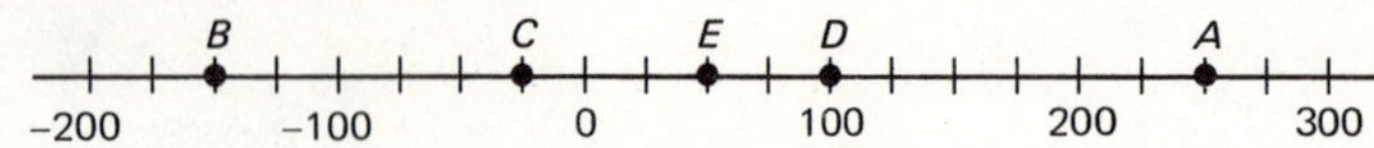

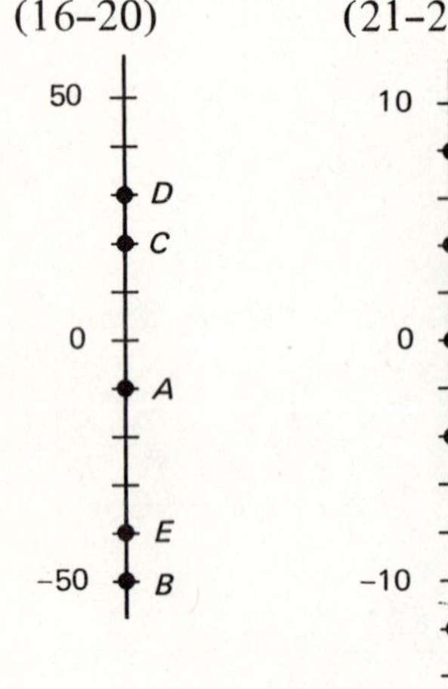

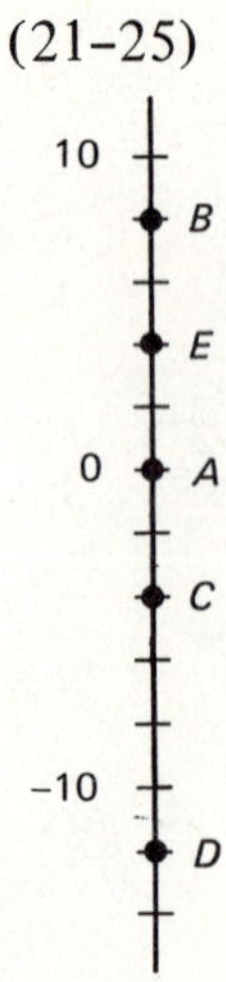

Exercises 4.3 A

1. 6
2. 7
3. 0
4. 4
5. 8
6. 9
7. 18
8. 3
9. 7
10. 3
11. 7
12. 4
13. 12
14. 14
15. 20
16. 0
17. -5
18. 0
19. 3
20. 8
21. 1

Exercises 4.3 B

1. 9
2. 20
3. 17
4. 7
5. 13
6. 7
7. 16
8. 20
9. 60
10. 40
11. 5
12. 8
13. 7
14. 1
15. 6

Exercises 4.4 A

1. $+4 < +7$, +4 is less than +7, point +4 is left of point +7
2. $+8 > +3$, +8 is greater than +3, point +8 is right of point +3
3. $-4 > -7$, -4 is greater than -7, point -4 is right of point -7
4. $-8 < -3$, -8 is less than -3, point -8 is left of point -3
5. $-2 < +5$, -2 is less than +5, point -2 is left of point +5
6. $+2 > -5$, +2 is greater than -5, point +2 is right of point -5
7. $0 > -4$, 0 is greater than -4, point 0 is right of point -4
8. $-6 = -6$, -6 equals -6, point -6 is the same as point -6
9. $6 > -3$, 6 is greater than -3, point 6 is right of point -3
10. $-6 < 0$, -6 is less than 0, point -6 is left of point 0
11. $x + 7 > 12$
12. $5y < 30$
13. $n + t < 0$
14. $x + 4 > 10$
15. $y - 5 < 8$
16. $|x| > 0$

Exercises 4.4 B

1. $2 > -3$, 2 is greater than -3, point 2 is right of point -3
2. $4 > 0$, 4 is greater than 0, point 4 is right of point 0
3. $0 < 5$, 0 is less than 5, point 0 is left of point 5
4. $7 > 5$, 7 is greater than 5, point 7 is right of point 5
5. $-12 > -14$, -12 is greater than -14, point -12 is right of point -14
6. $-3 = -3$, -3 equals -3, point -3 is the same as point -3
7. $-12 < 0$, -12 is less than 0, point -12 is left of point 0
8. $7 = 7$, 7 equals 7, point 7 is the same as point 7
9. $-8 < 4$, -8 is less than 4, point -8 is left of point 4
10. $-4 > -9$, -4 is greater than -9, point -4 is right of point -9
11. $6t < 24$
12. $x + y > xy$
13. $4y < 0$
14. $|ab| > 0$
15. $x - 7 > y$
16. $x + 8 < y$

Exercises 4.5 A

1. +9
2. +2
3. −9
4. +3
5. −4
6. −4
7. +8
8. −6
9. +10
10. −15
11. −9
12. +4
13. −4
14. 0

Exercises 4.5 B

1. +80
2. +75
3. −10
4. −42
5. −150
6. +24
7. +9
8. −5
9. −3
10. +7
11. −4
12. −17

Exercises 4.6 A

1. 8
2. 11
3. −10
4. −94
5. +24
6. −7
7. −14
8. −26

9. 70
10. -100
11. 1
12. -3
13. -7
14. -5
15. 7
16. -5
17. -5
18. 3
19. 0
20. 0
21. -5
22. -4

Exercises 4.6 B

1. -1
2. -21
3. 0
4. 13
5. 1
6. -1
7. -13
8. 1
9. -1
10. 0
11. 1
12. -1
13. 0
14. 12
15. 100
16. -100
17. -1
18. -5
19. 0
20. 0
21. -15
22. 50
23. 32
24. -32

Exercises 4.7 A

1. 2
2. 21
3. -2
4. 2
5. 10
6. -4
7. -7
8. -10
9. 16
10. 10
11. 2
12. -6
13. -2
14. 14
15. -12
16. 0
17. 0
18. -7
19. 6
20. 0
21. 8
22. 7
23. 20
24. 15
25. 7
26. 262
27. 34
28. 101

Exercises 4.7 B

1. -10
2. 7
3. 0
4. -10
5. 10
6. -4
7. -3
8. 2
9. 0
10. -9
11. 5
12. 5
13. -16
14. 1
15. 2
16. 10
17. 6
18. -6
19. 10
20. -2
21. 27
22. 47
23. $13\frac{3}{4}$
24. $4\frac{1}{4}$
25. 14,777
26. 856
27. 29,028
28. 74

Exercises 4.8 A

1. -10
2. 10
3. -10
4. 10
5. 27
6. -40
7. 4
8. -4
9. 1
10. 36
11. -56
12. -100
13. 9
14. 100
15. 1000
16. -1000
17. -64
18. -9
19. -1000
20. -10,000
21. -30
22. -100
23. -42
24. -18
25. 81
26. 3600
27. 100
28. 400
29. 0
30. 0
31. 150, counterclockwise
32. -100, clockwise
33. -360, clockwise
34. 480, counterclockwise
35. 0, no torque
36. 1000, east
37. -1050, west
38. 800, east
39. -150, west
40. 0, at airport

Exercises 4.8 B

1. 0
2. 0
3. 108
4. -1000
5. 100
6. 0
7. 0
8. 0
9. 0
10. -15
11. -25
12. 10
13. -4600
14. 58,000
15. 36
16. 64
17. 49
18. 81
19. -600
20. -3000
21. -150
22. -280
23. 7200
24. 144
25. -18
26. 36
27. 0
28. 0
29. -64
30. 216
31. -80
32. -12
33. 15
34. 84
35. 0
36. -1
37. -3
38. -4
39. +1
40. -1

Exercises 4.9 A

1. 3
2. 5
3. -5
4. 3
5. -2
6. -10
7. 5
8. -3
9. -4
10. -2
11. -6
12. -100
13. 10
14. 1
15. -1
16. -1
17. 1
18. 0
19. 0
20. -25

Exercises 4.9 B

1. −32
2. −80
3. 0
4. 7
5. 1000
6. −9
7. 0
8. −13
9. −4
10. 14
11. −1000
12. 100
13. −1
14. 1
15. −19
16. −27
17. 16
18. −4
19. 125
20. −625

Exercises 4.10 A

1. 5
2. 0
3. 0
4. 0
5. 0
6. −5
7. −1
8. −1
9. −4
10. 6
11. −27
12. 625
13. 20
14. 13
15. 6
16. 0
17. −5
18. −9
19. 5
20. 11

Exercises 4.10 B

1. 32
2. 4
3. −8
4. 2
5. −2
6. 1
7. −3
8. −14
9. 7
10. −7
11. −24
12. 6
13. 200
14. −200
15. −64
16. 64
17. 16
18. −1125
19. −7
20. 12

Exercises 4.11 A

1. −10
2. 56
3. 3
4. −20
5. −21, nearsighted
6. −2
7. −300
8. 60

Exercises 4.11 B

1. −2
2. 63
3. −6
4. 23
5. −3, nearsighted
6. −1250
7. $\frac{1}{2}$
8. −6

UNIT 4 POSTTEST

1. a latitude of 45° south of the equator; a latitude at the equator
2. A (15), B (−30), C (0)
3. P (80), Q (0), R (−40)
4. 7
5. 14
6. 15
7. 11
8. $y - 6 > 5$
9. $y + 9 < 0$
10. $+4 < +8$, +4 is less than +8
11. $+4 > -8$, +4 is greater than −8
12. $-4 > -8$, −4 is greater than −8
13. $-8 < -4$, −8 is less than −4
14. 30
15. −95
16. 15
17. 0
18. −57
19. −13
20. 13
21. −12
22. −25
23. 0
24. −20
25. −25
26. −12
27. 0
28. −45
29. 0
30. −1000
31. 150
32. −621
33. 35
34. 0
35. −17
36. 6
37. −9
38. −13
39. 0
40. −3800
41. −4
42. 44
43. −48
44. 2000
45. 15

UNIT 5 PRETEST

1. $9x$
2. $6y$
3. $7y$
4. x
5. $4x - 5$
6. $-3x - 7y$
7. $x^2 - 2x - 15$
8. $2x^2 - 5xy - 3y^2$
9. $4x + 24$
10. $-2x - 14$
11. $-6y + 12$
12. $-y + 8$
13. $-5x - 7$
14. $-x - 2y + 4$
15. $2x - 18$
16. $16 - x$
17. −38
18. $2x^2 + 3xy - 20y^2$
19. 7
20. 7
21. 16
22. 32
23. 7
24. −8
25. 10
26. −8
27. 11
28. −2
29. $\emptyset$ (no solution)
30. I (any integer is a solution)
31. $y = \dfrac{20 - 2x}{5}$
32. $h = \dfrac{2A - r^2}{r}$

Exercises 5.1 A

1. $7x$
2. $3y$
3. $2x$
4. y
5. 0
6. $-6z$
7. $5(r-s)$
8. $10x$
9. y
10. $8z$
11. $-n$
12. $2x$

Exercises 5.1 B

1. $6x$
2. $6y$
3. $-3y$
4. $-x$
5. 0
6. $7n$
7. $2(x+y)$
8. x
9. 0
10. $10k$
11. $-10d$
12. $-7x$

Exercises 5.2 A

1. $2x+5$
2. $x-2$
3. $y-6$
4. $-x+5$
5. $4x+10$
6. x^2-x-6
7. $3y^2-7y+2$
8. $5x+3y$
9. $s+t$
10. y
11. $9xy$
12. $-5xy$
13. x^2-xy+y^2
14. $9xy+xz$
15. $5a^2b+7a^2b^2$

Exercises 5.2 B

1. $8x-2$
2. $2y+5$
3. -2
4. $2x-2$
5. $-5t+6$
6. y^2+y-6
7. $5k^2-3k-8$
8. $-2x+5y$
9. $2s$
10. $3m+9n$
11. xy
12. $-5xy$
13. $r^2-8rs-s^2$
14. $ab-ac$
15. $-3xyz$

Exercises 5.3 A

1. $2x+8$
2. $5x-15$
3. $-3x-18$
4. $-4x+4$
5. $12x+18$
6. $21x+28$
7. $12y-20$
8. y^2-4y
9. $-6x-7$
10. $-18r-9$
11. $-8t+16$
12. $-4y+5$
13. $-6y+9$
14. $-5t-2$
15. $5x+5y-15$
16. $8x-12y+8$
17. $3x+15y+21$
18. $50x-60y-80$
19. $-3x-12y+3$
20. $-14x+35y-42$
21. $-x-y-1$
22. $-x+y-1$
23. $-2x+y+2$
24. $-4x-9y+8$
25. $6x^3-4x^2-2x$

Exercises 5.3 B

1. $3y + 18$
2. $-4x - 28$
3. $7x - 63$
4. $-6x + 48$
5. $2x^2 - 10x$
6. $-12y + 3y^2$
7. $-5x - 8$
8. $-6y + 9$
9. $10x - 15y + 30$
10. $-32x - 40y + 16$
11. $-r + s + t$
12. $-r + 2s - 1$
13. $2x^3 - 2x^2 - 2x$
14. $-8y - 4y^2 + 4y^3$
15. $-a^2 - ab - ac$
16. $a^3b - a^2b^2 - ab^3$
17. $x^3y - 5x^2y^2 - 6xy^3$
18. $-2x^2y - 14xy^2 + 2xy$

Exercises 5.4 A

1. $3x + 4$
2. $-2x + 4$
3. $3y - 5$
4. $-6t + 7$
5. $-x + 36$
6. $2y - 12$
7. $-3z + 5$
8. $8n - 2$
9. $x + 5$
10. $-6x + 18$
11. $-2x + 7y$
12. $y^2 - 3y + 6$
13. $9a + b$
14. y
15. $6r + 8s$
16. $x - y$
17. $2x$
18. $-3r - 3s + 6t$
19. $-7y + 4$
20. x

Exercises 5.4 B

1. $4x - 10$
2. $4y - 8$
3. $-5x - 2$
4. $2t - 7$
5. $x + 6$
6. $2y - 4$
7. $-4n - 3$
8. $x - 3$
9. $x - 6$
10. $5x + 6y$
11. $x^2 + 3x - 6$
12. $x + 5y$
13. $2r - 19s$
14. x
15. $2c$
16. $x - 9$
17. $-2x + 2y - 2z$
18. $4x - y + 12$
19. $x - 3$
20. $8x - 7y - 9$

Exercises 5.5 A

1. 12, addition
2. 4, subtraction
3. 3, division
4. 36, multiplication
5. −5, subtraction
6. −2, addition
7. −8, division
8. 50, multiplication
9. 0, subtraction
10. 4, multiplication by −1
11. −6, division
12. −12, multiplication
13. −13, subtraction
14. −2, addition
15. 0, addition
16. 4, division
17. 32, multiplication
18. 0, division
19. −50, multiplication
20. −400, division

Exercises 5.5 B

1. 7, subtraction
2. 15, addition
3. 75, multiplication
4. 6, division
5. −5, addition
6. −7, division
7. −4, subtraction
8. −80, multiplication
9. 24, multiplication
10. −1, subtraction
11. 0, subtraction
12. 8, division
13. 0, division
14. 30, addition
15. −2, division
16. −200, multiplication
17. 2, division
18. −10, subtraction
19. 0, addition
20. −6, multiplication

Exercises 5.6 A

1. 4
2. 3
3. 12
4. 5
5. 2
6. 3
7. 45
8. 10
9. −6
10. 4
11. −2
12. 8
13. 2
14. 0
15. 0

Exercises 5.6 B

1. 0
2. 16
3. −4
4. 5
5. 2
6. 5
7. 9
8. −3
9. −5
10. −1
11. 6
12. −9
13. 11
14. −1
15. 0

Exercises 5.7 A

1. 3
2. 7
3. −6
4. 2
5. −3
6. 9
7. 7
8. 8
9. *I*, the set of integers
10. 13
11. 13
12. 2
13. ∅
14. ∅
15. 0

Exercises 5.7 B

1. -14
2. -4
3. 5
4. -6
5. $\emptyset$
6. -9
7. 10
8. -30
9. 11
10. -7
11. I
12. 0
13. -5
14. 12
15. $\emptyset$

Exercises 5.8 A

1. $y = 10 - 2x$
2. $y = 4x - 5$
3. $y = \frac{12 - 3x}{2}$
4. $y = \frac{5x - 15}{3}$
5. $y = \frac{-ax - c}{b}$
6. $t = \frac{d}{r}$
7. $c = p - a - b$
8. $h = \frac{2A}{b}$
9. $t = \frac{A - P}{Pr}$
10. $L = 2s - A$
11. $n = \frac{L - a + d}{d}$
12. $a = \frac{mg - T}{m}$
13. $A = SW + W$
14. $S = C - Rn$
15. $f = \frac{ab}{a + b}$

Exercises 5.8 B

1. $y = 7 - 3x$
2. $y = 2x - 6$
3. $x = \frac{-3y - 12}{2}$
4. $x = \frac{2y + 10}{5}$
5. $y = mx + b$
6. $I = \frac{E}{R}$
7. $W = \frac{P - 2L}{2}$
8. $b = \frac{2A}{h} - a$
9. $d = \frac{L - a}{n - 1}$
10. $R = \frac{PV}{T}$
11. $C = S - P$
12. $c = \frac{L - a}{at}$
13. $A = \frac{2L}{cdv^2}$
14. $M = \frac{QC}{100}$
15. $c = \frac{K - aK}{a^2}$

UNIT 5 POSTTEST

1. $8x$
2. y
3. $7x$
4. $-2y$
5. $-5x - 2$
6. $2x^2 - 3x - 5$
7. $x^2 - 2xy - 8y^2$
8. $x^2y + 4xy^2$
9. $6x - 18$
10. $-5x - 35$
11. $-4x - 9$
12. $6x - 8$
13. $4x - 8y - 24$
14. $-2x^2 - 2xy + 2x$
15. $10 - 4x$
16. $4x - 4$
17. $6x^2 - 11x + 4$
18. $6x^2 - 11xy - 2y^2$
19. 100
20. -4
21. -8
22. -225
23. 40
24. -2
25. 9
26. 7
27. 6
28. -5
29. I
30. $\emptyset$
31. $c = 2s - a - b$
32. $y = \frac{9x - 18}{2}$

UNIT 6 PRETEST

1. $4x - 7 = x + 8; x = 5$
2. $2x - 5 = 3(x - 5); x = 10$ years, $2x = 20$ years
3. $2x + 2(x - 6) = 180$; length = 48 ft., width = 42 ft.
4. $450x + 375(20 - x) = 8400$; 12 beef, 8 fish
5. $2x + 2(x + 50) = 2500$; 600 m.p.h., 650 m.p.h.
6. $200x = 40(x + 2); \frac{1}{2}$ hr.
7. $640x = 576(x + 1)$; 9 hr., 10 hr., 5760 mi.

Exercises 6.1 A

1. $x + 5 = 5(x - 3); x = 5$
2. $4x + 1 = 2x - 3; x = -2$
3. $4(x - 4) = 5(x - 6); x = 14$
4. $10x - 2(x - 1) = 58; x = 7$
5. $5 + 4x = 17 - 2x; x = 2$

Exercises 6.1 B

1. $3(7 - x) + 12 = 8x; x = 3$
2. $\frac{7 + x}{2} = x + 1; x = 5$
3. $3x + 5 = 5x - 3; x = 4$
4. $4 = x + 2(3x - 5); x = 2$
5. $6(x - 3) = 5(x - 5); x = -7$

Exercises 6.2 A

1. $30 - 5$
2. $x - 5$
3. $x - 3$
4. $25 + 12$
5. $x + 12$
6. $3x + 12$
7. $40 - 7$
8. $40 - x$
9. $x - 7$
10. $8 + 10$
11. $y + 10$
12. $8 + x$
13. $x - 3$
14. $2x + 6$

Exercises 6.2 B

1.

Age 3 yr. ago	Age 7 yr. from now
$x - 3$	$x + 7$
$x - 8$	$x + 2$
$4x - 3$	$4x + 7$

2.

Age 10 yr. ago	Age 50 yr. from now
$x - 10$	$x + 50$
$x + 10$	$x + 70$
$x - 40$	$x + 20$

3.

Age now	Age 6 yr. ago	Age 9 yr. from now
14	8	23
$14 - x$	$8 - x$	$23 - x$
$14 + y$	$8 + y$	$23 + y$

4.

Age now	Age 8 yr. ago	Age 4 yr. from now
x	$x - 8$	$x + 4$
$2x$	$2x - 8$	$2x + 4$
$x + 5$	$x - 3$	$x + 9$

Exercises 6.3 A

1. $x + 20 = 3(x - 8)$; 22 years
2. $\frac{x}{2} = x - 9$; 18 years
3. $x - 14 = \frac{x + 8}{3}$; 25 years
4. $x + 20 = 2x$, present age is 20; $20 + y = 3(20)$, 40 years
5. $3x + 14 = 2(x + 14)$; 42 years

Exercises 6.3 B

1. $(x + 25) + (x + 2) + 25 = 100$; 26 years and 24 years
2. $(x - 2) - 5 + (x - 5) = (x + 5) - 5$; 12 years
3. $x + \frac{x}{4} = 20$; 16 years

4. $\frac{x}{2} + x + 2x = 2(2x) - 2$; 4 years

5. $x + 5 = 4(x - 54 + 5)$; $x = 67$, $100 - 67 = 33$ years

Exercises 6.4 A

1. $4s = 32$, $s = 8$, area $= 8^2 = 64$ sq. in.
2. $2W + 2(2W) = 54$; $W = 9$ in., $L = 18$ in.
3. $2x + 2(x + 10) = 180$; 40 ft. by 50 ft.
4. $x + 2x + 3x = 180$; 30°, 60°, 90°
5. $x + (x + 6) + (x + 3) = 36$; 9 cm., 12 cm., 15 cm.
6. $\frac{x(x + 5)}{2} - \frac{(x - 5)x}{2} = 900$; 180 ft. by 185 ft.

Exercises 6.4 B

1. $2(48 + 2x) + 2(36 + 2x) = 176$, 1 in.
2. window area $= \frac{44(20)}{8} = 110$
   $(4 \cdot 4)x + 2(5 \cdot 3) = 110$; 5 windows
3. $x + x + 2x = 60$; 15 ft. by 30 ft.
4. $x + 2x + 2x = 160$; 32 cm., 64 cm., 64 cm.
5. $\frac{(x + 3)5}{2} = 2\left(\frac{15}{2}\right)$; 3 ft.
6. $x + 2x + 60 = 180$; 40°, 80°

Exercises 6.5 A

1. 100
2. $25x$
3. 0.05 (2000)
4. 0.065 (5000)
5. $0.05d$
6. $0.05(x + 500)$
7. 5 (15)
8. $15y$
9. $15(20 - x)$
10. $5c$
11. 0.05 (40)
12. $0.05x$
13. $0.05(x + 15)$
14. $0.05(50 - x)$

Exercises 6.5 B

1. 40 (3.10)
2. $3.10h$
3. $40x$
4. 0.6 (20)
5. $0.6y$
6. $0.6(x + 20)$
7. $7x$
8. $60y$
9. $60(60h)$
10. $3s$
11. $50x$

Exercises 6.6 A

1. $5 + 3$
2. $3 + x$
3. $12 - 3$
4. $12 - x$
5. $2000 + 3000$
6. $2000 + d$
7. $5000 - 3000$
8. $5000 - x$
9. $20 + 40$
10. $40 + x$
11. $75 - 50$
12. $x - 50$
13. $5 + y$
14. $15 - x$
15. $x + 2x$

Exercises 6.6 B

1. $25 + x$
2. $x - 60$
3. $100 - x$
4. $40 - x$
5. $x - 15$
6. $y + (y + 6)$
7. $d + 3d + 5d$
8. $x + (x - 8)$
9. $d + 2d + (d + 2000)$
10. $x + 3x + (3x - 5)$

Exercises 6.7 A

1. $30(40) + 55x = 45(30 + x)$; 15 lb.
2. $150x + 50(450 - x) = 54{,}500$; 320 adult tickets
3. $0.05(x - 3000) + 0.07x = 630$; 6500 dollars
4. $x + 0.16(18 - x) = 0.30(18)$; 3 qt.
5. $0.03(60 - x) = 0.01(60)$; 40 gal.
6. $50x + 20(20) = 30(x + 20)$; 10 lb.
7. $45x + 120(10 - x) = 600$; 8 lb.
8. $0.05x + 0.07(2x) = 760$; \$4000 bonds, \$8000 stocks

Exercises 6.7 B

1. $2x + 11(32) = 400$; 24 two-cent stamps
2. $5(3x) + 4x = 304$; $x = 16$, $3x = 48$ oranges
3. $100x + 50(2x) + 75(60 - 3x) = 4200$; 12 lb.
4. $0.06x + 0.09x + 0.07(100{,}000 - 2x) = 7300$; $x = \$30{,}000$ (bonds), $100{,}000 - 2x = \$40{,}000$ (mortgages)
5. $x + 0.75(30) = 0.90(x + 30)$; 45 lb.
6. $0.35x + 0.04(155 - x) = 0.12(155)$; 40 qt.
7. $0.03(900 - x) = 0.02(900)$; 300 gal.
8. $20x + 5x = 400$; 16 lb.

Exercises 6.8 A

1. $40(3)$ mi.
2. $40h$ mi.
3. $40(h+2)$ mi.
4. $5x$ mi.
5. $5(x-2)$ mi.
6. $6(35)$ m.p.h.
7. $6x$ m.p.h.
8. $4(x-2)$ m.p.h.

9a. $(30+20)$ m.p.h.
9b. $(30+x)$ m.p.h.
9c. $(30-5)$ m.p.h.
9d. $(30-x)$ m.p.h.
10a. $(x+20)$ m.p.h.
10b. $(x-5)$ m.p.h.

Exercises 6.8 B

1. $3(450)$ mi.
2. $450x$ mi.
3. $450(x-2)$ mi.
4. $(450-r)$ m.p.h.
5. $(450+x)$ m.p.h.
6. $6x$ knots
7. $9x$ knots
8. $(x+12)$ knots
9. $(x-12)$ knots
10. $3x$ knots

Exercises 6.9 A

1. $520x+460x=1470$; $1\frac{1}{2}$ hr.
2. $6x+4(x+50)=550$; 35 m.p.h. (freight), 85 m.p.h. (passenger)
3. $40x+35x=450$; 6 hr.
4. $55x+55(x+2)=330$; 1:00 P.M.

Exercises 6.9 B

1. $4x+4(x+20)=440$; 45 m.p.h. (northbound), 65 m.p.h. (southbound)
2. $60x+50(x-3)=400$; $x=5$
   distance from Sacramento = $60x$ = 300 mi.
   clock time = 6 A.M. + 5 hr. = 11 A.M.
3. $650x+700x=4050$; 9:00 A.M.
4. $4x+4(x+5)=380$; 45 m.p.h., 50 m.p.h.

Exercises 6.10 A

1. $3(x+3)=39x$; $x=\frac{1}{4}$, 11:00 + 0.15 = 11:15 A.M.
2. $90x=80x+5$; $x=\frac{1}{2}$ hr. or 30 min.
3. $60x=30(x+1)$; 1 hr.
4. $25x=9x+16$; 1 hr.

Exercises 6.10 B

1. $165(x-2) = 55x$; 3 hr.
2. $60x = 40x + 50$; $2\frac{1}{2}$ hr.
3. $75x = 35x + 120$; 12:00 noon
4. $15t + \frac{1}{3} = 1$; $tx = 1$; $\frac{2x}{45} = 1$, $x = 22\frac{1}{2}$ m.p.h.

Exercises 6.11 A

1. $6x = 7(x-25)$; 175 m.p.h.
2. $240x = 180(7-x)$; $x = 3$ hr., $3(240) = 720$ mi.
3. $36(x+1) = 48x + 18$; $x = 1.5$ hr.
   distance $= 48x = 72$ mi.

Exercises 6.11 B

1. $60x = 20(2-x)$; $x = \frac{1}{2}$ hr., $d = 30$ mi.
2. $10x = 40(5-x)$; 4 hr.
3. $4x = 3(x+5)$; 15 m.p.h.

Exercises 6.12

1. $195x + 49(2x-5) + 76 = 1589$; 6 yd.
2. $8x = 90 + 5x$; 30 sec.
3. $0.20(x + 500) = 0.25(500)$; 125 cc.
4. rectangle: $2w + 2w = 48$, $w = 12$, $l = 24$, $A = 288$ sq. ft.
   square: $3s = 48$, $s = 16$, $A = (16)^2 = 256$ sq. ft.
   the rectangle has the greater area by 32 sq. ft.
5. $x + 5 = 2(x - 45 + 5)$; $x = 85$, $100 - 85 = 15$ yr.
6. $90x + 40(x-2) = 310$; $x = 3$, $90x = 270$ mi.
7. $x + (x+10) + (x+20) = 180$; $50°$, $60°$, $70°$
8. $60x = 40(5-x)$; $x = 2$ hr., $d = 120$ mi.
9. $0.40x + 0.10(100-x) = 0.16(100)$; 20 cc.

UNIT 6 POSTTEST

1. $6(x+5) = x - 10; -8$
2. $(x+3) + (x+4+3) = 3[(x+3-50) + (x+4+3-50)]; x = 70$, age of Kay; $x + 4 = 74$, age of Jay
3. $2(x + x - 4) - 3 = 45$; 14 ft. by 10 ft.
4. $0.75(60 - x) = 0.50(60)$; 20 liters
5. $3x + 3(x + 15) = 285$; 40 m.p.h. (truck), 55 m.p.h. (car)
6. $70(x - 4) = 30x$; 7 hr., 210 mi.
7. $12x = 3(5 - x)$; 1 hr.

UNIT 7 PRETEST

1. $-24x^5$
2. $56x^5y^5$
3. $x^3 - 2x^2 + 7x - 5$
4. $-x^3y + x^2y^4 - x + 15$
5. $x^2 - 5x + 3$
6. $7x - 5$
7. $x^3 - 12x^2 + 48x - 64$
8. $-8x^2 + 16xy + 24y^2$
9. $18x^4 - 21x^3 + 3x^2$
10. $-24x^3y + 18x^2y^2 + 30xy^3$
11. $2x^2 + x - 21$
12. $25x^2 - 36y^2$
13. $6xy - 15x - 8y + 20$
14. $x^3 + 15x^2 + 75x + 125$
15. $27x^3 - 64y^3$
16. $4x^2 + 20x + 25$
17. $49x^2 - 42xy + 9y^2$
18. $9x^2 - 49y^2$
19. $81x^4 - 16$
20. $x^2 - 3x - 70$
21. $y^4 + 7y^2 - 144$
22. $20x^2 - 23x + 6$
23. $18x^2 - 27xy - 35y^2$
24. $xy - x + 9y - 9$
25. $x^3 + 10x^2 - 36x - 360$

Exercises 7.1 A

1. $20x^4$
2. $-42y^6$
3. $40y^5$
4. $100x^7$
5. $-36x^6$
6. $49x^4$
7. $64x^6$
8. $36y^6$
9. $81y^4$
10. x^3y^3
11. $-24x^4y^4$
12. $8x^3$
13. $-27x^6$
14. $-25x^2y^2$
15. $25x^2y^2$
16. $-64x^6$
17. $-4x^6$
18. $75x^5$
19. $80x^5$
20. $56x^7$

Exercises 7.1 B

1. $21x^6$
2. $-16y^6$
3. $30t^8$
4. $-16x^3y^3$
5. $81x^4$
6. $100x^4$
7. $25y^6$
8. $144y^6$
9. $30r^4s^4$
10. $-64t^6$
11. $-216t^6$
12. $-8x^3y^3$

13. $9x^8$
14. $-3x^8$
15. $24x^9$
16. $10x^4y^6$
17. $-21x^3y^5$
18. $-48x^8y^7$
19. $24x^6y^8$
20. $72x^9y^{11}z^4$

Exercises 7.2 A

1. $3x^2 - 5x + 8$
2. $x^3 - x^2 + x - 1$
3. $-x^4 + 14x^2 - 49$
4. $5x^4 + 7x^3 + 6x^2 + 4x$
5. $x^2 + 2x - y^2 - 4y - 3$
6. $-8x^3y + xy^2 + y^4$
7. $-4x^2 + 8x + 9y^2 - 6y - 10$
8. $-x^3 - x - y^3 - y + 1$
9. $-x^2 + 6x + 36y^2 - 72y + 5$
10. $2x^3y + 4x^2y^2 - 7xy^3$

Exercises 7.2 B

1. $3x^2 - 2x + 5$
2. $x^3 - 6x + 1$
3. $-x^4 + 10x^2 - 25$
4. $x^5 - 2x^3 + 4x - 10$
5. $x^4 - 4x^3 - x + 4$
6. $x^2 + 4xy + y^2 + 6y$
7. $-x^3 + 5x - 2$
8. $x^4 + x^2y^2 + y^4$
9. $-x^3 + 3x^2y - 3xy^2 + y^3$
10. $-3x^2 - 2xy + y^2$

Exercises 7.3 A

1. $8x^2 - 9x + 1$
2. $7x - 3y + 3$
3. $-5x + 9$
4. $2y - 12$
5. $10x - 2$
6. $x^3 + 1$
7. $y - 7$
8. $-8x - 5y + 8z$
9. $y^4 - y^2 + 4$
10. $x^3 - 3x^2y + 3xy^2 - y^3$

Exercises 7.3 B

1. $6x^2 + x - 9$
2. $-5y^2 - 2y$
3. $x^3 + 3x^2 + 3x + 1$
4. $5x - 15$
5. $y^3 - 8$
6. $2y - 3$
7. $-x + y - 9$
8. $-x^2 + 2y^2$
9. $x^3 + 27y^3$
10. $x^4 + 4$

Exercises 7.4 A

1. $5x^2 + 4x + 3$
2. $x^2 - 3x - 4$
3. $-x - 9y + 5$
4. $-10x + 4y + 15$
5. $x^4 - 28x^2 + 8x + 35$
6. $-x^2 - 14x - 24$
7. $2y - 2z$
8. $5x^2y - 2x^2y^2 - 10xy^2$
9. $-12y^2 + 14$
10. $x^2 + x - 2xy - y + y^2$

Exercises 7.4 B

1. $3x - 3y - 25$
2. $11x^2 + 10xy - y^2$
3. $x^3 + 343$
4. $x^2 - xy - 7y^2$
5. $4xy - 2y^2$
6. $x^2y^2 - x^2 - 36y^2 + 81$
7. $1 - xy - xz - yz + tx$
8. $10 - 2x - 2y$
9. $x^3 - 9x^2 + 27x - 27$
10. $x + y - z - 7$

Exercises 7.5 A

1. $14x^3 - 6x^2 - 8x$
2. $-20x^2 + 8xy - 28x$
3. $5x^4 - 10x^3y + 5x^2y^2$
4. $12y^4 - 18xy^3 - 6x^2y^2$
5. $6x^3y - 24x^2y^2 + 18xy^3$
6. $7a^3 - 35a^2b + 7a$
7. $-9b^4 + 9b^3 + 54b$
8. $32x^3y^2 - 40x^2y^3 - 48x^2y^2$
9. $-60a^4b^2 + 60a^3b^2 - 15a^2b^2$
10. $5x^4 + 150x^3 - 425x^2$

Exercises 7.5 B

1. $-48r^3s + 48r^2s^2 - 12rs$
2. $-15st^3 + 75st^2 - 60st$
3. $-r^4 + r^3 - r^2 + r$
4. $-c^4d - c^3d + c^2d^2 + c^2d^3$
5. $8x^6 + 4x^5 - 8x^4 - 4x^3$
6. $8y^3 - 8y^4 - 8y^5 - 8y^6$
7. $10y^4 - 6y^3 + 2y^2$
8. $-xy + x^2y - xy^2 + x^2y^2$
9. $7s^3t - 7s^2t - 7st^2 + 7st^3$
10. $5a^2b^2c^3 + 10\,ab^3c^3 - 10ab^2c^4 + 25ab^2c^3$

Exercises 7.6 A

1. $x^2 + 6x + 8$
2. $x^2 - 6x + 5$
3. $y^2 - y - 30$
4. $y^2 + 4y - 21$
5. $2x^2 - 13x + 20$
6. $12x^2 - 7x - 10$
7. $x^2 - 36$
8. $4x^2 - 49$
9. $25x^2 - 9y^2$

10. $16x^2 + 8xy + y^2$
11. $36x^2 - 60xy + 25y^2$
12. $xy + 5x + 2y + 10$
13. $20xy + 4x - 30y - 6$
14. $x^3 - x^2 + x - 1$
15. $5x^3 - 2x^2 - 20x + 8$
16. $x^3 + 7x^2 + 17x + 15$
17. $x^3 - 2x^2y - xy^2 + 2y^3$
18. $x^3 - 216$
19. $y^3 + 6y^2 + 12y + 8$
20. $x^4 - 1$
21. $x^2 + 2xy + 2xz + y^2 + 2yz + z^2$
22. $x^4 - 5x^2 + 4$
23. $y^4 - 26y^2 + 25$
24. $y^4 + y^3 - 3y^2 - y + 2$
25. $4x^2 + y^2 + 9z^2 + 4xy - 12xz - 6yz$

Exercises 7.6 B

1. $3x^2 + 2x - 8$
2. $3x^2 - 2x - 8$
3. $49x^2 - 4y^2$
4. $49x^2 - 28xy + 4y^2$
5. $25x^2 - 4$
6. $x^3 - x^2y + xy^2 - y^3$
7. $2x^2 - 3xy - 2y^2$
8. $-25x^2 + 40xy - 16y^2$
9. $x^2 + 4xy + 4y^2$
10. $49x^2 - 84xy + 36y^2$
11. $a^4 - b^4$
12. $x^4 - 16$
13. $x^6 - 1$
14. $64x^6 - 729$
15. $4x^2 + 24x - 7xy - 42y$
16. $2y^3 - 7y^2 + 7y - 2$
17. $-x^3 - 9x^2y - 19xy^2 + 4y^3$
18. $x^3 + 125$
19. $y^3 - 12y^2 + 48y - 64$
20. $r^2 + 4rs + 4s^2 - 25$
21. $x^4 + 4$
22. $4y^4 - 20y^3 + 2y^2 - 11y + 5$
23. $x^4 - 3x^2y + x^2 - 6y - 2$
24. $x^2 + y^2 + z^2 - 2xy - 2xz + 2yz$
25. $9a^2 + 25b^2 + c^2 - 30ab + 6ac - 10bc$

Exercises 7.7 A

1. $6x^2 + 23x + 20$
2. $20x^2 - 19x + 3$
3. $12y^2 - 16y - 35$
4. $12y^2 + 16y - 35$
5. $12y^2 - 44y + 35$
6. $12y^2 + 44y + 35$
7. $15x^2 + 14xy - 8y^2$
8. $15x^2 + 2xy - 8y^2$
9. $15x^2 - 14xy - 8y^2$
10. $15x^2 - 2xy - 8y^2$
11. $4x^4 + 3x^2 - 1$
12. $36y^4 - 25y^2 + 4$
13. $xy + 2x + 3y + 6$
14. $xy - 5x - 4y + 20$
15. $xy - x + 7y - 7$
16. $xy + 4x - 6y - 24$
17. $x^3 - x^2 - x + 1$
18. $y^3 - 6y^2 + 5y - 30$
19. $x^4 - 5x^3 + x - 5$
20. $y^5 + 4y^3 - 2y^2 - 8$

Exercises 7.7 B

1. $10x^2 - 13x - 30$
2. $18x^2 - 45x + 28$
3. $2x^2 + 3x - 2$
4. $8x^2 - 23xy - 3y^2$
5. $36x^2 - 63xy + 20y^2$
6. $36r^4 - 13r^2 + 1$
7. $4x^4 - 37x^2y^2 + 9y^4$
8. $4x^4 - 24x^2 + 35$
9. $9r^2s^2 - 6rs - 35$
10. $16x^2y^2 - 32xy + 15$
11. $xy + 6x - 7y - 42$
12. $xy - 8x - 5y + 40$
13. $3r^2 + 2rs - 15rt - 10st$
14. $ac + ad + bc + bd$
15. $x^3 + 6x^2 - 4x - 24$
16. $x^3 - 8x^2 - 9x + 72$
17. $y^3 - 7y^2 + y - 7$
18. $y^5 - y^3 - y^2 + 1$
19. $x^5 - 3x^3 + 27x^2 - 81$
20. $5x^2y - 2xy^2 + 25x - 10y$

Exercises 7.8 A

1. $x^2 + 8x + 16$
2. $x^2 - 6x + 9$
3. $4x^2 - 20x + 25$
4. $4x^2 - 20x + 25$
5. $25x^2 - 20x + 4$
6. $x^2 - 14xy + 49y^2$
7. $x^2 - 2xy + y^2$
8. $x^4 + 12x^2 + 36$
9. $16y^6 + 8y^3 + 1$
10. $x^2y^2 + 16xy + 64$

Exercises 7.8 B

1. $x^2 + 10x + 25$
2. $y^2 - 12y + 36$
3. $100x^2 - 60xy + 9y^2$
4. $x^2y^2 + 8xy + 16$
5. $64x^4 - 16x^2 + 1$
6. $25x^2 + 10xy + y^2$
7. $81a^2 - 72ab + 16b^2$
8. $81a^2 - 72ab + 16b^2$
9. $81a^2 + 72ab + 16b^2$
10. $16a^2 - 72ab + 81b^2$

Exercises 7.9 A

1. $x^2 - 9$
2. $x^2 - 25$
3. $y^2 - 4$
4. $y^2 - 16$
5. $25x^2 - 1$
6. $16x^2 - 49$
7. $x^2 - y^2$
8. $s^2 - r^2$
9. $36r^2 - s^2$
10. $x^4 - 36$
11. $x^4 - 64y^4$
12. $9y^4 - 100$
13. $b^4 - a^4$
14. $49a^2 - 4b^2$
15. $1 - x^2y^2$

Exercises 7.9 B

1. $x^2 - 36y^2$
2. $4a^2 - 49b^2$
3. $16 - x^4$
4. $r^2s^2 - 1$
5. $x^2y^2 - z^2$
6. $9x^6 - 64$
7. $81r^4 - s^4$
8. $y^2 - x^2$
9. $b^2y^2 - a^2x^2$
10. $y^4 - x^4$

Exercises 7.10 A

1. $x^2 + 7x + 10$
2. $x^2 + 11x + 28$
3. $x^2 - 9x + 18$
4. $x^2 - 6x + 8$
5. $x^2 + x - 42$
6. $x^2 - x - 72$
7. $x^2 - 4x - 5$
8. $x^2 + 6x - 16$
9. $y^2 + 5y + 6$
10. $y^2 - y - 6$
11. $x^2 - 5xy + 6y^2$
12. $x^2 + xy - 6y^2$
13. $x^2y^2 - 5xy - 6$
14. $x^2y^2 - 2xy - 24$
15. $x^4 - 10x^2 + 16$
16. $x^4 + 5x^2y^2 - 36y^4$
17. $t^4 + t^2 - 30$
18. $t^4 - 11t^2 + 28$
19. $y^4 + y^2 - 20$
20. $y^4 - 8y^2 - 9$

Exercises 7.10 B

1. $x^2 + 14x + 48$
2. $x^2 - 14x + 48$
3. $x^2 + 2x - 48$
4. $x^2 - 2x - 48$
5. $x^2 + 2xy - 8y^2$
6. $4x^2 - 4xy - 15y^2$
7. $x^4 - 3x^2 - 4$
8. $x^4 + 15x^2y^2 - 16y^4$
9. $r^4 - 13r^2s^2 + 36s^4$
10. $4xy - 10x - 10y + 25$
11. $x^2y^2 + 5xyz - 50z^2$
12. $u^6 + u^3v^3 - 2v^6$
13. $x^4 + 3x^2 - 54$
14. $a^2b^2c^2 - abc - 2$
15. $r^2x^2 - 7rsx + 12s^2$

UNIT 7 POSTTEST

1. $-63x^7$
2. $100x^4y^6$
3. $x^4 - x^2 + 18x - 81$
4. $-2x^3y^2 + 5x^2y^3 - 10xy^4$
5. $x^2 - 3$
6. $5x - 7$
7. $x^4 - 25x^2 + 12x + 60$
8. $9x^2 + y^2$
9. $-14x^5 + 2x^3 - 6x^2$
10. $10x^5y^2 - 75x^4y^3 + 20x^3y^4$
11. $20x^2 - 7x - 6$
12. $18xy - 3x - 42y + 7$
13. $81x^2 - 64y^2$
14. $x^3 - 20x + 16$
15. $12x^3 - 8x^2y - 21xy^2 - 5y^3$
16. $16x^2 + 72xy + 81y^2$

17. $36x^2 - 84x + 49$
18. $64x^4 - y^4$
19. $36x^2 - 25$
20. $y^2 + y - 72$
21. $x^4 - 21x^2 - 100$
22. $16x^2 - 38x - 63$
23. $18x^2 - 9xy - 20y^2$
24. $x^3 - 6x^2 + 6x - 36$
25. $xy - 5x - 4y + 20$

UNIT 8 PRETEST

1. $5x^2 (4x + 3)$
2. $12xy (2x - y^2)$
3. $(x + 6)(x - 6)$
4. $(x^2 + 9y)(x^2 - 9y)$
5. $(x + 5)(x + 7)$
6. $(y - 4)(y - 6)$
7. $(x - 4)(x + 7)$
8. $(y - 8)(y + 2)$
9. $(x - 5)^2$
10. not possible
11. not possible
12. $(6x + y)^2$
13. $(x - 6)(3x - 1)$
14. $(2x + y)(3x - 5y)$
15. $4x (x - 5)(x + 5)$
16. $3 (y - 8)(y + 5)$
17. $(x - 4)(x + 4)(x^2 + 1)$
18. $2 (5t - 2)^2$
19. $2y (y + 1)(7y - 6)$

Exercises 8.1 A

1. $6x^2 (2x + 1)$
2. $6y^3 (3y - 2)$
3. $5xy (2 + 3xy)$
4. $7xy (2y - 3x)$
5. $8x (2x^2 + 1)$
6. $-9x^2 (2x^2 + 1)$
7. $8x^2y^2 (2x - 5y)$
8. $9x^2y^3 (6x + y^2)$
9. $-15at^2 (t^2 + 5)$
10. $6c^2s (4s - 7c)$
11. $7 (x + y + z)$
12. $4 (x - y + 1)$
13. $15a (x - 3y - 1)$
14. $4a^2 (x^2 + 2ax - 1)$
15. $-9y^2 (y^2 + y - 3)$
16. $50y (3y^2 - 4y - 5)$
17. $35t^2 (2t^2 - 3t + 4)$
18. $abc (ab + bc + ac)$
19. $r^2s^2t^2 (r^2 - s^2 - t^2)$
20. $24xyz (x^2 + 2y^2 - 3z^2)$

Exercises 8.1 B

1. $10xy (4x - 5y)$
2. $25x^3 (x - 4)$
3. $4x^4 (x^2 + 16)$
4. $-15y^2 (x^2 + y^2)$
5. $6x^2 (7x + 1)$
6. $-8xy^2 (5x - 4y)$

7. $21(x - 2y + z)$
8. $x^2y^2(x^2 - y^2 - 1)$
9. $12(4r - 7s + 2)$
10. $6(x^2 - 3x + 5)$
11. $4x(x^2 + 4x - 1)$
12. $5xy(2x^2 - xy - 2y^2)$
13. $-8y^2(3y^2 - 5y + 4)$
14. $rst(r + s + t)$
15. $-3a^2b^2c^2(a^2 + 2b^2 - c^2)$

Exercises 8.2 A

1. $(x + 5)(x - 5)$
2. $(x + 4)(x - 4)$
3. $(y + 6)(y - 6)$
4. $(y + 7)(y - 7)$
5. $(8x + 1)(8x - 1)$
6. $(3y + 1)(3y - 1)$
7. $(9x + 5)(9x - 5)$
8. $(10y + 7)(10y - 7)$
9. $(5x + 8y)(5x - 8y)$
10. $(2x + 3y)(2x - 3y)$
11. $(4x^2 + 5y)(4x^2 - 5y)$
12. $(6y + x^2)(6y - x^2)$
13. $(xy^2 + z)(xy^2 - z)$
14. $(20ab + c)(20ab - c)$
15. $(x^3 + y)(x^3 - y)$

Exercises 8.2 B

1. $(x + 1)(x - 1)$
2. $(4x + y)(4x - y)$
3. $(6x + 7)(6x - 7)$
4. $(8y + 5)(8y - 5)$
5. $(x + 10y)(x - 10y)$
6. $(3xy + 2)(3xy - 2)$
7. $(9 + k)(9 - k)$
8. $(xy + z^2)(xy - z^2)$
9. $(30a^2 + 1)(30a^2 - 1)$
10. $(a^3 + bc)(a^3 - bc)$
11. $(2 + 9y^3)(2 - 9y^3)$
12. $(c^3 + a^2)(c^3 - a^2)$
13. $(11x^3 + y^2)(11x^3 - y^2)$
14. $(y^3 + 12x^2)(y^3 - 12x^2)$
15. $(x^3 + 13y^3)(x^3 - 13y^3)$

Exercises 8.3 A

1. $(x + 7)(x + 2)$
2. $(x - 2)(x - 7)$
3. $(x + 2)(x - 7)$
4. $(x - 2)(x + 7)$
5. $(x - 1)(x - 14)$
6. $(x + 1)(x - 14)$
7. $(x - 1)(x + 14)$
8. $(x + 6)(x + 5)$
9. $(x - 4)(x - 1)$
10. $(x - 3)(x - 2)$
11. $(x - 7)(x + 3)$
12. $(x - 3)(x + 7)$
13. $(y - 4)(y + 5)$
14. $(y - 5)(y + 4)$
15. $(y - 15)(y + 1)$
16. prime
17. $(x + 13)(x + 2)$
18. $(x + 13)(x - 2)$
19. $(x - 13)(x + 2)$
20. $(t - 13)(t - 2)$
21. $(x + 6y)(x - 2y)$
22. $(x - 6y)(x + 2y)$
23. $(r - 12s)(r + s)$
24. $(s - 12t)(s - t)$
25. prime

Exercises 8.3 B

1. $(x + 4)(x + 2)$
2. $(x - 5)(x - 6)$
3. $(x - 6)(x + 4)$
4. $(x + 4)(x - 2)$
5. prime
6. $(y - 5)(y + 1)$
7. $(y + 5)(y - 1)$
8. $(y - 1)(y - 4)$
9. prime
10. $(x + 2y)(x + 9y)$
11. $(x + 8y)(x - 3y)$
12. $(x - 7y)(x + 5y)$
13. prime
14. $(u - 2v)(u + v)$
15. prime
16. $(u + 8v)(u - 7v)$
17. $(r - 9s)(r + 6s)$
18. $(x^2 - 3y^2)(x^2 + y^2)$
19. $(x^2 - 5)(x^2 - 10)$
20. $(x^2 - 8)^2$

Exercises 8.4 A

1. $(x + 4)^2$
2. $(y + 1)^2$
3. $(y - 2)^2$
4. $(y - 3)^2$
5. not a perfect square
6. not a perfect square
7. $(2x + 5)^2$
8. $(2x - 5)^2$
9. not a perfect square
10. not a perfect square
11. not a perfect square
12. $(x - 8y)^2$
13. $(x + 9y)^2$
14. not a perfect square
15. $(x - 10y)^2$

Exercises 8.4 B

1. $(3x - 7y)^2$
2. $(x^2 - 5)^2$
3. $(y^2 + 2)^2$
4. $(5x^2 + 6y^2)^2$
5. not a perfect square
6. not a perfect square
7. $(5x + 1)^2$
8. $(7y - 1)^2$
9. not a perfect square
10. $(6x - 1)^2$
11. $(10t^2 + 1)^2$
12. $(8t^2 - 1)^2$
13. $(2x - y)^2$
14. not a perfect square
15. $(3x^2 - 1)^2$

Exercises 8.5 A

1. $3x + 4$
2. $2x + 1$
3. $y - 9$
4. $2y - 3$
5. $2x - 3$
6. $x + 2$
7. $x - 10$
8. $5x - 3$
9. $2t + 3$
10. $6t - 7$
11. $(6x - 7)(x + 1)$
12. $(6x + 7)(x - 1)$
13. $(6x - 7)(x - 1)$
14. $(6x + 7)(x + 1)$
15. $(3y + 1)(3y + 2)$
16. $(3y - 1)(3y + 2)$
17. $(3y + 1)(3y - 2)$
18. $(3y - 1)(3y - 2)$
19. $(5x + 1)(x + 2)$
20. $(x - 7)(3x - 1)$
21. $(x - 3)(2x + 1)$

22. $(x + 2)(7x - 1)$
23. $(y - 1)(3y + 5)$
24. $(y + 1)(2y - 5)$
25. $(t - 1)(7t + 5)$
26. prime
27. $(x + y)(11x - 2y)$
28. $(3x - y)(11x + y)$
29. $(x - 2y)(2x - 3y)$
30. $(x - 2y)(2x + 3y)$
31. $(y + 2)(5y - 4)$
32. $(y - 2)(5y + 4)$
33. $(2x - 5)(3x - 1)$
34. prime

Exercises 8.5 B

1. $3r - 2$
2. $6r - 7$
3. $13x + 10y$
4. $2x - y$
5. $5x - 2y$
6. $2x + 3y$
7. $x - 4y$
8. $3x - 2$
9. $3x + 2$
10. $3x - 2$
11. $(4s - 5t)(2s + t)$
12. $(8s + 3t)(5s - 4t)$
13. $(12x - y)(x - 8y)$
14. $(4x - 25y)(5x + 4y)$
15. $(2a + 7b)(5a - 2b)$
16. $(2x + 3y)(15x - 7y)$
17. $(7a - 4b)(3a - 2b)$
18. $(7a - 4b)(3a + 2b)$
19. $(3y + 1)(5y - 2)$
20. $(3y - 1)(5y + 2)$
21. $(3x - y)(5x - 2y)$
22. $(x - y)(15x - 2y)$
23. $(x + 2)(15x - 1)$
24. prime
25. $(2t - 1)(3t + 2)$
26. $(3a + 5)(9a + 1)$
27. $(2b + 1)(5b - 2)$
28. $(x - 5)(8x + 3)$
29. $(2x + 5)(4x - 3)$
30. $(3x - 2y)(5x - 4y)$
31. prime
32. $(3r + 4s)(6r - 7s)$
33. $(3r - 4s)(6r + 7s)$
34. $(2r + 7s)(9r - 4s)$

Exercises 8.6 A

1. e
2. b
3. f
4. h
5. a
6. g
7. d
8. c
9. $(3x - 5)(3x + 5)$
10. $x(9x - 25)$
11. $9(x^2 + 4)$
12. prime
13. $(x - 6)^2$
14. $(3x - 5)^2$
15. $(x - 9)(x + 4)$
16. $(9x + 1)(x + 2)$
17. $9(x^2 + 2x + 3)$
18. $(7x - 1)(7x + 1)$
19. $(x + 8)(x - 4)$
20. $(5x + 3)(x - 1)$
21. $(x^2 - 10)(x^2 + 10)$
22. $x^3(x - 100)$
23. $(x^2 + 10)^2$
24. $x^2(x^2 - x + 1)$
25. $(x^2 - y)(x^2 + 2y)$
26. $(x^2 + 5)(x^2 + 4)$

Exercises 8.6 B

1. e
2. g
3. f
4. h
5. b
6. a
7. d
8. c
9. $(x + 6)(x - 6)$
10. $(x^2 + 6)(x^2 - 6)$
11. $x^2(x^2 + 36)$
12. $(x - 7)^2$
13. $(x^2 - 7)^2$
14. $(x - 2)(x + 8)$
15. $(x^2 - 2)(x^2 + 8)$
16. $(7x - 3)(x + 1)$
17. $(7x^2 - 3)(x^2 + 1)$
18. $(x^2 - 8y)(x^2 + 8y)$
19. $x^2(x^2 + 64)$
20. $(6x^2 + 1)^2$

Exercises 8.7 A

1. $5x(x - 2)(x - 5)$
2. $(2x - 3y)(2x + 3y)(4x^2 + 9y^2)$
3. $4(5x - y)(5x + y)$
4. $4x(25x - 1)$
5. $16y^2(y^2 + 4)$
6. $6x^2(x - 7)(x + 3)$
7. $9x^2(x - 4)^2$
8. $25(x^2 - 2x + 4)$
9. $(4x - 5y)(4x + 5y)(16x^2 + 25y^2)$
10. $15(x + 2)^2$
11. $12(x - 1)(x + 6)$
12. $36x^2(x - 2)(x + 2)$
13. $21x(x^2 - x - 1)$
14. $x(x - 1)(3x - 5)$
15. $3(2x - 1)(7x + 2)$
16. $4(6x - 7)(3x + 1)$
17. $(x - 1)(x + 1)(x - 3)(x + 3)$
18. $4(2x + 1)(2x - 1)(x + 2)(x - 2)$
19. $25(5x - 1)(5x + 1)(x^3 + 1)$
20. $(6x - 1)(6x + 1)(2x^2 + 5)$

Exercises 8.7 B

1. $9(x - 2)(x + 1)$
2. $9y(x + 2)(x - 1)$
3. $x(x - 6)(3x + 1)$
4. $4xy(6x - 5y)(6x + 5y)$
5. $4(3x - y)(3x + y)(9x^2 + y^2)$
6. $9t(3t^2 + 4t + 7)$
7. $x(7 - x)(7 + x)$
8. $y^4(y - 2)(y + 2)(y^2 + 4)$
9. $5(x - 5)(x + 2)$
10. $(2y - 7)^2$
11. $2(x - 2)(3x - 4)$
12. $2c^2(5c - 1)^2$
13. $y(y - 1)(y + 1)(y^2 + 1)$
14. $6(x - 2)(x - 7)$
15. $-3x(x^2 + 9)$
16. $5a(a - 2)(a + 2)(a^2 + 16)$
17. $b^2(3b - 4)(2b + 3)$
18. $x^2(x - 4)(x + 4)$
19. $(x - 2)(x + 2)(x - 3)(x + 3)$
20. $4(a^2 - 2b^2)(a^2 + 2b^2)$
21. $2(xy - 2)(xy + 2)(x^2y^2 + 4)$
22. $t^2(t - 2)(t + 2)(t^2 + 9)$

23. $4x^2 (x - 6)^2$

24. $(x - 5)(x + 5)(x^2 + 1)$

25. $4(2u^2 - 3v^2)(2u^2 + 3v^2)$

26. $-x^2 (x - 5)(x + 1)$

UNIT 8 POSTTEST

1. $3ax^2 (6x + 1)$
2. $9x(x^2 - 2x + 4)$
3. $(7t + 6)(7t - 6)$
4. $(5x^2 + 8y)(5x^2 - 8y)$
5. $(x + 3)(x + 4)$
6. $(x - 4y)(x - 5y)$
7. $(y + 9)(y - 3)$
8. $(x - 5y)(x + 3y)$
9. $(x - 7)^2$
10. not possible
11. $(y^2 + 8)^2$
12. not possible
13. $(x - 1)(2x - 1)$
14. $(x + 2)(10x - 7)$
15. $(x + 2)(x - 2)(x^2 + 4)$
16. $3(x - 7)(x + 5)$
17. $10x(6x - 5)^2$
18. $x^2 (x - 5)(2x + 3)$
19. $(y + 3)^2 (y - 3)^2$

UNIT 9 PRETEST

1. $\frac{3x}{2y}$
2. $\frac{1}{2}$
3. $\frac{y + 3}{y - 3}$
4. $\frac{x - 2}{x + 3}$
5. $\frac{4x}{x - 2}$
6. $\frac{-2y^2}{7x}$
7. -1
8. $\frac{-x}{x + 1}$
9. $40xy^2$
10. -1
11. $3x^2 + 9x$
12. $x^2 - x - 30$
13. $\frac{11}{30x}$
14. $\frac{n - 9}{20(n + 2)}$
15. $\frac{y^2 + 15y + 25}{5y(y + 5)}$
16. $\frac{2x}{(x - 1)(x + 1)(x - 5)}$
17. $\frac{3x}{7y}$
18. $y - 2$
19. $\frac{2rt}{5x}$
20. $\frac{2y}{y + 4}$
21. $\frac{1}{x - 2}$
22. $\frac{x - 2}{x - 5}$
23. $x^2 - 4x + 5 + \frac{7}{x - 3}$
24. $x^2 - 5x - 2$

Exercises 9.1 A

1. $\frac{2}{3}$
2. $\frac{7}{8}$
3. $\frac{1}{9}$
4. $\frac{1}{25}$
5. $\frac{3}{5}$
6. $\frac{1}{8}$
7. $\frac{2}{3x}$
8. $\frac{4}{9xy}$

9. $4x$

10. $\frac{x - 2}{x + 1}$

11. $\frac{x^2 - 1}{x^2 + 1}$

12. $\frac{x}{x - 5}$

13. $\frac{x + 1}{x + 2}$

14. $\frac{4x}{x + 1}$

15. $\frac{x + 7}{x + 4}$

16. $\frac{2x + 1}{x + 2}$

17. $\frac{x - 7}{x + 7}$

18. $\frac{2y + 3}{2y - 3}$

Exercises 9.1 B

1. $\frac{3}{4}$

2. $\frac{1}{4}$

3. $\frac{9}{16}$

4. $\frac{1}{4}$

5. $\frac{4}{5}$

6. $\frac{2}{125}$

7. $\frac{2x}{3y}$

8. $\frac{2x}{3}$

9. $\frac{1}{5y}$

10. $\frac{4}{x - 2}$

11. $\frac{x^2 + 9}{2(x^2 - 9)}$

12. $\frac{x + 6}{x}$

13. $\frac{x - 2}{x - 3}$

14. $\frac{x + 3}{x}$

15. $\frac{x - 5}{x - 11}$

16. $\frac{x + 3}{x - 3}$

17. $\frac{y + 6}{y - 6}$

18. $\frac{5x + 2y}{5x - 2y}$

Exercises 9.2 A

1. $\frac{-1}{3x}$

2. $\frac{-2x}{9y}$

3. $\frac{-1}{3}$

4. -1

5. -1

6. -1

7. $\frac{-3}{x}$

8. -1

9. 1

10. 1

Exercises 9.2 B

1. $\frac{-2}{5x}$

2. $\frac{-2}{3x}$

3. $\frac{-3y^2}{5x^2}$

4. -1

5. -1

6. -1

7. $\frac{-5}{x}$

8. -1

9. 1

10. $\frac{b - c}{a - c}$

Exercises 9.3 A

1. 15
2. 28
3. 60
4. 40
5. 25
6. 64
7. 16
8. $10x$
9. $5x$
10. $-4x$
11. $-x$
12. -1
13. -4
14. $2x + 8$
15. $6x - 18$
16. $4x - 20$
17. $10x - 40$
18. $x^2 + 2x + 1$
19. $x^2 - 16$
20. $x^2 + 3x - 4$
21. $5x^2 + 26x + 5$
22. $x(x + 8)^2$

Exercises 9.3 B

1. 12
2. 20
3. 48
4. 35
5. 125
6. 8
7. 16
8. $63x^2$
9. 10
10. $-8y$
11. $-y$
12. -2
13. -1
14. $5x + 30$
15. $3x + 12$
16. $2x + 14$
17. $40x + 80$
18. $x^2 - 10x + 25$
19. $x^2 - 9$
20. $x^2 - 4x + 4$
21. $4x^2 + x - 3$
22. $5(x - 9)^2$

Exercises 9.4 A

1. $\frac{4}{9}$
2. $\frac{1}{6}$
3. $\frac{41}{24}$
4. $\frac{1}{6}$
5. $\frac{5}{3}$
6. 1
7. $\frac{-3}{35}$
8. $\frac{1}{28}$
9. $\frac{2}{x}$
10. $\frac{x + 6}{x + 3}$
11. 1
12. $\frac{-6}{x - 9}$
13. 0
14. $\frac{5x + 6}{3x^2}$
15. $\frac{y^2 - 5}{y^2}$
16. $\frac{21 - 2x}{9x^2}$
17. $\frac{5x + 8}{x^2 - 49}$
18. $\frac{-x}{(x + 4)^2}$
19. $\frac{2x^2}{(x - 7)(x + 7)}$
20. $\frac{-8}{(x + 4)^2 (x - 4)}$
21. $\frac{2x + 2}{x(x + 2)}$
22. $\frac{4}{15(y - 3)}$
23. $\frac{-27}{(t + 1)(t - 1)}$
24. $\frac{1}{6}$
25. $\frac{6x}{(x - 6)(x - 6)(x + 6)}$

Exercises 9.4 B

1. $\frac{2}{3}$
2. $\frac{-1}{14}$
3. $\frac{3}{16}$
4. $\frac{7}{12}$
5. $\frac{7}{5}$
6. $\frac{1}{3}$
7. $\frac{-1}{12}$
8. $\frac{1}{10}$
9. $\frac{1}{3x}$
10. $\frac{x+5}{x-5}$
11. 1
12. $\frac{6}{y-6}$
13. $\frac{2}{x-10}$
14. $\frac{2y+4}{y}$
15. $\frac{8x-15}{10x^2}$
16. $\frac{-1}{24x^2}$
17. $\frac{2x}{x^2-25}$
18. $\frac{-2x+15}{(x-6)^2}$
19. $\frac{50}{(x-5)(x+5)}$
20. $\frac{60x}{(5x+3)(5x-3)^2}$
21. $\frac{y+10}{5(y+5)}$
22. $\frac{1}{3y}$
23. $\frac{4}{4k^2-1}$
24. $\frac{n-6}{2n-6}$
25. $\frac{-8}{(2x+1)(2x-1)(2x-1)}$

Exercises 9.5 A

1. $\frac{1}{5}$
2. $\frac{-2}{7}$
3. 12
4. $\frac{1}{2}$
5. -10
6. 0
7. $\frac{2x^2}{7}$
8. $\frac{4y}{3}$
9. $-14x$
10. $\frac{1}{7a}$
11. $\frac{3}{t+3}$
12. $\frac{8}{2x+3y}$
13. $\frac{x^2}{3(x-7)}$
14. $\frac{2(x+1)}{x^2(x+2)}$
15. -1
16. $\frac{t^2-4}{t^2-1}$

Exercises 9.5 B

1. $\frac{8}{15}$
2. $\frac{-7}{33}$
3. 27
4. 2
5. $\frac{-9}{2}$
6. 1
7. $\frac{y^2}{4}$
8. 1

9. $4x$

10. $x - y$

11. $\frac{1}{2(x+8)}$

12. $\frac{7(x-10)}{9}$

13. $\frac{4(x+3y)}{3(x-2y)}$

14. $\frac{y(x-1)}{x(y+1)}$

15. $\frac{b-3}{b+2}$

16. $\frac{3r+s}{4(r-3s)}$

Exercises 9.6 A

1. $\frac{2y}{5x}$

2. $\frac{4x}{3}$

3. $\frac{t^3}{3}$

4. $\frac{3}{50bc^4}$

5. $\frac{(x+y)^2}{4(x-y)^2}$

6. 1

7. $(5a+6b)^2$

8. $\frac{-2}{3x}$

9. $\frac{a-1}{a+1}$

10. $\frac{x(x+2y)}{5y(x-2y)}$

Exercises 9.6 B

1. $\frac{y}{3}$

2. $5x$

3. $\frac{5s^3}{6}$

4. $\frac{1}{25t^2}$

5. $\frac{2(a+2b)}{a-2b}$

6. -1

7. $\frac{1}{(7r-2s)^2}$

8. $\frac{2(y-6)}{y(y+6)}$

9. $\frac{2(t+5)}{5(t-5)}$

10. $\frac{b^3}{a^3}$

Exercises 9.7 A

1. 2

2. $\frac{11}{8}$

3. $\frac{x-1}{x^2}$

4. $2(x-5)$, or $2x-10$

5. $\frac{5}{12}$

6. $\frac{10x^2}{x+5}$

7. $\frac{3x}{x+3}$

8. $\frac{1}{4}$

9. $\frac{t^2+5t}{t^2+25}$

10. $\frac{x+1}{x-1}$

11. 48 m.p.h.

12. $\frac{x-3}{3x+1}$

13. $\frac{1}{r(r+s)} + \frac{1}{s(r+s)} = \frac{s+r}{rs(r+s)} = \frac{1}{rs} = \frac{1}{c}$; $c = rs$

Exercises 9.7 B

1. $\frac{2x}{27y}$
2. $\frac{y-x}{2y}$
3. $7x+49$
4. $\frac{x-1}{x+1}$
5. $\frac{5}{4}$
6. $\frac{32}{y-4}$
7. $\frac{x^2+25}{x^2-25}$
8. $\frac{2xy}{x+y}$
9. $\frac{t}{7-t}$ or $\frac{-t}{t-7}$
10. $\frac{x-4}{x+4}$

11a. $\frac{40x}{x+40}$

11b. 15 ohms

12a. $\frac{-1}{x(x+h)}$

12b. $\frac{-1}{x^2}$

13. $\frac{k(b^2-a^2)}{(a^2+b^2)^2}$

Exercises 9.8 A

1. $x-9+\frac{28}{x+2}$
2. $y-1$
3. $3x-1+\frac{-5}{2x-1}$
4. $4x-3$
5. $2y^2+2y-3+\frac{-2}{y+1}$
6. x^2+4x+6
7. $y^2-5y+25$
8. $2x-3+\frac{18}{2x+3}$

Exercises 9.8 B

1. $x+1+\frac{-11}{x+3}$
2. $y-5$
3. $2x+\frac{-9}{4x+3}$
4. $x-4$
5. $4y^2+3y-1+\frac{-8}{2y-1}$
6. x^2+3x-6
7. y^2+3y+9
8. $5x-4y+\frac{32y^2}{5x+4y}$

UNIT 9 POSTTEST

1. $\frac{3}{8x}$
2. $\frac{4}{5y}$
3. $x+2$
4. $\frac{x+4}{x-5}$
5. $\frac{2x+7}{2x}$
6. $\frac{-3x}{2y}$
7. -1
8. $\frac{-x-2}{4}$
9. $40x^3y$
10. -1
11. $6t^3+36t^2$
12. x^2+x-12
13. $\frac{8x+15}{x(x+3)}$

14. $\dfrac{n + 24}{(n - 6)(n + 4)}$

15. $\dfrac{x^2 - 10}{(x + 3)^2}$

16. $\dfrac{-1}{2x - 1}$

17. $\dfrac{4x}{5}$

18. $\dfrac{2x}{x + 3}$

19. $\dfrac{y^3 (x - 4)}{x (x + 2)}$

20. $\dfrac{x (x + 4)}{3x + 2}$

21. $\dfrac{x - 1}{x + 1}$

22. $\dfrac{xy}{x + y}$

23. $4x^2 - 20x + 25$

24. $x^3 - 2x^2 + 6x - 21 + \dfrac{60}{x + 3}$

UNIT 10 PRETEST

1. −4
2. 3
3. ∅
4. 4
5. 2
6. −1
7. 8
8. 0
9. ∅
10. all values of x except 4 and −3
11. $62\frac{1}{2}$ m.p.h.
12. 45 ft. by 72 ft.
13. 40 m.p.h.
14. 15 min.

Exercises 10.1 A

1. 9
2. 8
3. 24
4. 30
5. 7
6. −4
7. all values of x except 0
8. 1
9. ∅
10. 6
11. ∅
12. 3
13. 16

Exercises 10.1 B

1. 14
2. 10
3. 40
4. −72
5. −2
6. 1
7. $\dfrac{-1}{2}$
8. ∅
9. −21
10. −9
11. 20
12. −8
13. $\dfrac{5}{6}$

Exercises 10.2 A

1. 10
2. −3
3. 1
4. ∅
5. 9
6. 2
7. ∅
8. 8
9. 2
10. 2

Exercises 10.2 B

1. 24
2. 10
3. ∅
4. 7
5. −3
6. −8
7. $\frac{1}{5}$
8. $\frac{-3}{2}$
9. −5
10. 1

Exercises 10.3 A

1. $7\frac{1}{2}$ inches, $10\frac{1}{2}$ inches
2. 40°, 60°, 80°
3. $21,250, $3,750
4. 16 inches
5. 1200 voters under 21

6a. 66 ft. per sec.

6b. 18,000 m.p.h.

7. about 3.21 meters
8. 4.77 gal.
9. 0.28 mole
10. $563.20

Exercises 10.3 B

1. 24 in., 24 in., 36 in.
2. 50 lb. pigment, 250 lb. vehicle
3. 8 A's
4. $50
5. large eggs

6a. 220 volts

6b. 2.75 amps

7. 8 ft.
8. 40.45 cm.
9. $52\frac{1}{2}$ palas
10. 33 days

Exercises 10.4 A

1. $3\frac{1}{2}$ hr.
2. 4 m.p.h.
3. 48 m.p.h., 32 m.p.h.
4. 55 m.p.h., 30 m.p.h.
5. 20 m.p.h.
6. 40 m.p.h.
7. 3 hr.

Exercises 10.4 B

1. 36 m.p.h.
2. 30 m.p.h., 75 m.p.h.
3. 15 m.p.h.
4. $2\frac{1}{2}$ hr.
5. 4 hr.
6. 175 m.p.h., 525 m.p.h.
7. 12 m.p.h.

Work Problems: Questions on Basic Ideas

1. $\frac{1}{40}$
2. $\frac{1}{5}$
3. $\frac{1}{x}$
4. $\frac{1}{x+20}$
5. $\frac{6}{10}=\frac{3}{5}$
6. $\frac{2}{x}$
7. $\frac{x}{36}+\frac{x}{45}$
8. $1-\frac{1}{4}=\frac{3}{4}$
9. $1-\frac{1}{3}=\frac{2}{3}$
10. 1

Exercises 10.5 A

1. 21 hr.
2. 40 days
3. 5 min.
4. 90 min.
5. 8 days

Exercises 10.5 B

1. 15 min.
2. 50 min.
3. 22 hr., 44 hr., 66 hr.
4. $4\frac{1}{2}$ hr.
5. 30 hr.

UNIT 10 POSTTEST

1. 20
2. 6
3. 5
4. $\emptyset$
5. -1
6. 2
7. $\emptyset$
8. any number except 4 and -3
9. 10
10. 9
11. 25 francs
12. $87\frac{1}{2}$ gm. sodium, $57\frac{1}{2}$ gm. chlorine
13. 40 m.p.h., 60 m.p.h.
14. 36 min., 180 min.

UNIT 11 PRETEST

1.

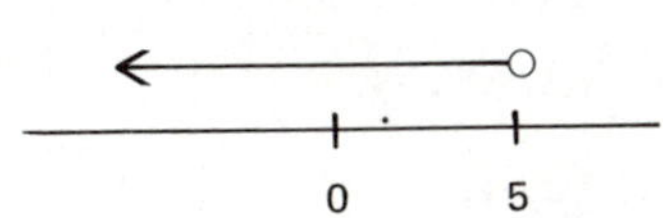

2.

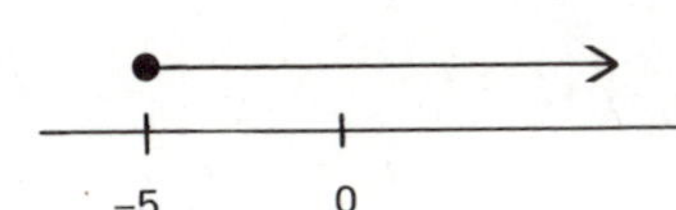

3.

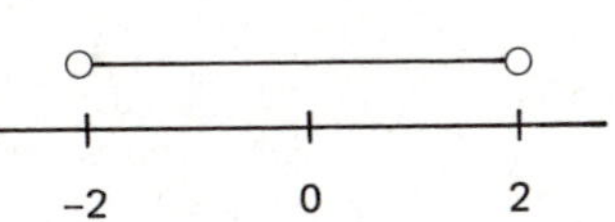

4.

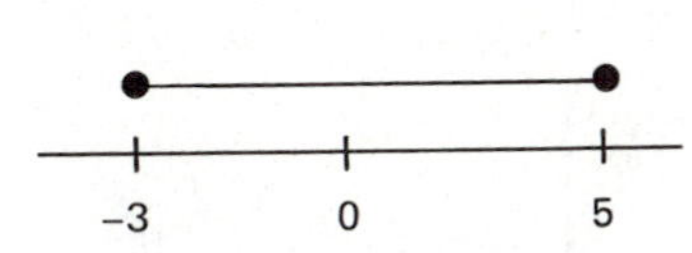

5.

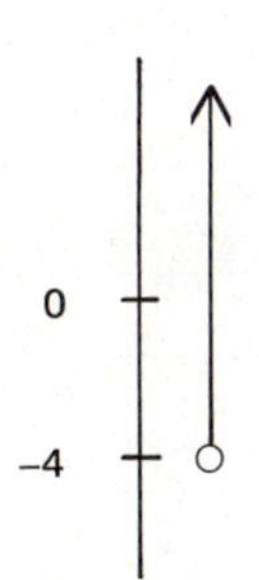

6.

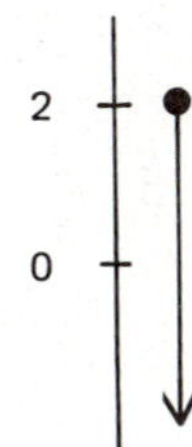

7.

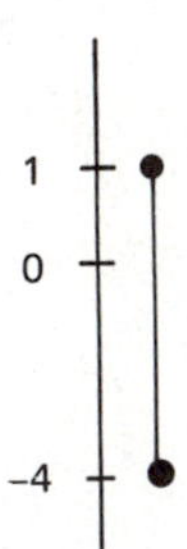

8.

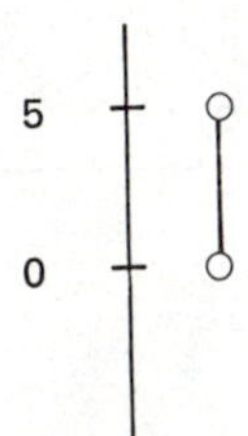

(9–13)

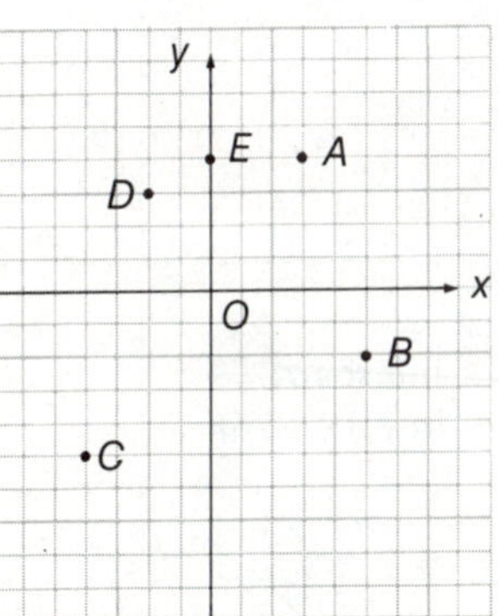

14. $P(-5, 2)$
15. $Q(5, -4)$
16. $R(5, 6)$
17. $S(6, 0)$
18. $T(-3, -4)$
19. yes
20. no
21. yes
22. no
23. yes
24. yes
25. (3, 5)
26. (2, 4)

27.

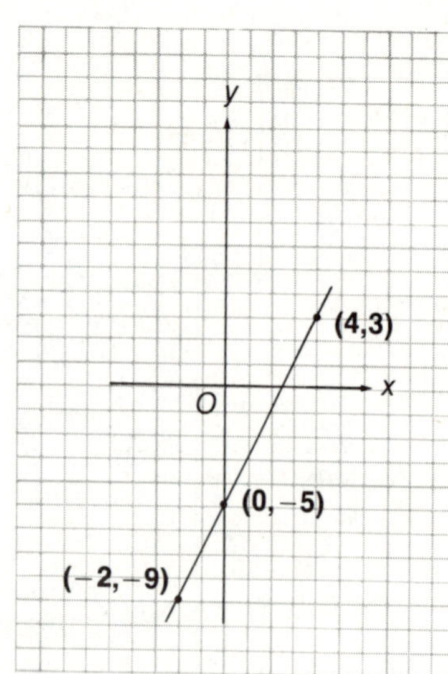

28.

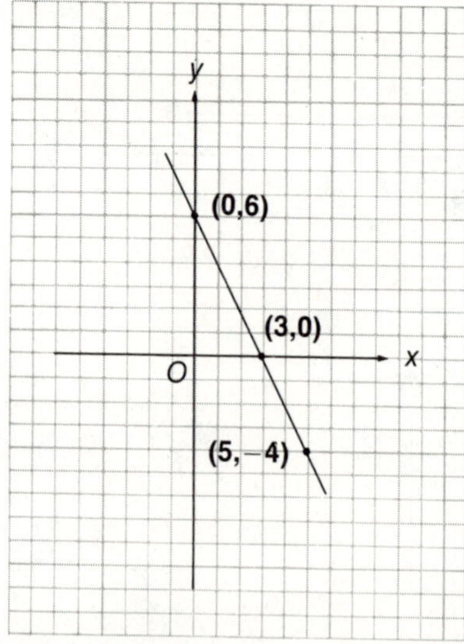

29.

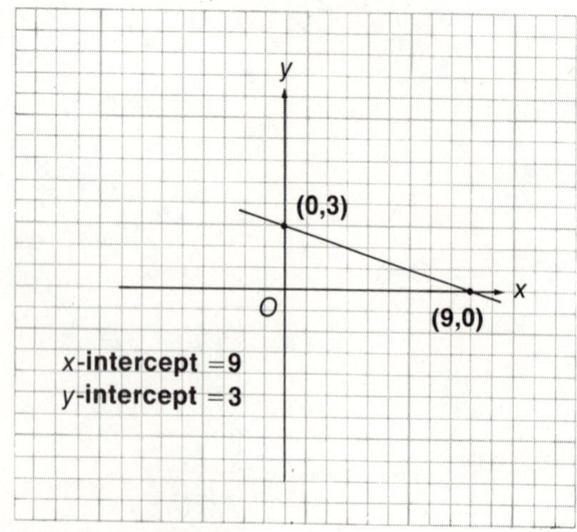

30.

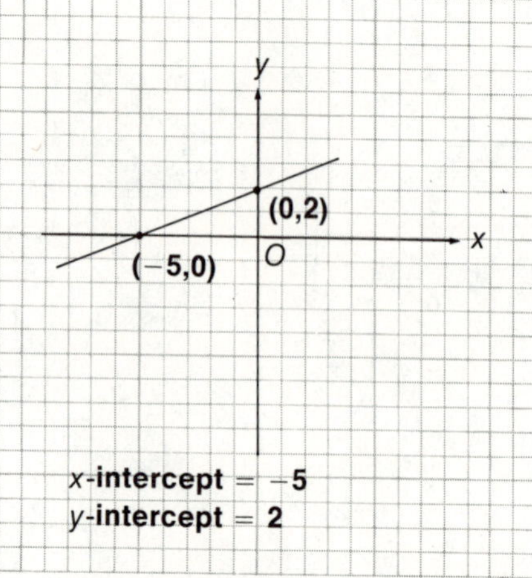

31. slope $= \frac{4}{3}$

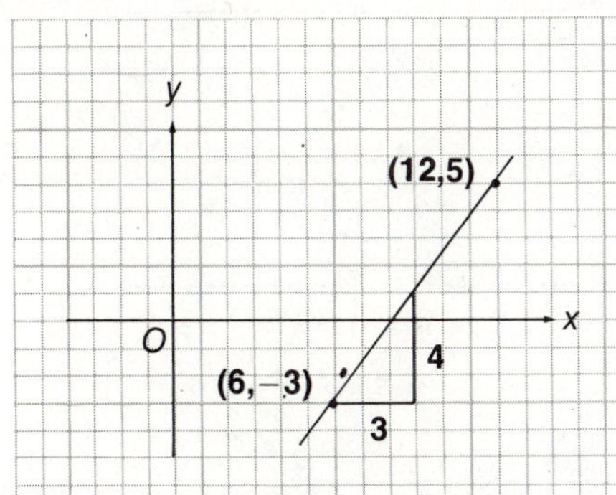

32. slope $= -3$

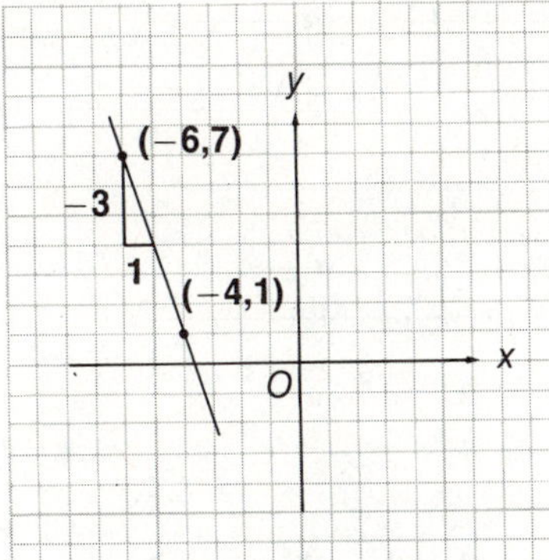

33. slope $= 4$; y-intercept $= -8$

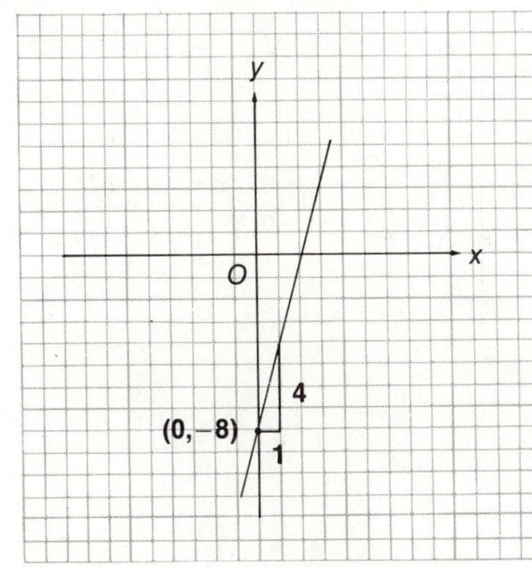

34. slope $= \frac{-5}{3}$; y-intercept $= 5$

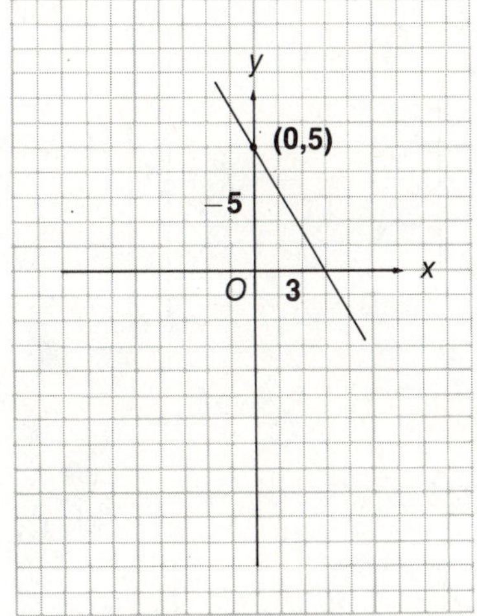

Exercises 11.1 A

1.

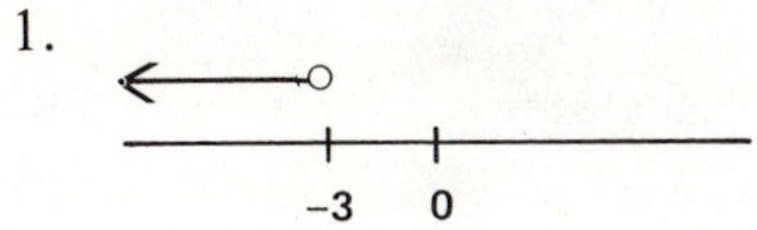

2.

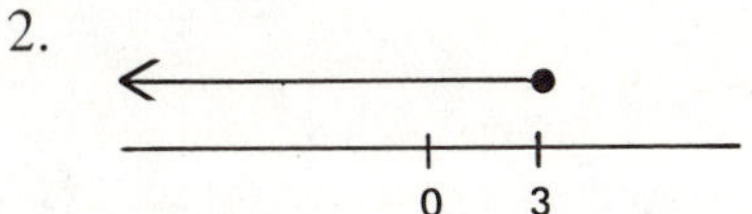

3.

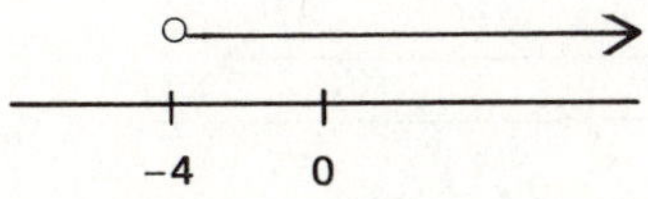

4.

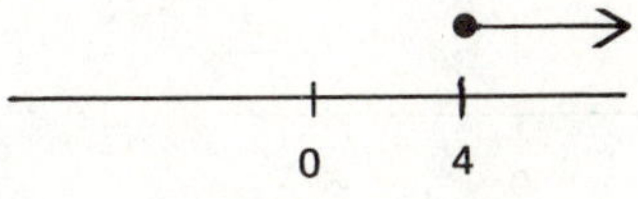

5.

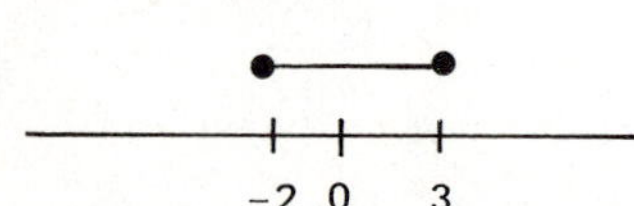

6.

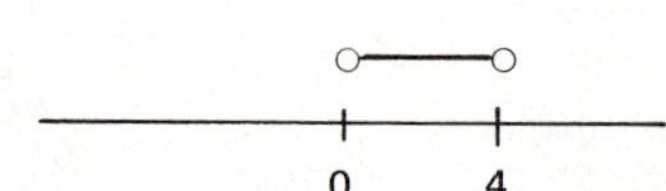

7.

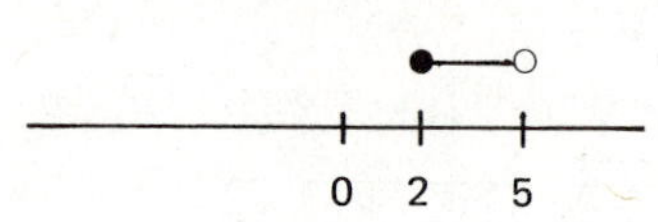

8.

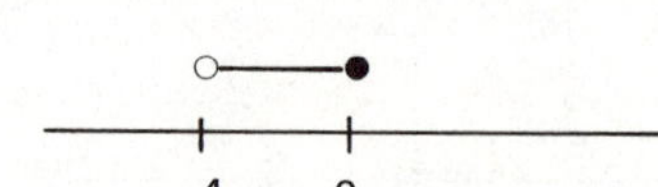

9.

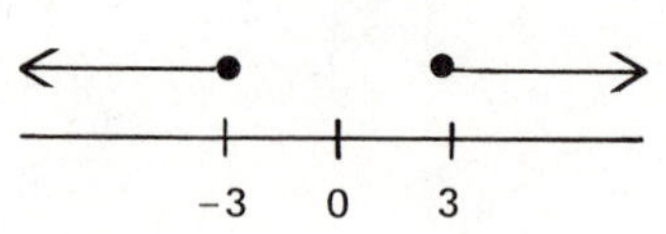

10.

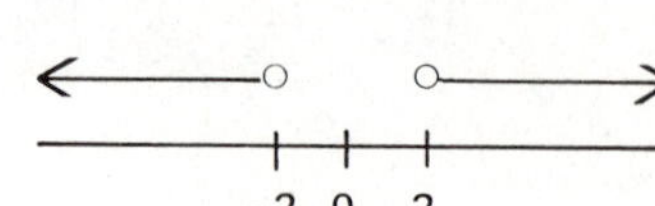

11.

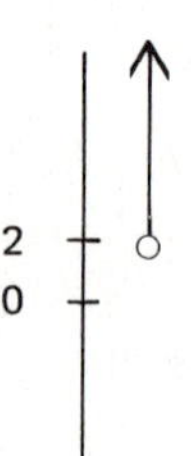

12.

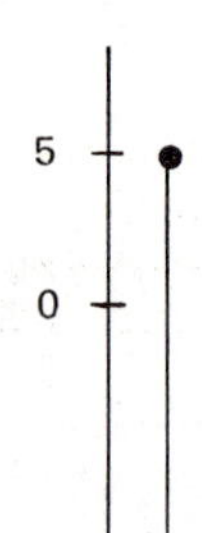

13.

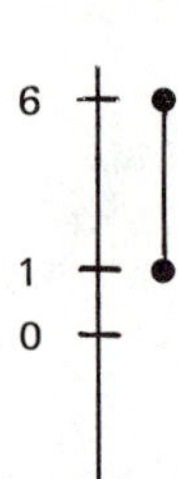

14.

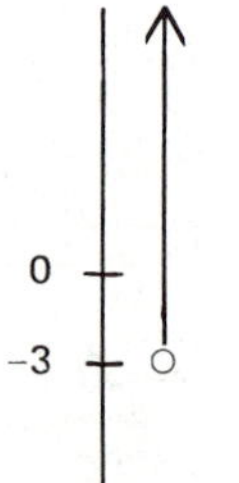

15.

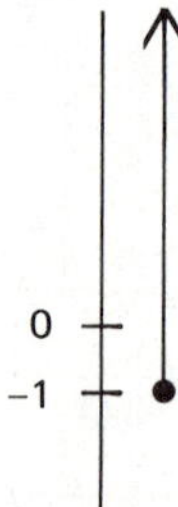

16.

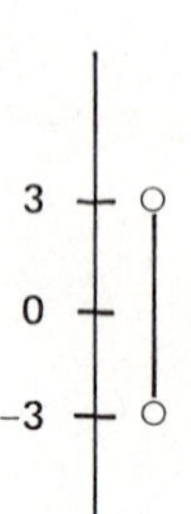

17.

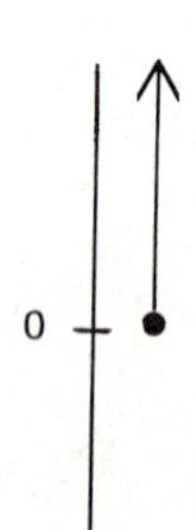

18.

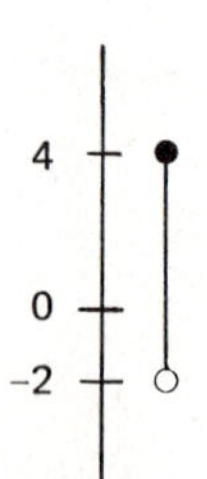

19.

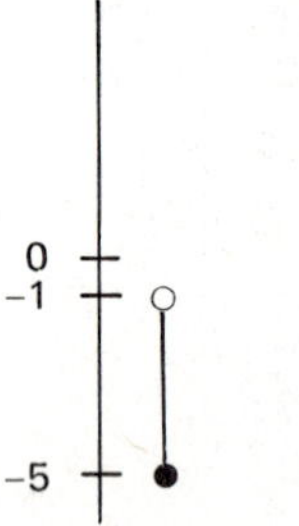

20.

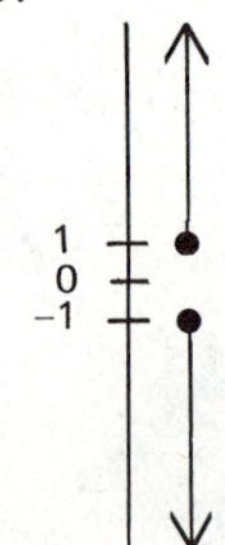

Exercises 11.1 B

1.

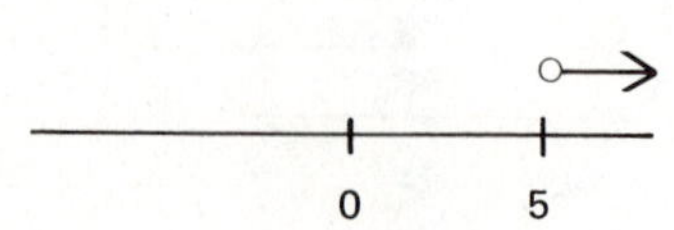

2.

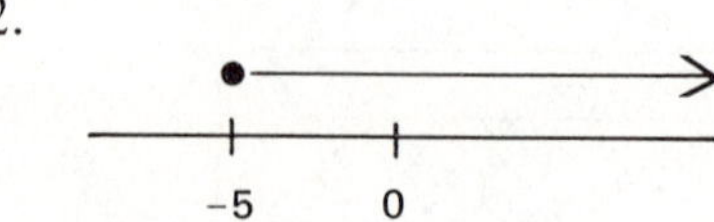

3.

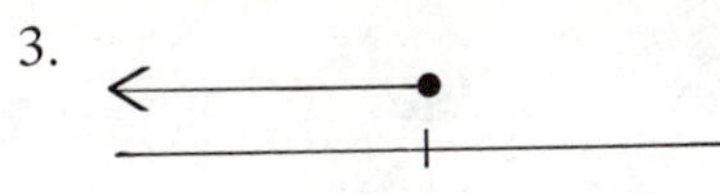

4.

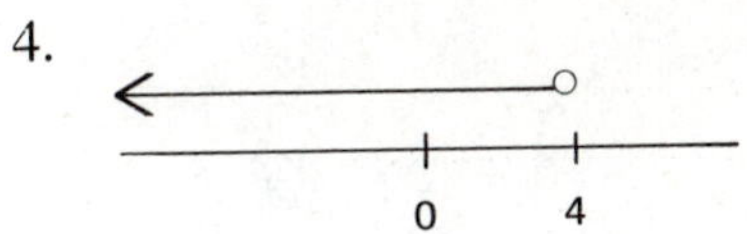

5.

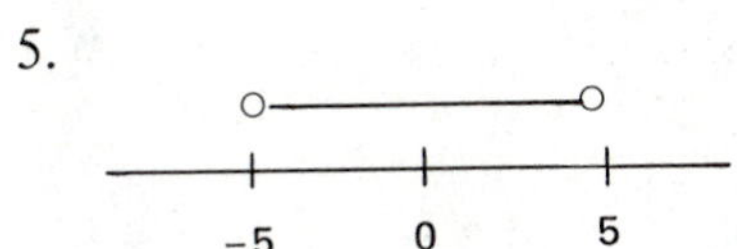

6.

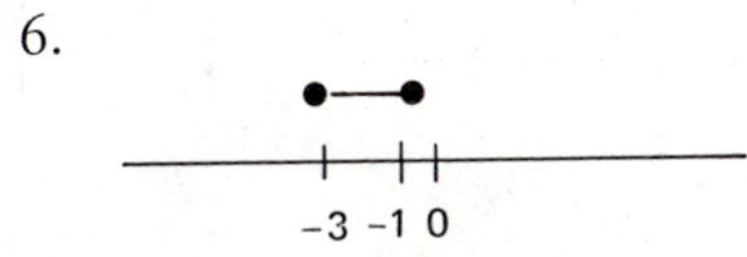

7.

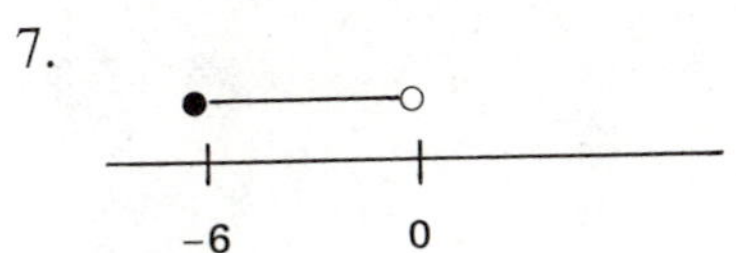

8.

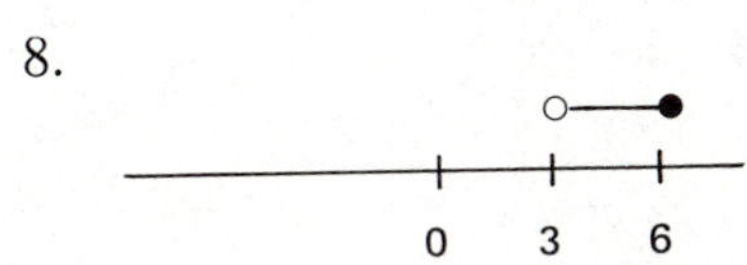

9.

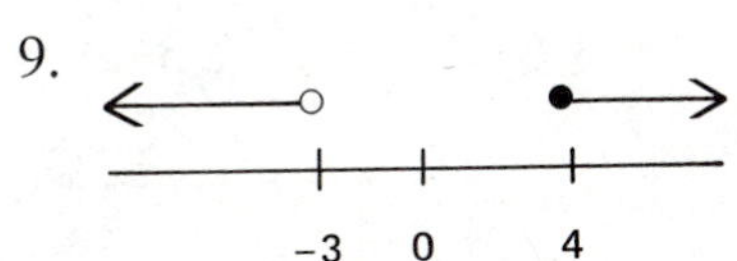

10.

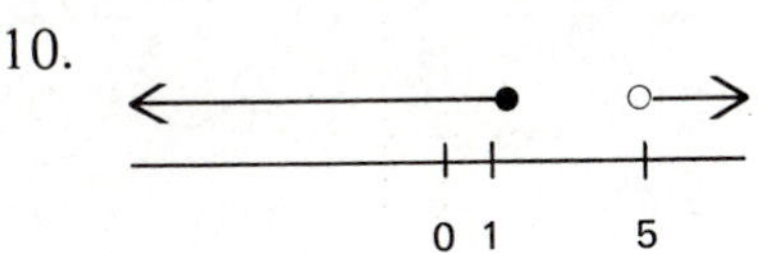

11. 12.

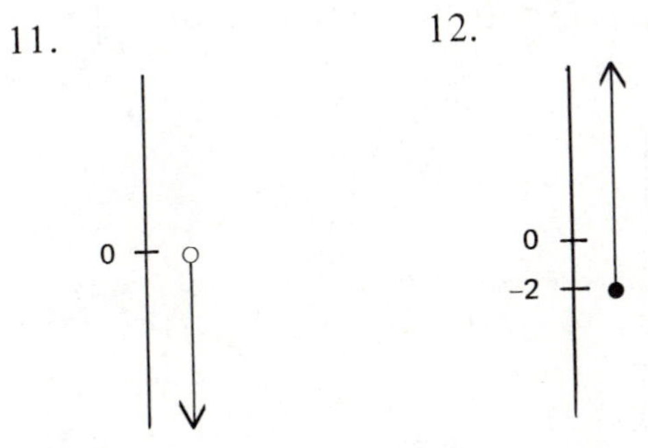

13.

0
−4

14.

3
0
−3

15. 16.

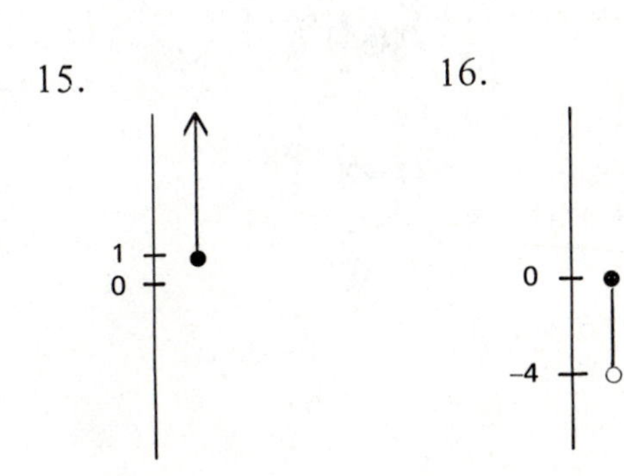

17.

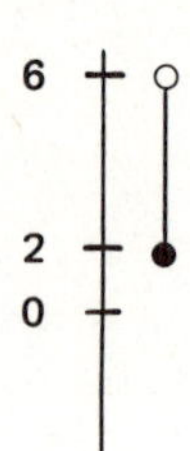

18.

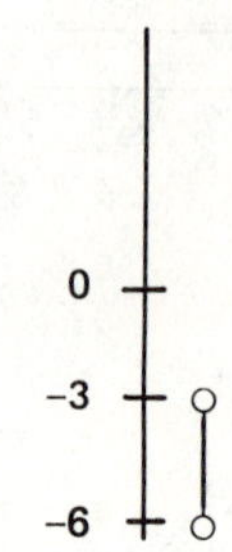

19.

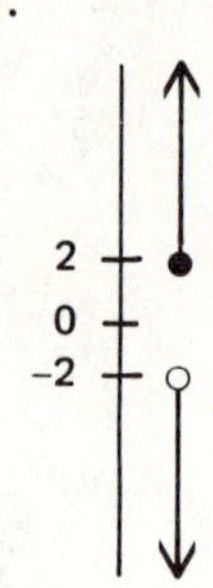

20.

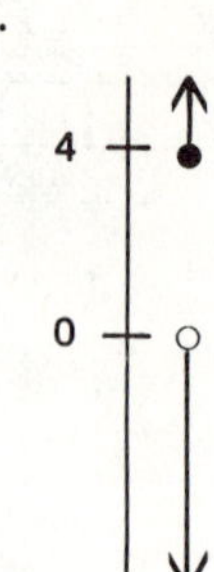

Exercises 11.2 A

(1–8)

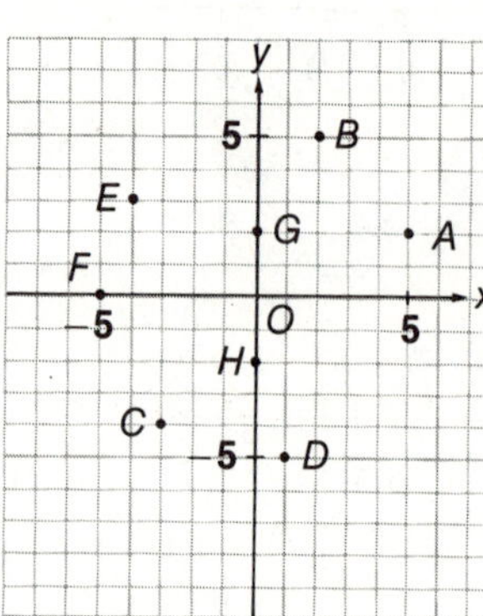

9. A (5, 3)
10. B (-4, 4)
11. C (-3, 0)
12. D (0, -2)
13. E (-5, -4)
14. F (4, -2)

15.

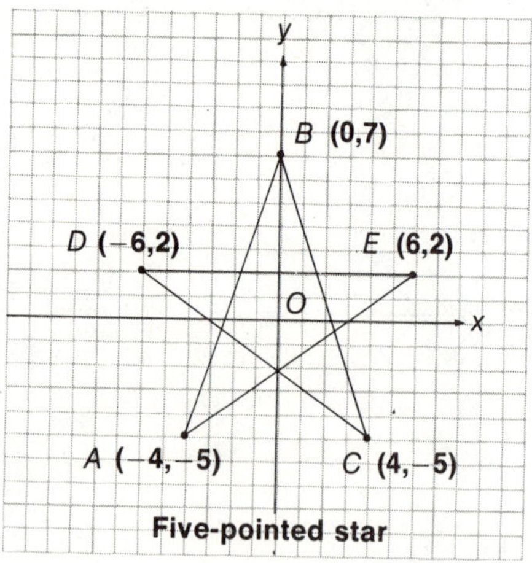

Exercises 11.2 B

(1–8)

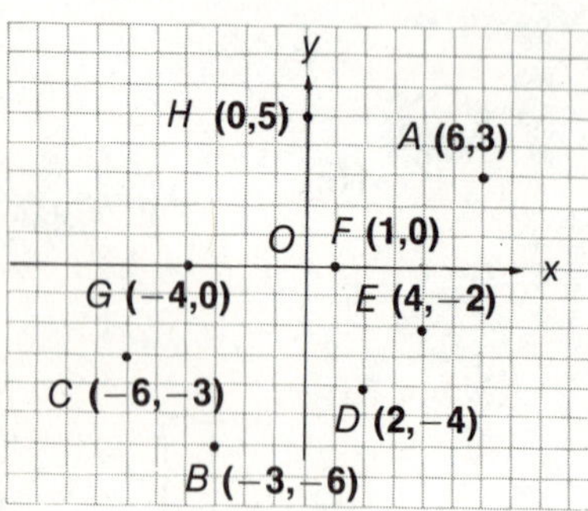

9. P (0, 4)
10. Q (0, 0)
11. R (4, -3)
12. S (-5, 0)
13. T (0, -4)
14. U (-3, -1)

15.

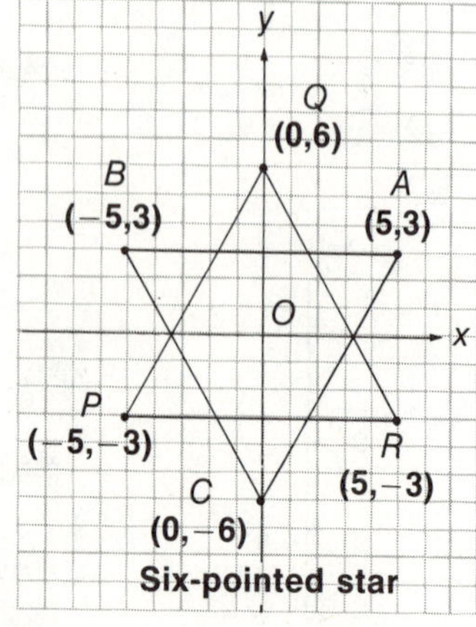

Exercises 11.3 A

1. yes
2. no
3. yes
4. yes
5. yes
6. no
7. no
8. yes
9. yes
10. no
11. yes
12. yes
13. no
14. yes
15. no
16. no
17. no
18. yes
19. yes
20. no
21. yes
22. yes
23. yes
24. yes
25. yes
26. yes
27. no
28. no
29. yes
30. yes

Exercises 11.3 B

1. yes, no, yes, no
2. yes, yes, yes, no
3. no, yes, yes, no
4. yes, no, no, yes
5. no, yes, no, yes
6. no, yes, yes, yes
7. yes, yes, yes, no
8. yes, yes, no, yes
9. yes, yes, yes, yes
10. yes, no, no, yes

Exercises 11.4 A

1. $(7, 5)$
2. $(14, -2)$
3. $(2, -3)$
4. $(-4, -9)$
5. $(1, 6)$
6. $(-2, 13)$
7. $(-1, -8)$
8. $(2, -2)$
9. $(0, -2)$
10. $(5, 0)$

Exercises 11.4 B

1. $(0, -3)$
2. $(20, -2)$
3. $(4, 1)$
4. $(3, 3)$
5. $(5, 24)$
6. $\left(\frac{-4}{3}, 5\right)$
7. $(1, -2)$
8. $\left(\frac{-5}{2}, 0\right)$
9. $(0, -2)$
10. $\left(\frac{11}{2}, 1\right)$

Exercises 11.5 A

1. $y = x + 4$

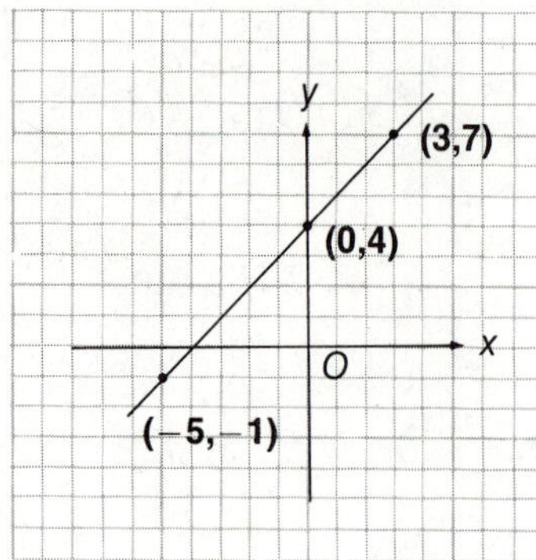

2. $y = 2x + 3$

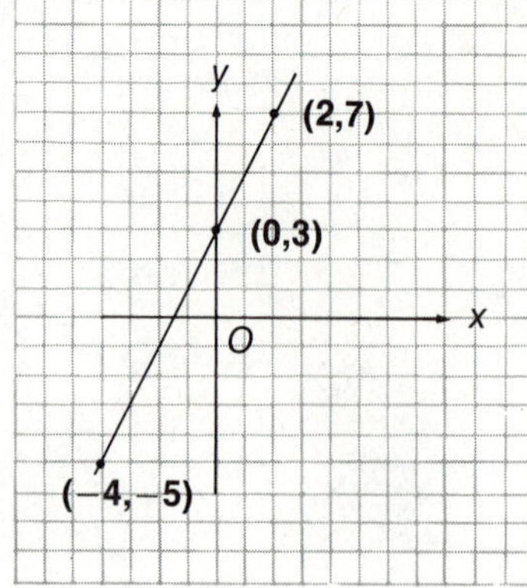

3. $y = 5 - x$

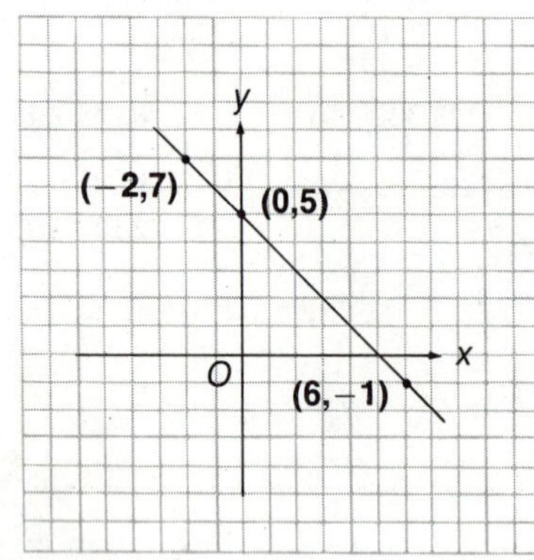

4. $y = 2x$

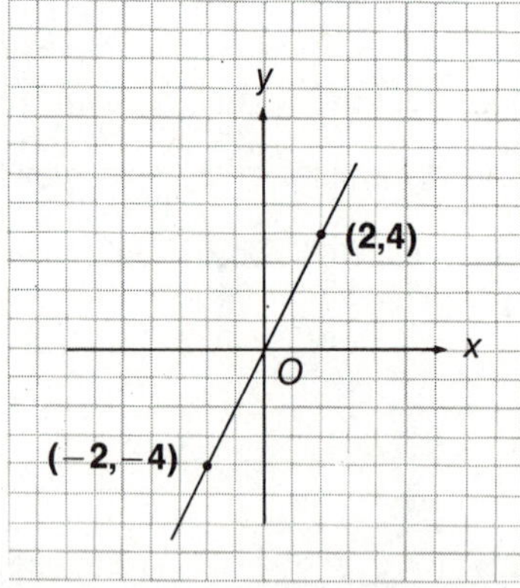

5. $y = \frac{-1}{2}x$

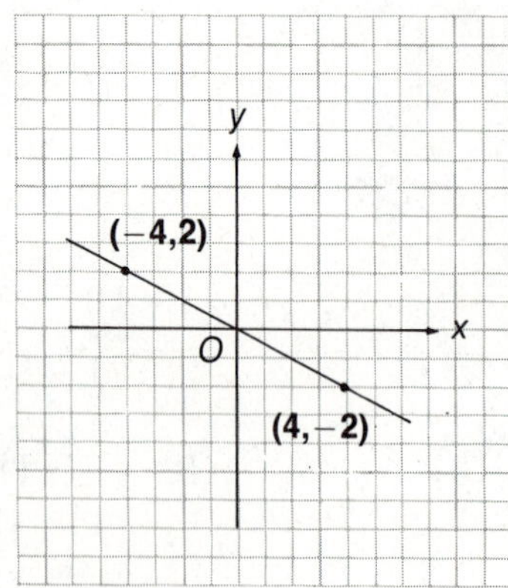

6. $y = 4 - x$

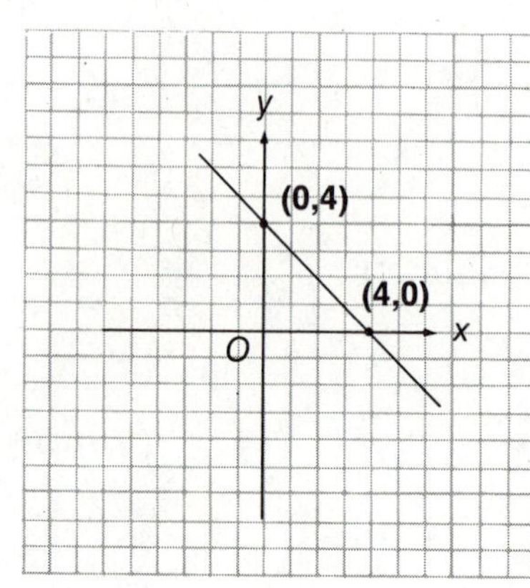

7. $y = 3x + 12$

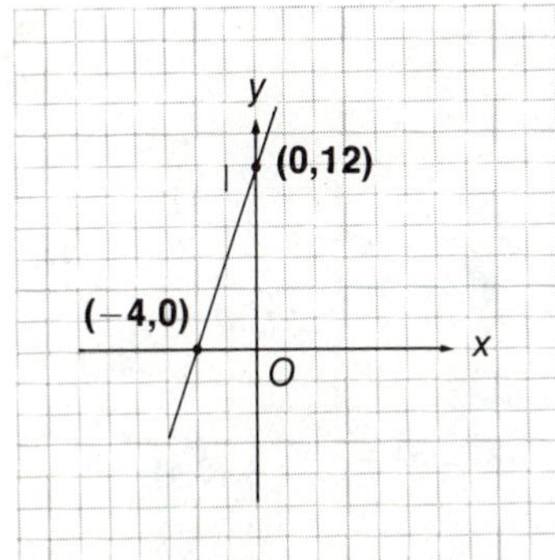

8. $y = 6 - 3x$

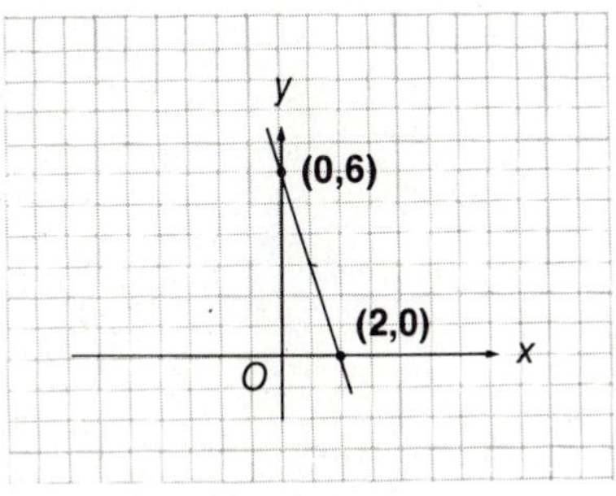

9. $c = \dfrac{5\,(f - 32)}{9}$

10. $y = 12 - \dfrac{3x}{4}$

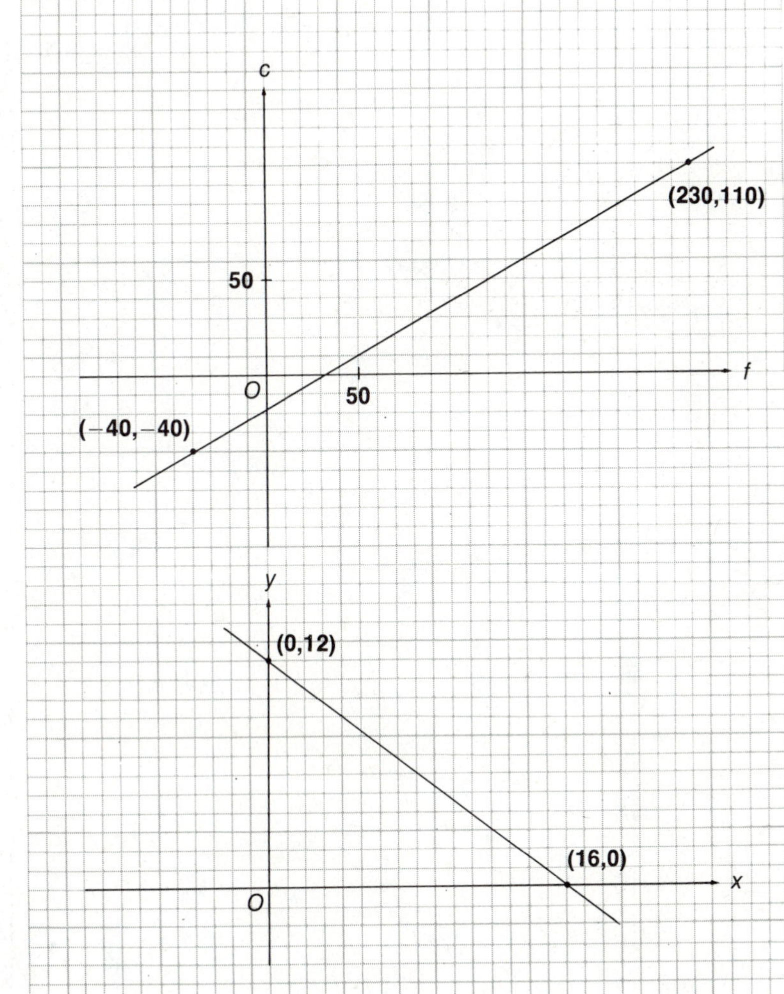

Exercises 11.5 B

1. $y = 3x - 12$

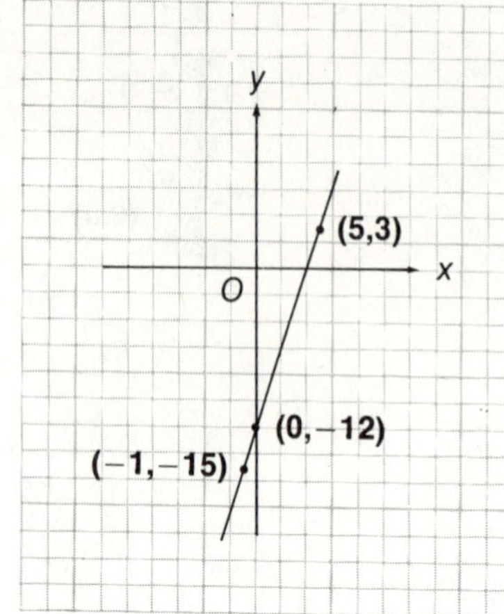

2. $y = 4 - 2x$

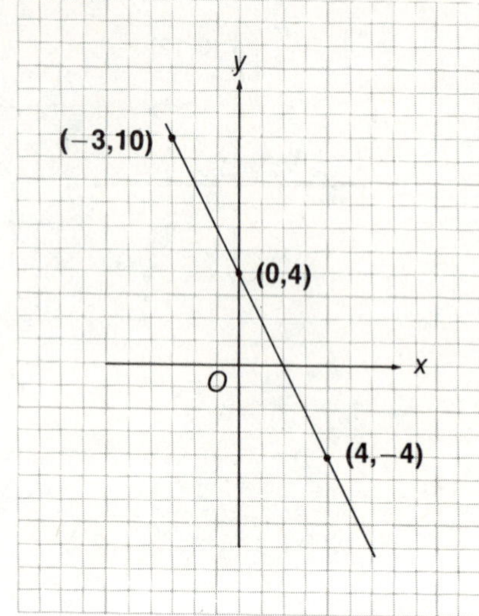

3. $y = \frac{-x}{3}$

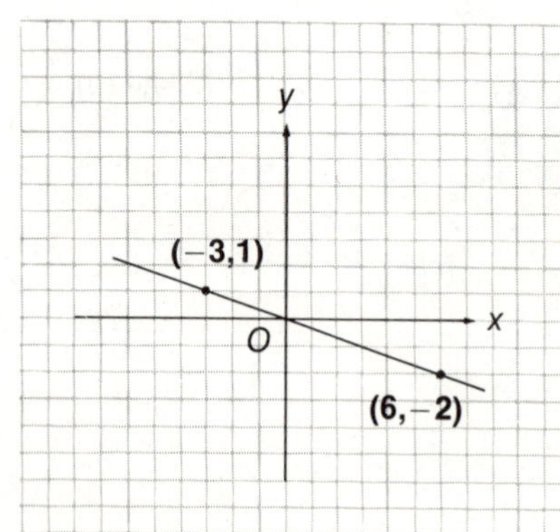

4. $y = x$

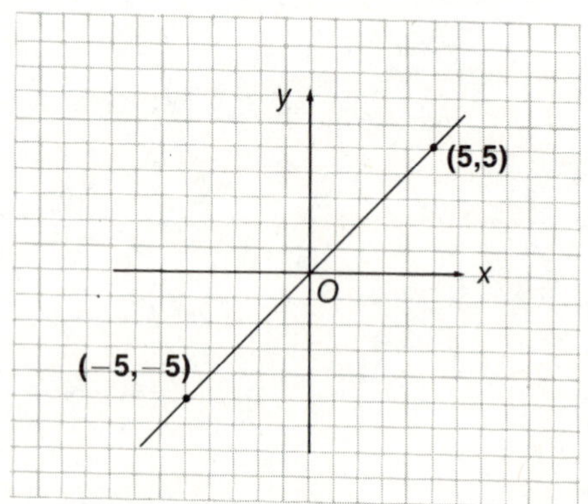

5. $y = x + 5$

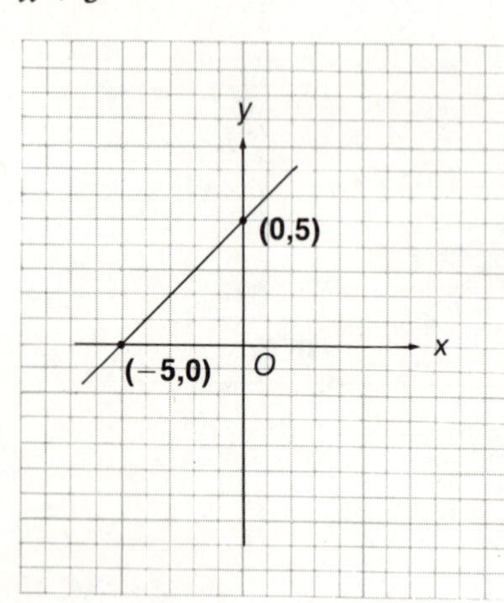

6. $y = 2x - 6$

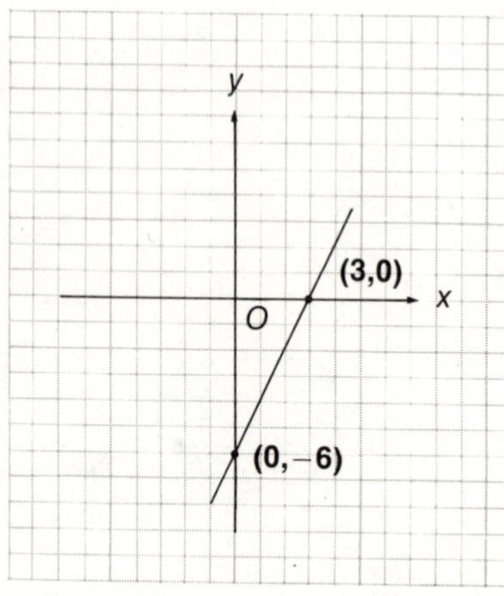

7. $y = 8 - 4x$

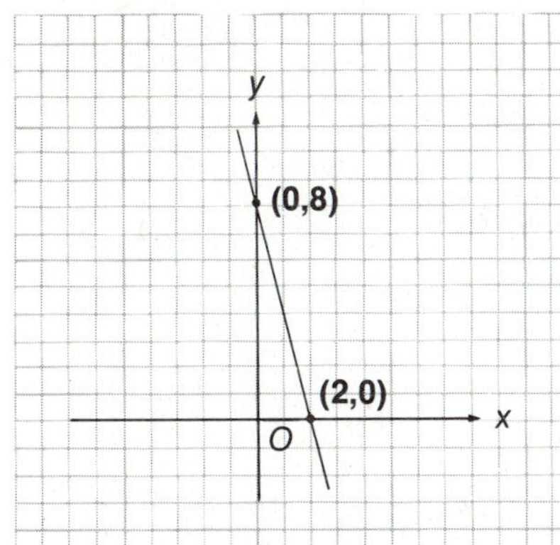

8. $y = \frac{x}{2} + 4$

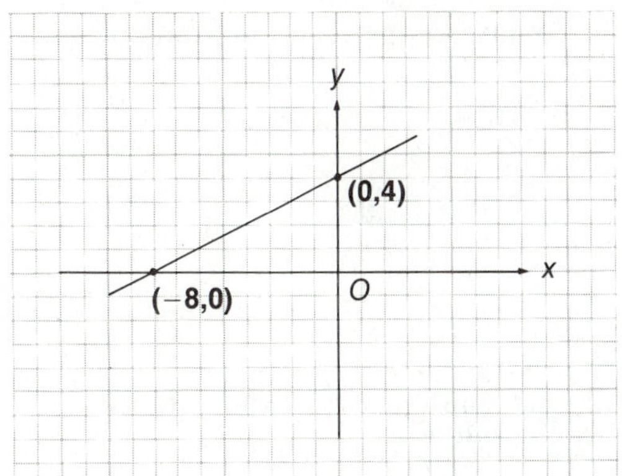

9.

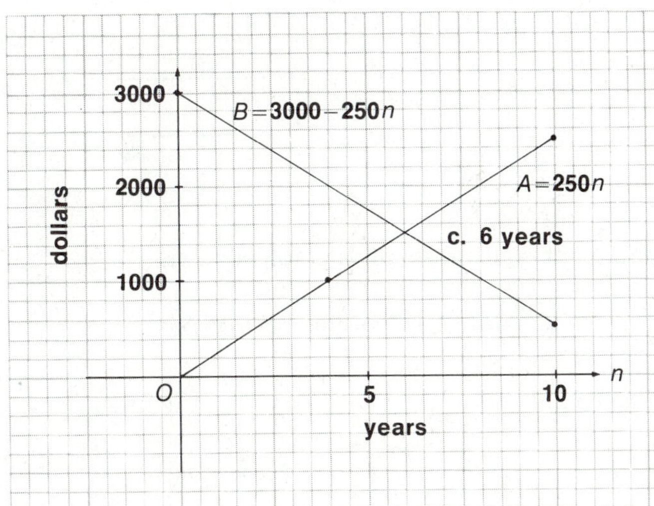

10.

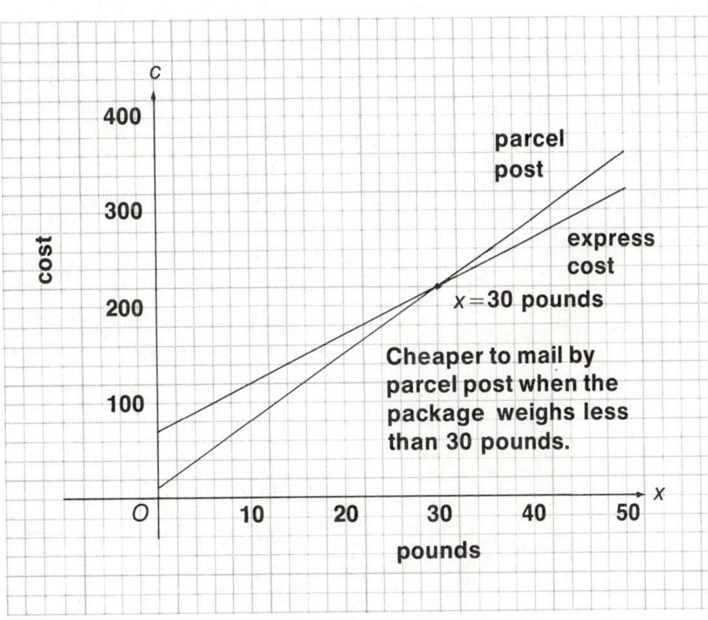

Exercises 11.6 A

1. x-intercept = 2; no y-intercept

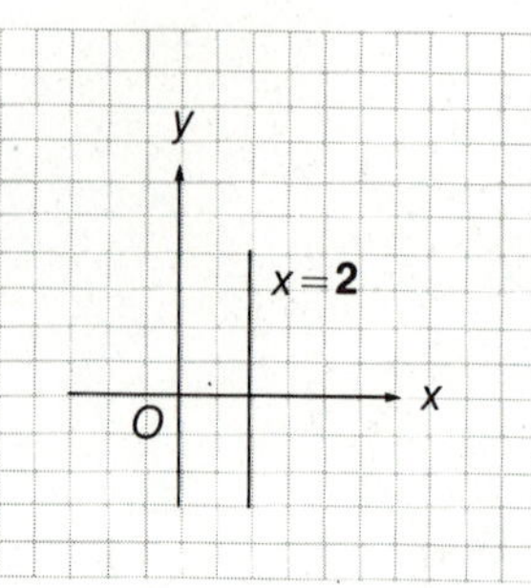

2. no x-intercept; y-intercept = −3

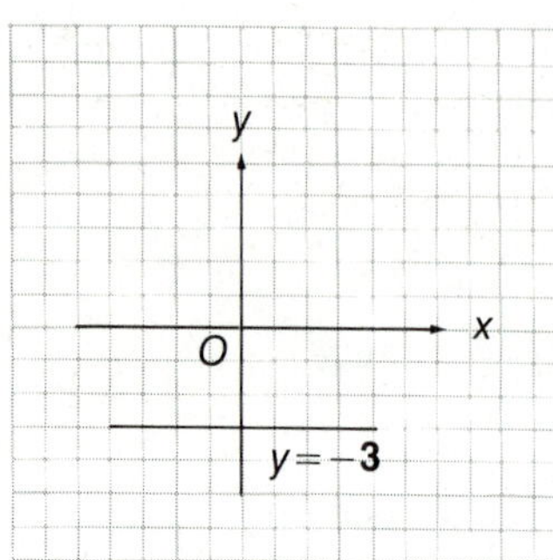

3. x-intercept = 0; y-intercept = all values of y; graph is y-axis

4. x-intercept = all values of x; y-intercept = 0; graph is x-axis

5. x-intercept = 2; y-intercept = 4

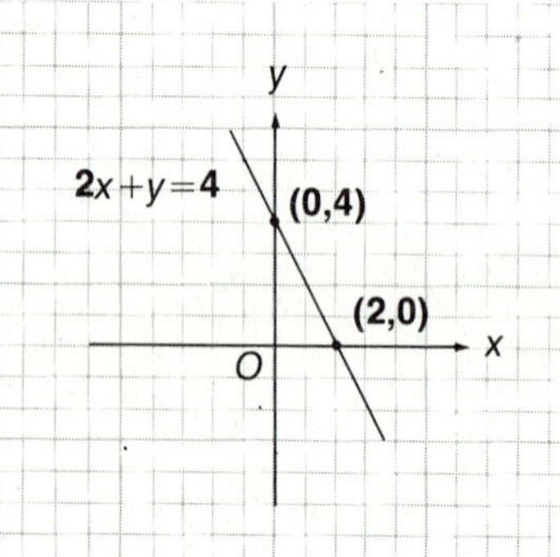

6. x-intercept = −3; y-intercept = 9

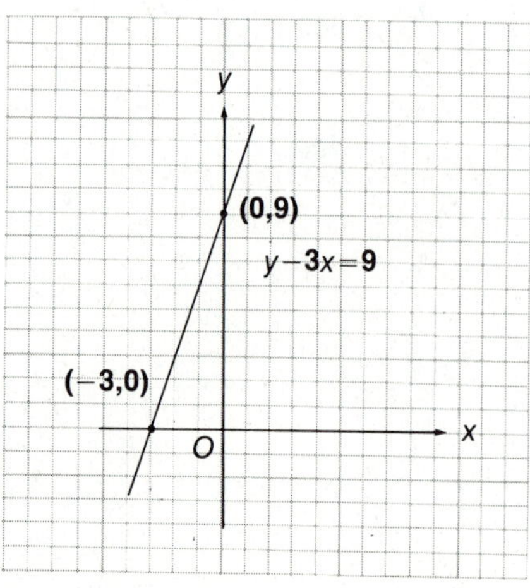

7. x-intercept = 5; y-intercept = -5

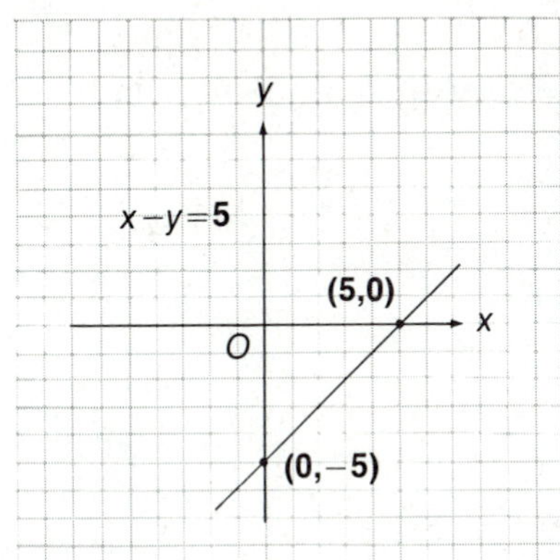

8. x-intercept = 2; y-intercept = 5

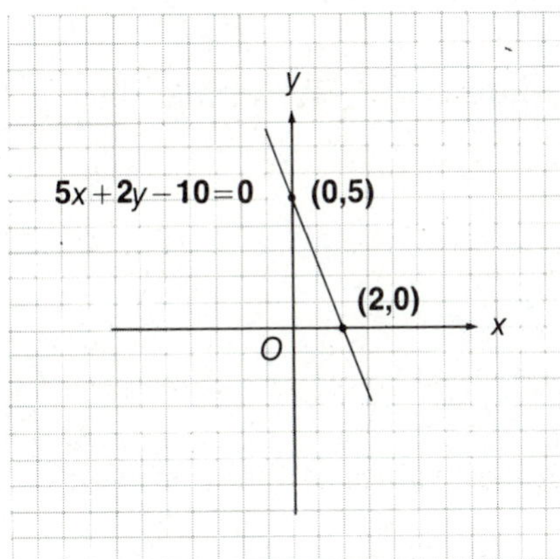

9. x-intercept = -4; y-intercept = -3

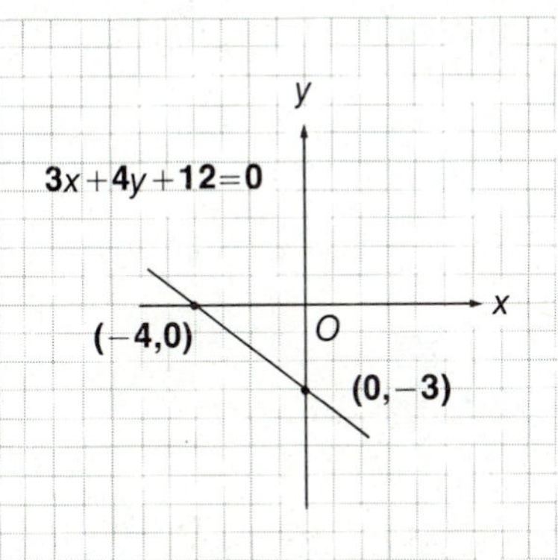

10. x-intercept = 5; y-intercept = -2

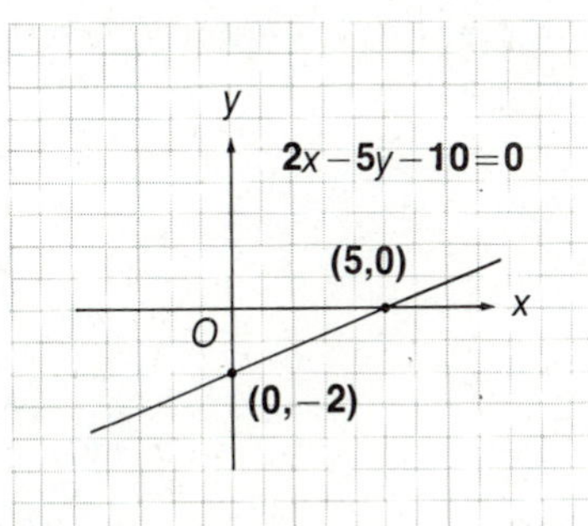

Exercises 11.6 B

1. $x = -5$; x-intercept $= -5$; no y-intercept

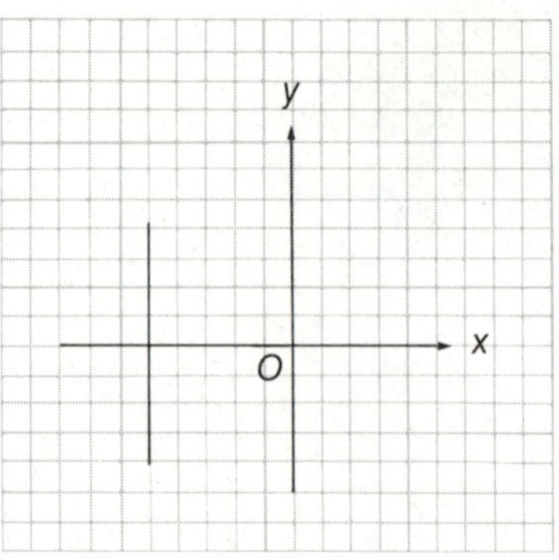

2. $y = 4$; no x-intercept; y-intercept $= 4$

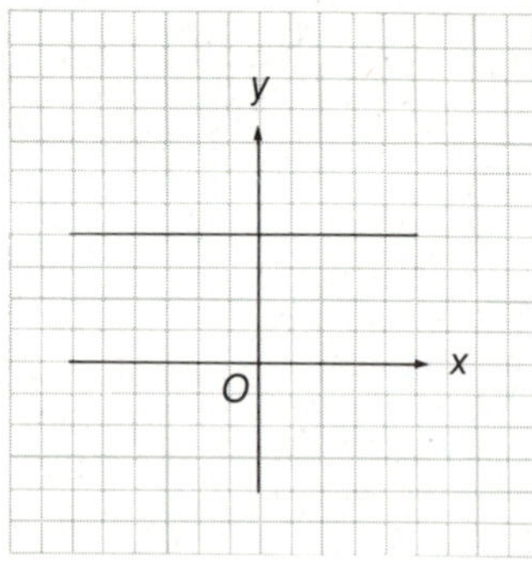

3. $x + 2y = 6$; x-intercept $= 6$; y-intercept $= 3$

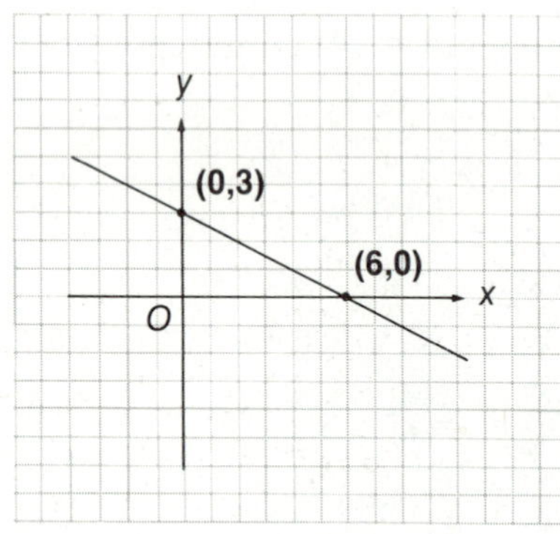

4. $x - 2y = 0$; x-intercept $= 0$; y-intercept $= 0$

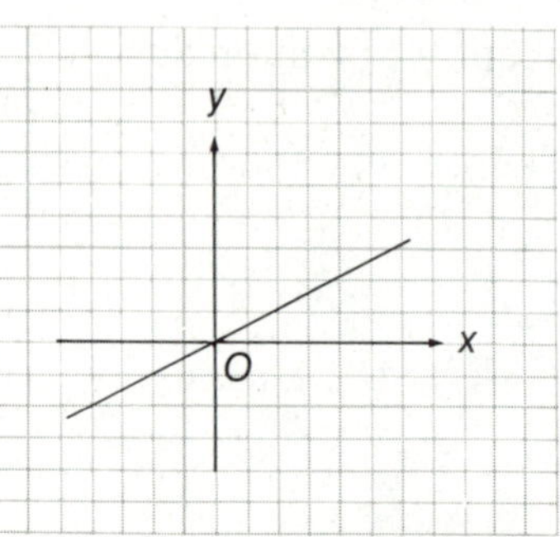

5. $3x - y = 6$; x-intercept $= 2$; y-intercept $= -6$

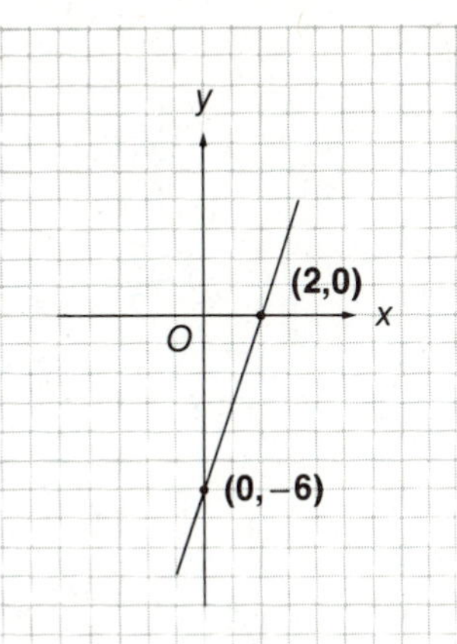

6. $5x - 2y + 10 = 0$; x-intercept $= -2$; y-intercept $= 5$

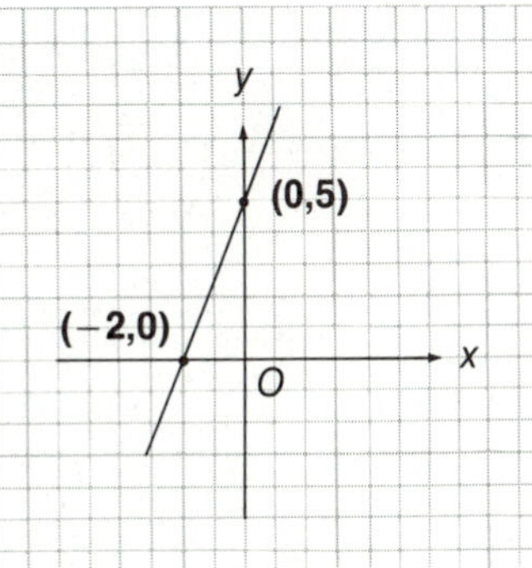

7. $5x - 6y - 30 = 0$; x-intercept $= 6$; y-intercept $= -5$

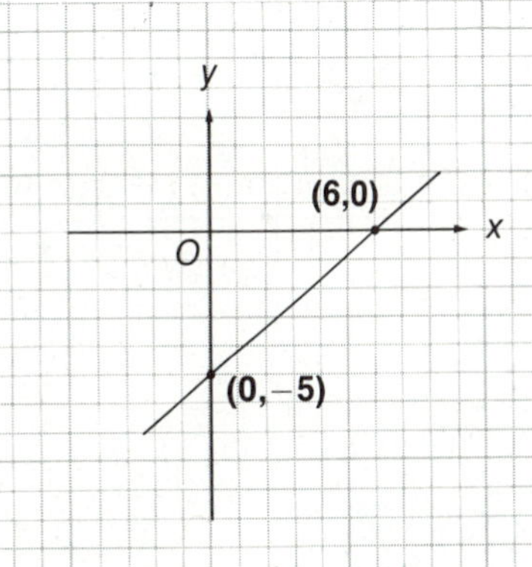

8. $2x + 7y + 14 = 0$; x-intercept $= -7$; y-intercept $= -2$

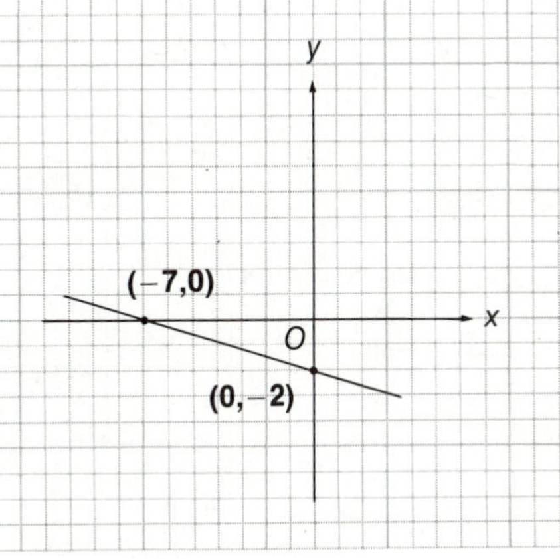

Exercises 11.7 A

1. 2
2. -4
3. $\frac{1}{3}$
4. $\frac{-3}{2}$
5. $\frac{5}{4}$
6. 0
7. $\frac{-2}{5}$
8. undefined

9.

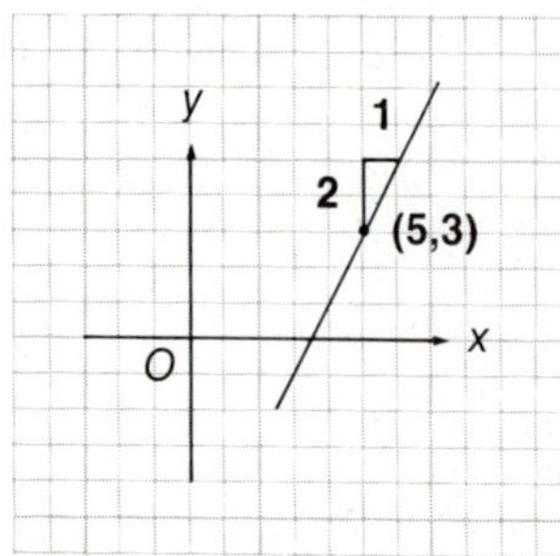

10.

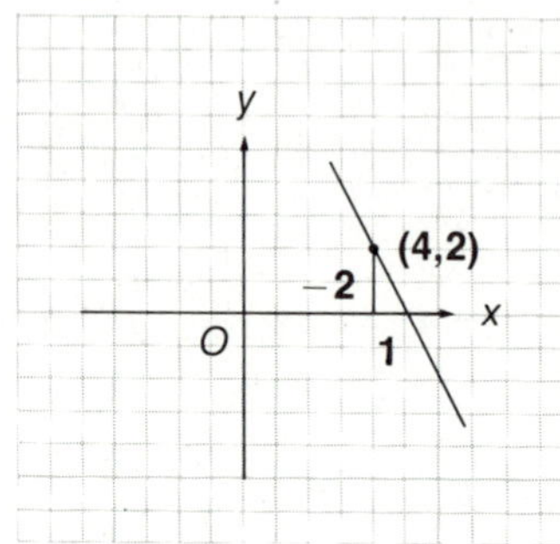

11.

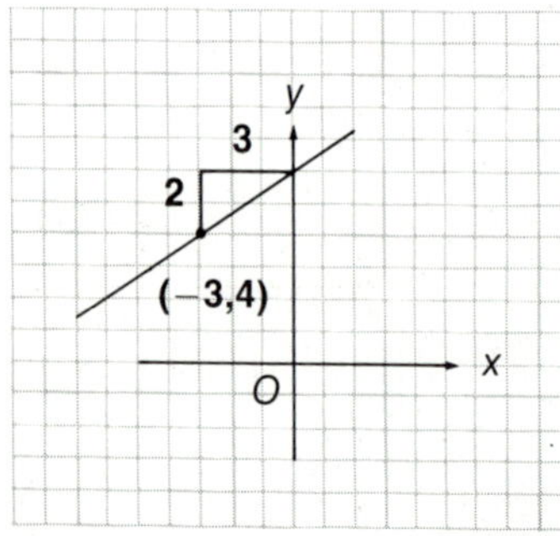

12.

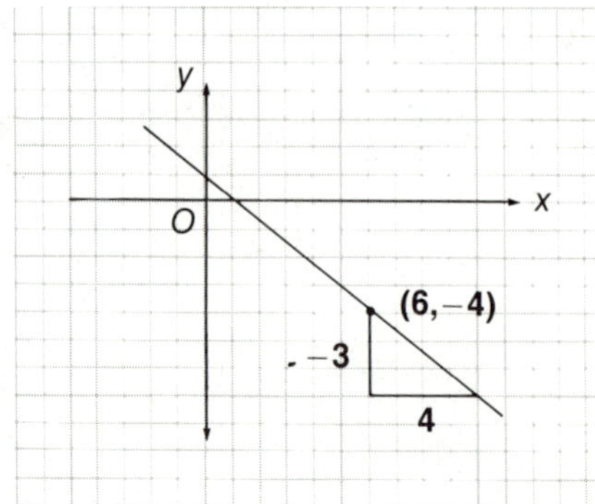

13.

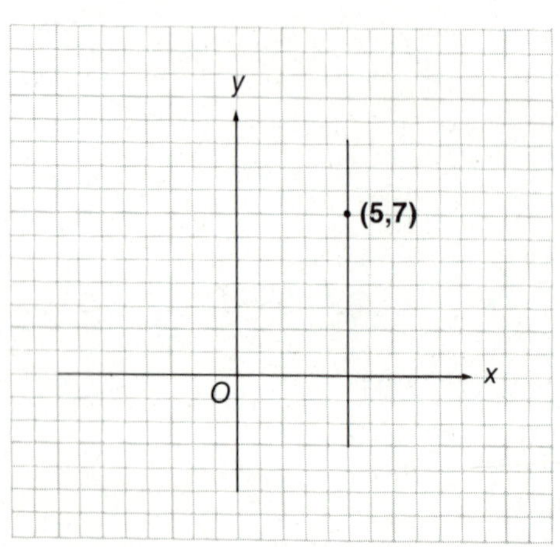

14.

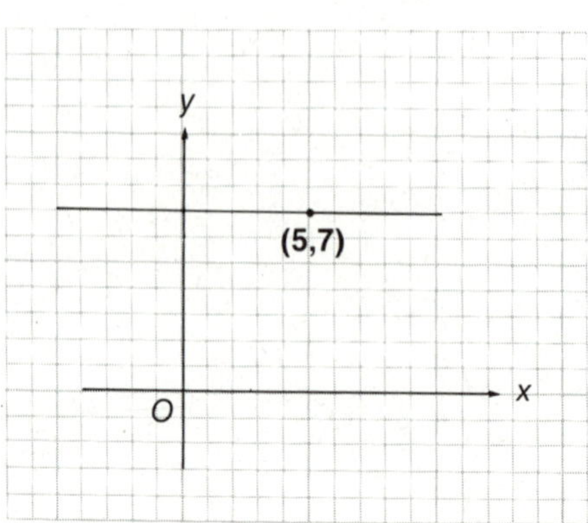

15. 92.4 ft.

16. 237.6 ft.

17. a and b

Exercises 11.7 B

1. 3
2. -2
3. $\frac{7}{6}$
4. $\frac{-5}{3}$
5. $\frac{-2}{3}$
6. $\frac{1}{6}$
7. undefined
8. 0

9.

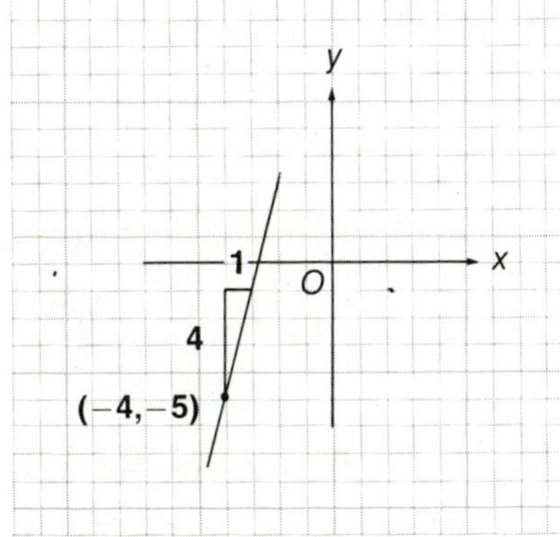

10.

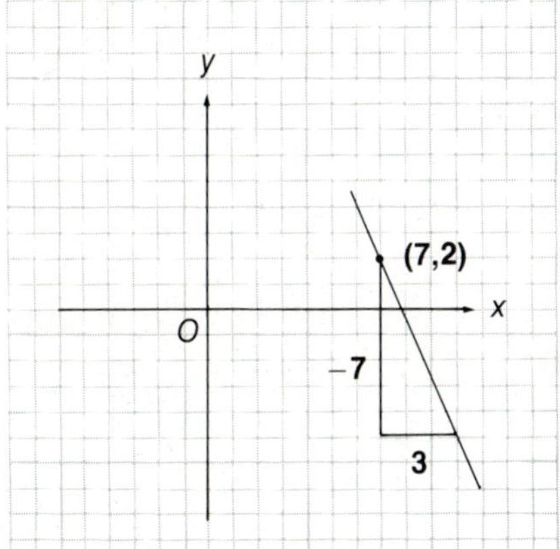

11.

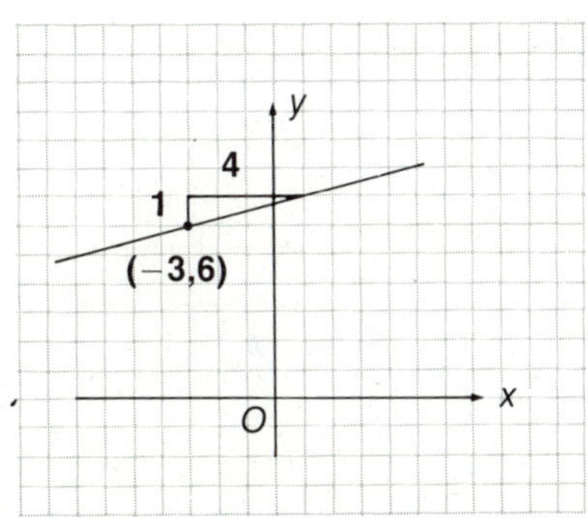

12.

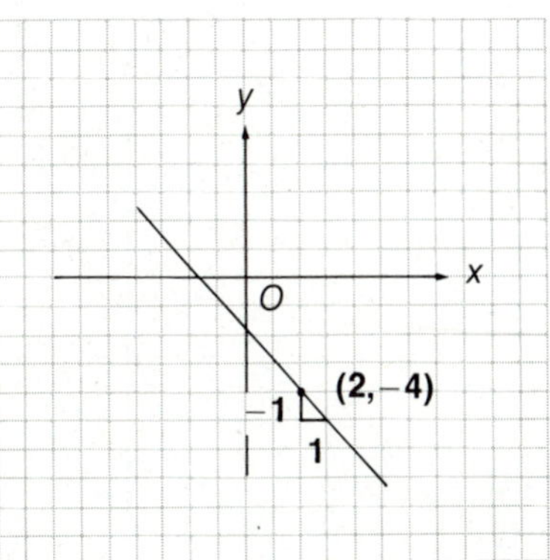

13.

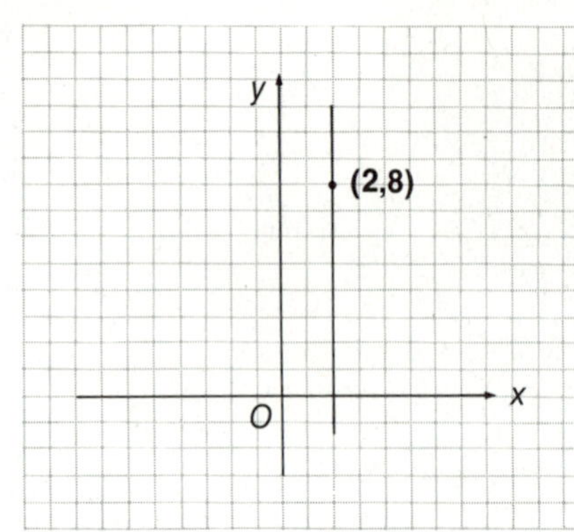

14.

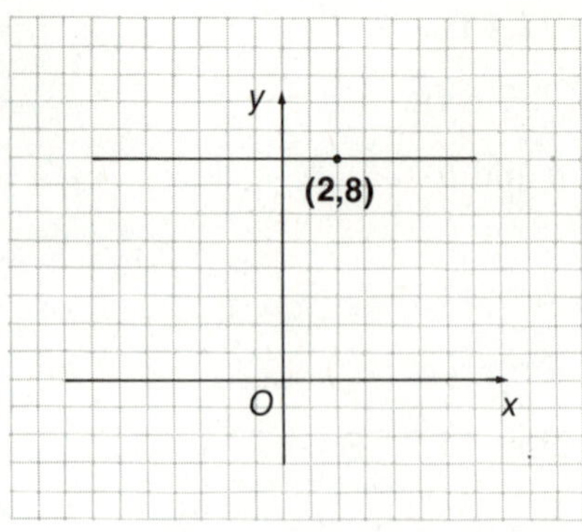

15. $4\frac{1}{2}$ ft.

16. 18 ft.

Exercises 11.8 A

1. $m = 2, b = 6$

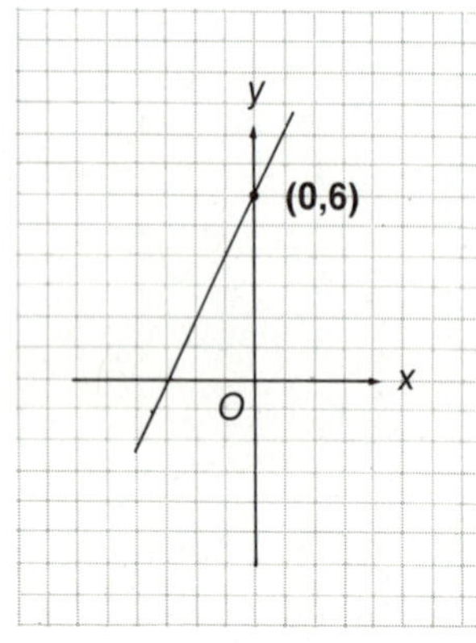

2. $m = -2, b = 6$

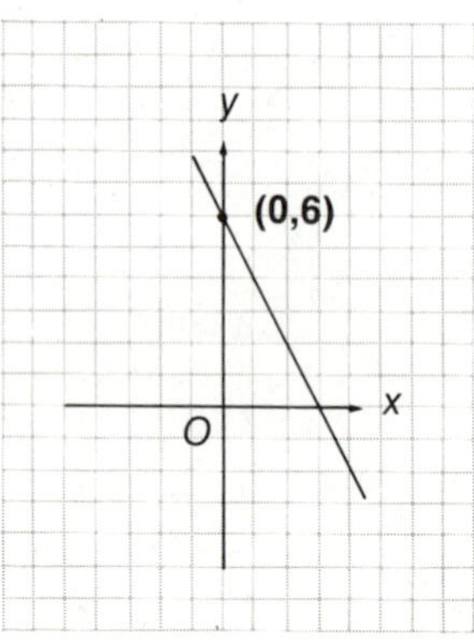

3. $m = 2, b = 0$

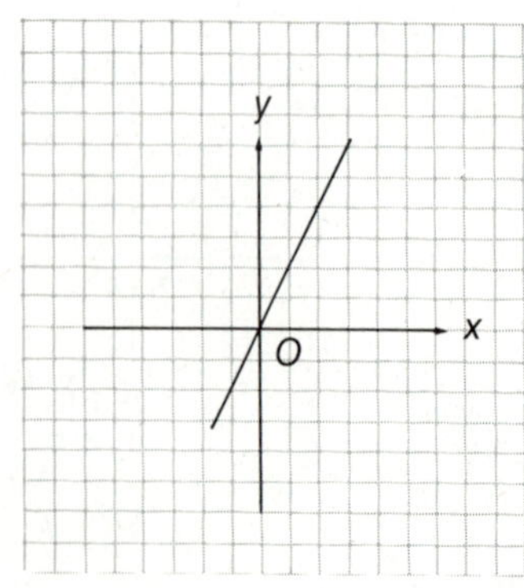

4. $m = 0, b = 2$

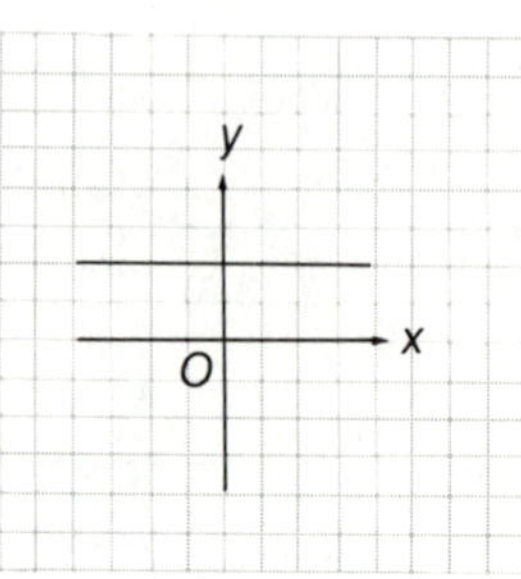

5. undefined

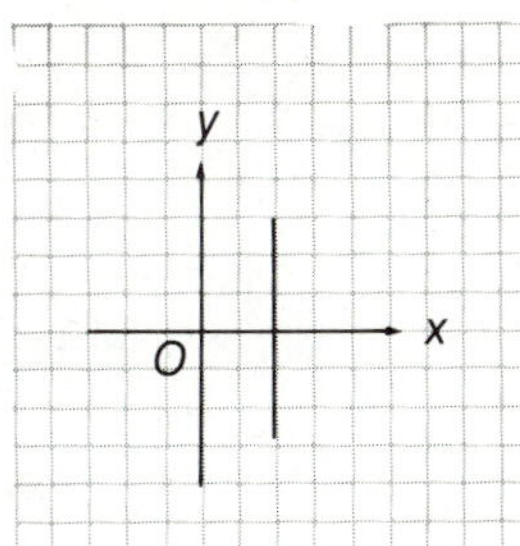

6. $m = \frac{1}{2}$, $b = 4$

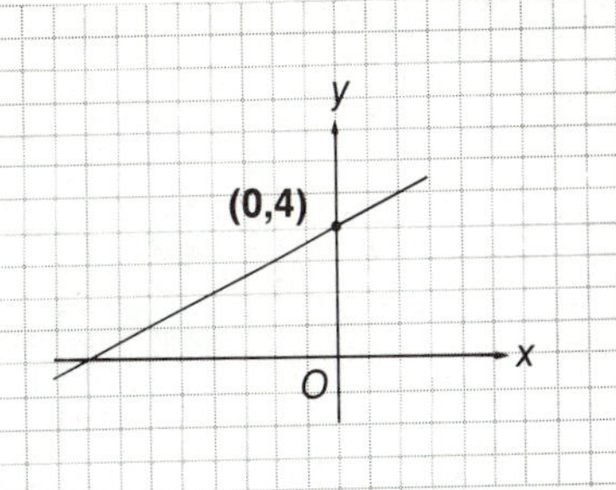

7. $m = \frac{-2}{3}$, $b = 2$

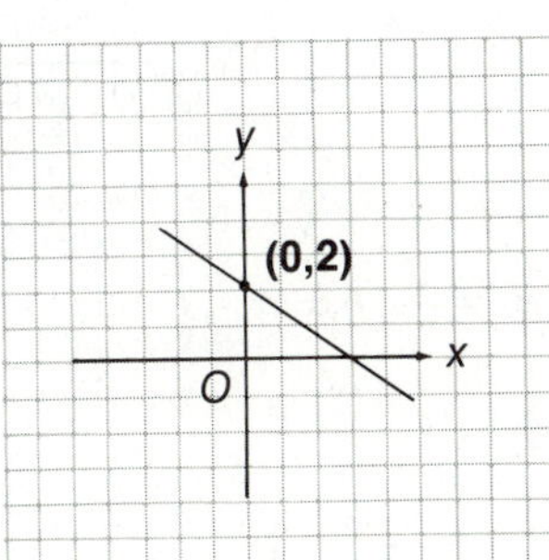

8. $m = \frac{2}{3}$, $b = 2$

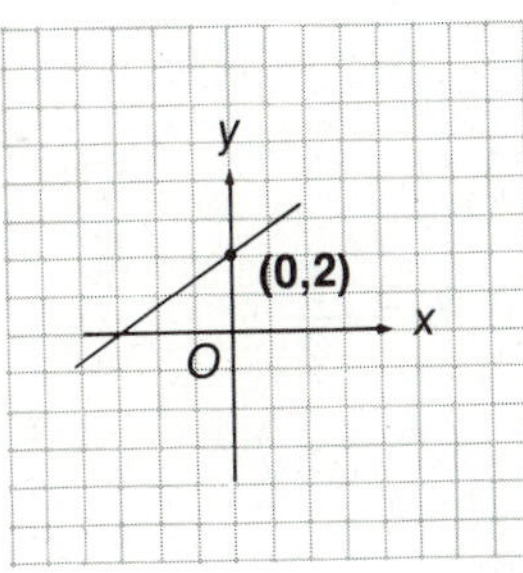

9. $m = \frac{-2}{3}$, $b = -2$

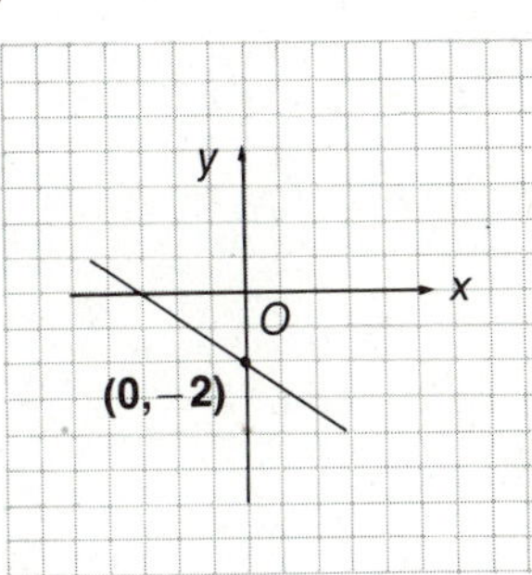

10. $m = \frac{2}{3}$, $b = 4$

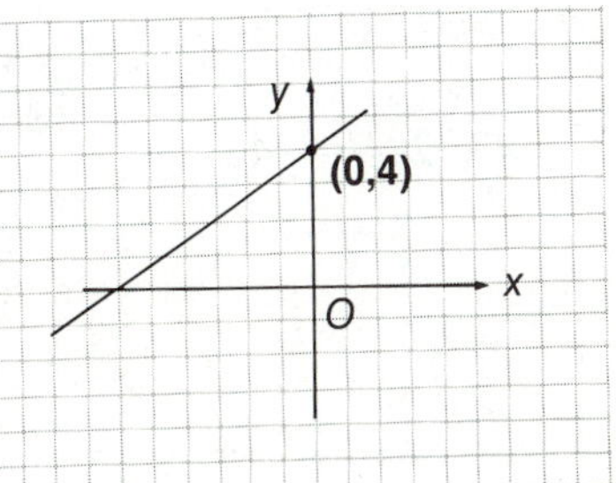

Exercises 11.8 B

1. $m = 3$, $b = -6$

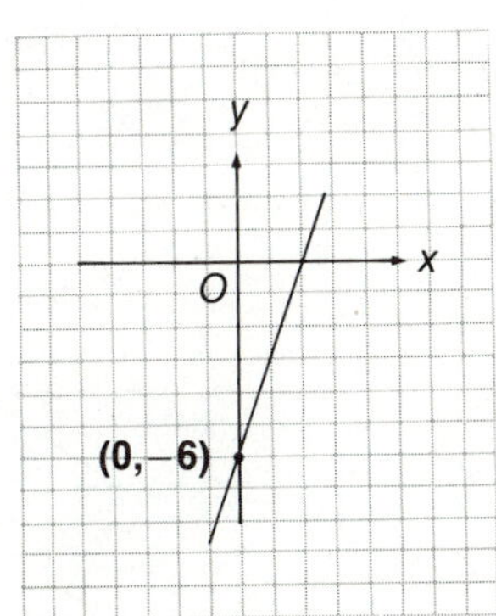

2. $m = -2$, $b = 8$

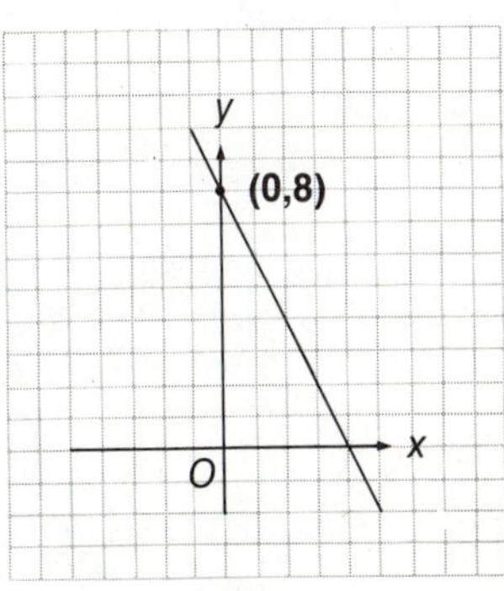

3. $m = 1, b = -6$

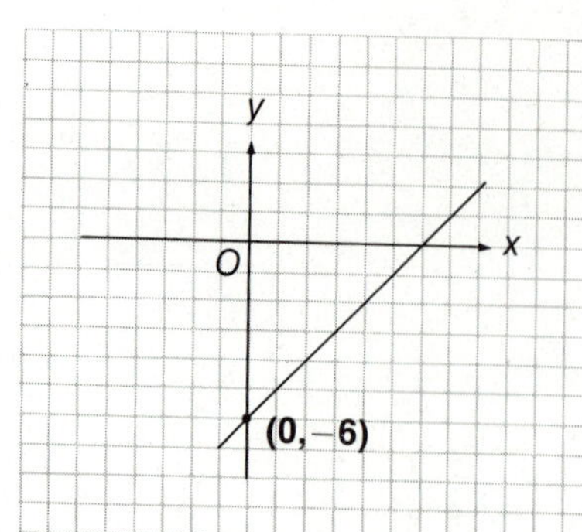

4. $m = -2, b = 12$

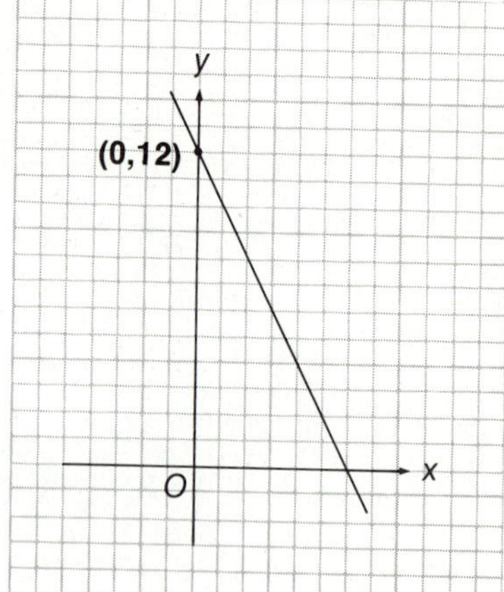

5. $m = 0, b = \frac{-5}{2}$

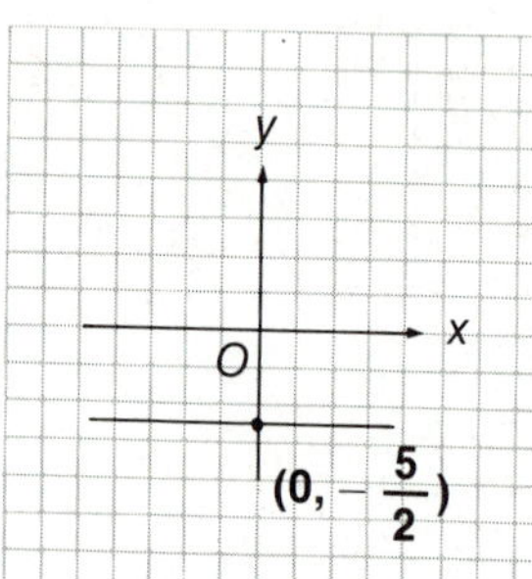

6. $m = \frac{1}{2}, b = 4$

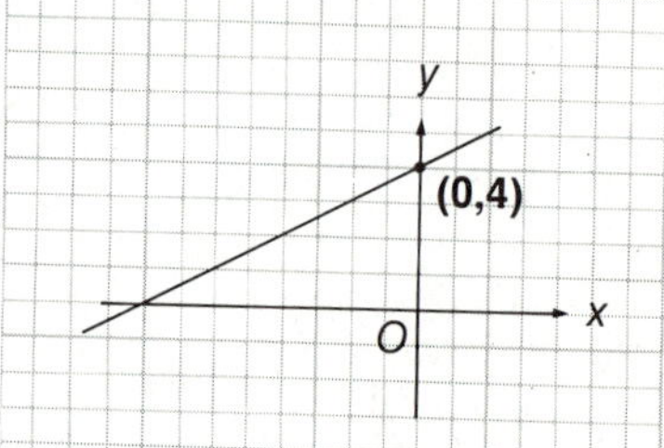

7. $m = 2, b = -4$

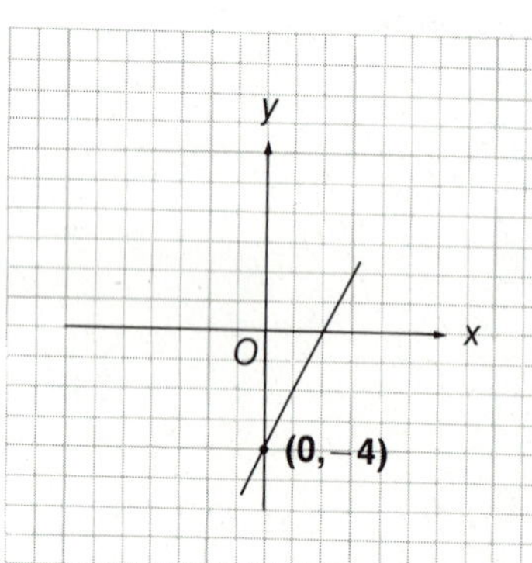

8. $m = 1, b = \frac{5}{2}$

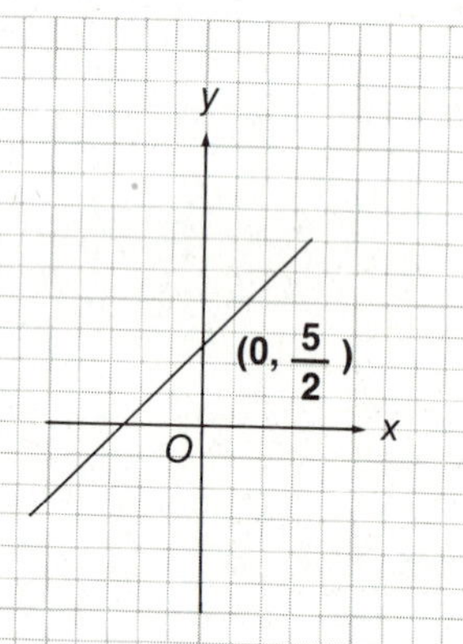

9. $m = \frac{-1}{2}, b = 0$

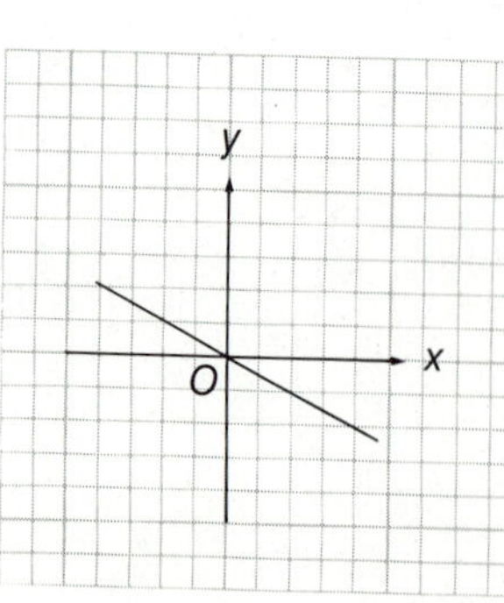

10. $m = \frac{2}{3}, b = 0$

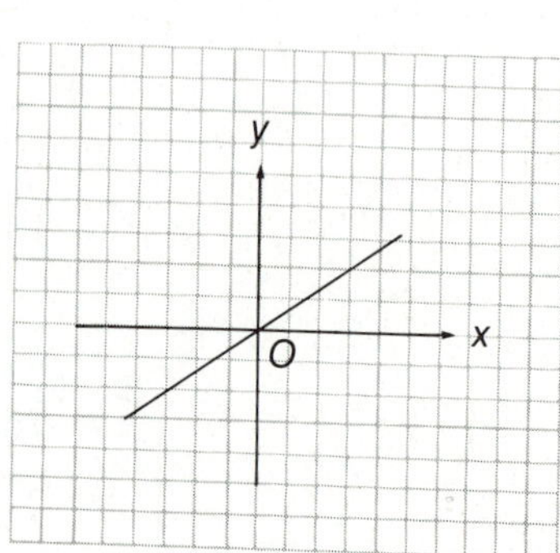

UNIT 11 POSTTEST

1.

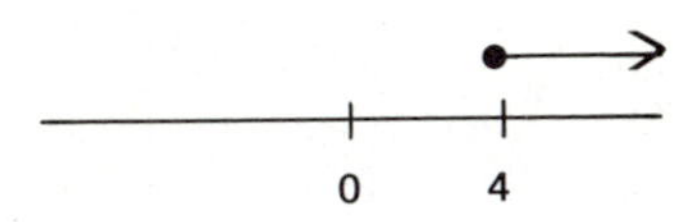

2.

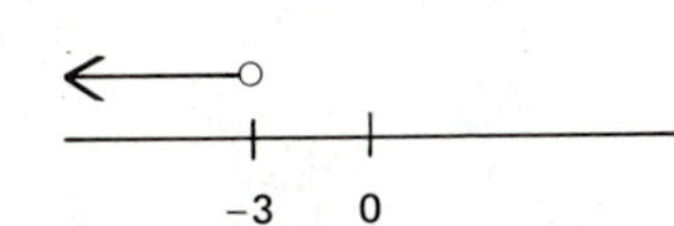

3.

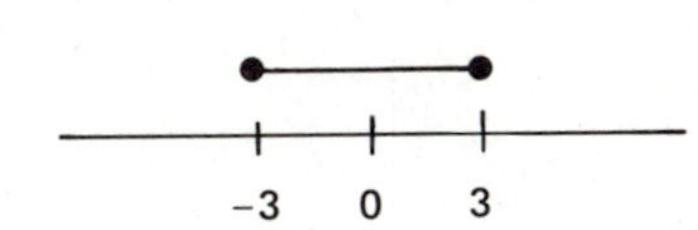

4.

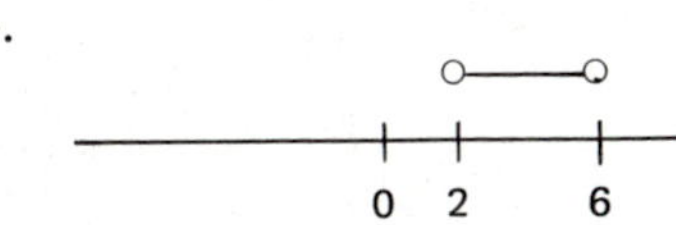

5.

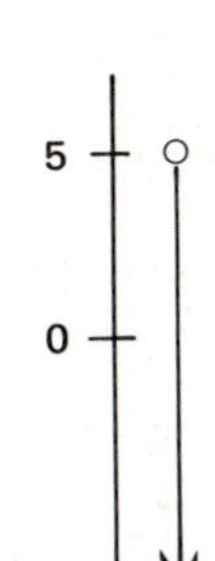

6.

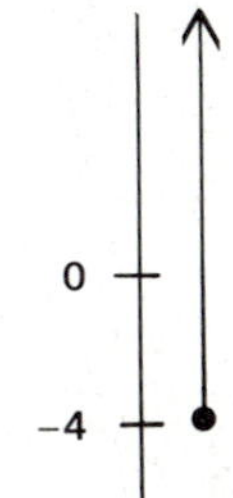

7.

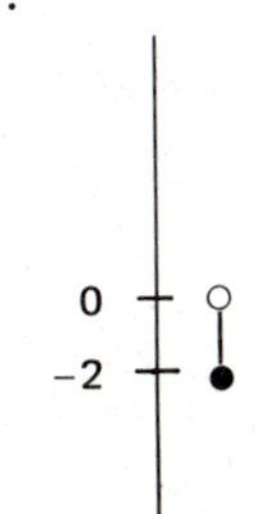

8.

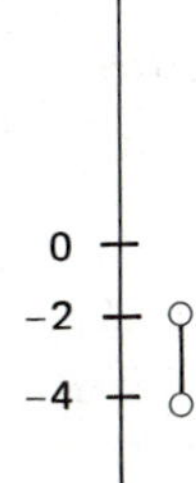

(9–13)

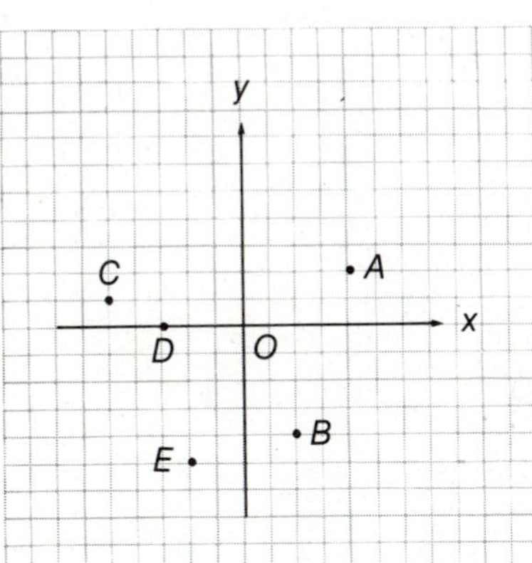

14. $P(4, -2)$
15. $Q(0, 3)$
16. $R(0, 0)$
17. $S(-5, -4)$
18. $T(5, 5)$
19. yes
20. no
21. no
22. yes
23. yes
24. yes
25. $(5, -6)$
26. $(4, 2)$

27.

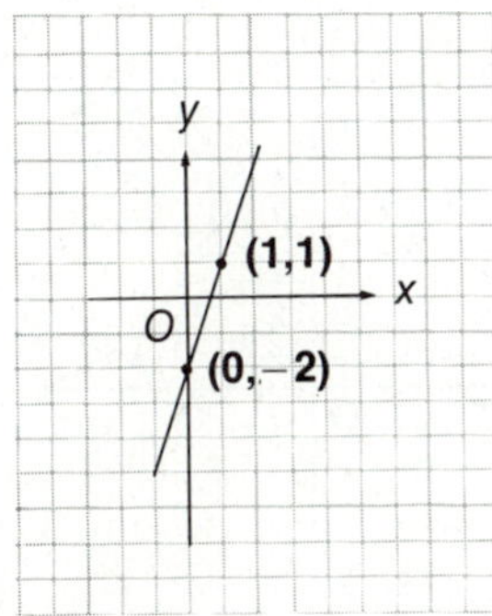

28.

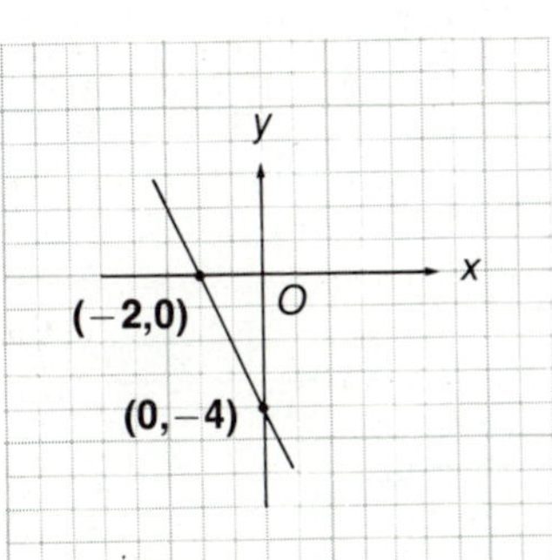

29. x-intercept = 8, y-intercept = −2

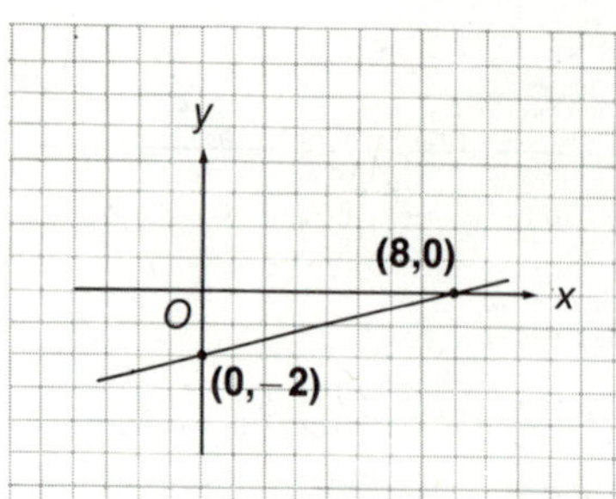

30. x-intercept = −2, y-intercept = −10

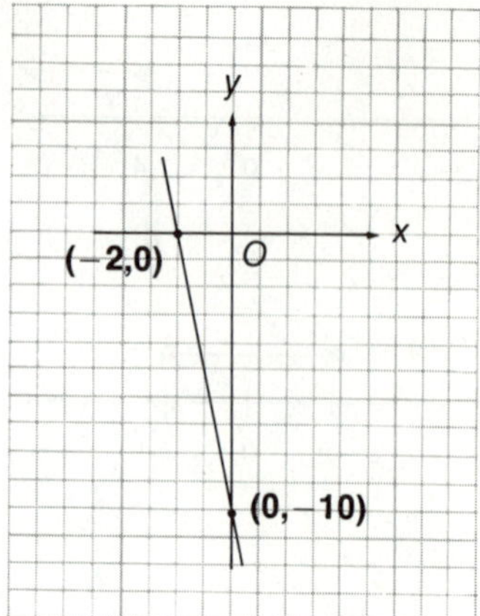

31. $m = \frac{-1}{2}$

32. $m = \frac{1}{3}$

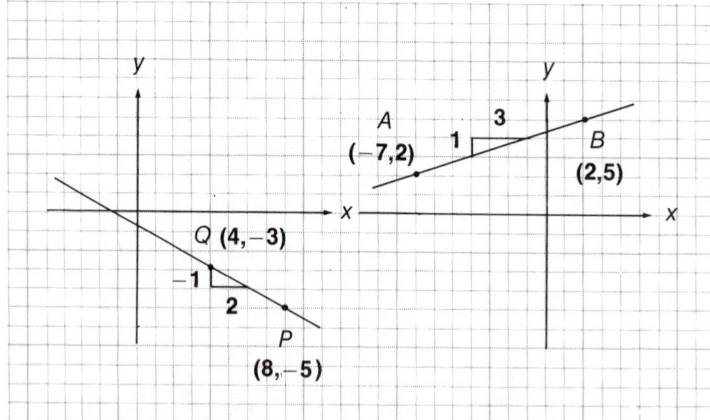

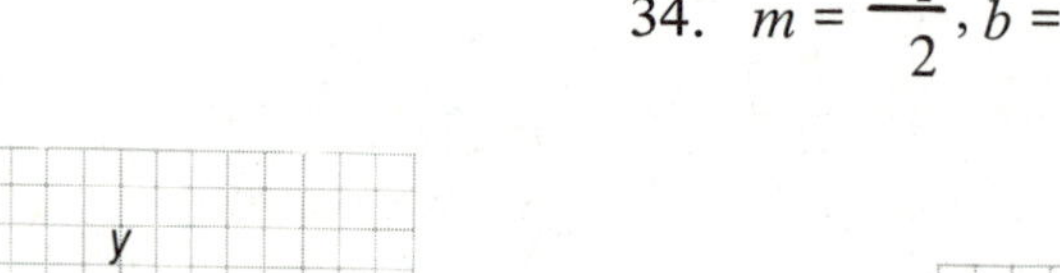

33. $m = 2$, $b = 6$

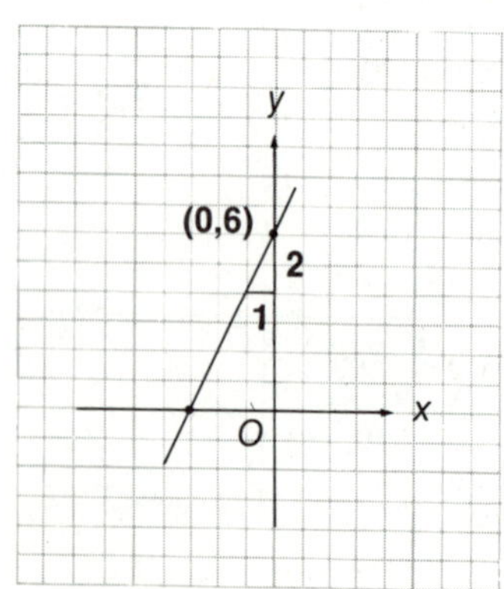

34. $m = \frac{-1}{2}$, $b = 4$

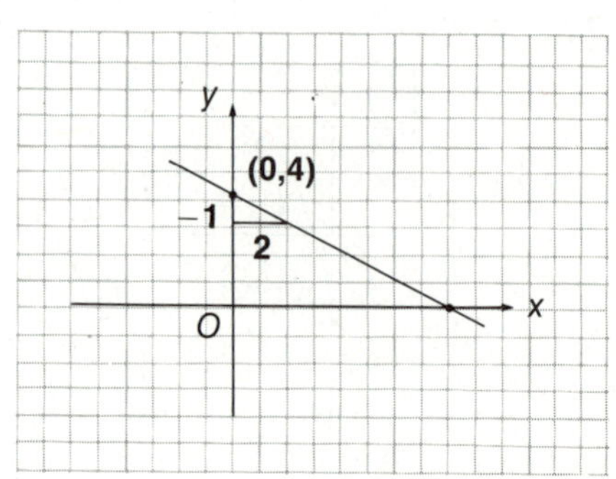

UNIT 12 PRETEST

1. intersecting
2. intersecting
3. parallel
4. coincident

5a. yes

5b. no

6a. no

6b. yes

7. (-4, 3)

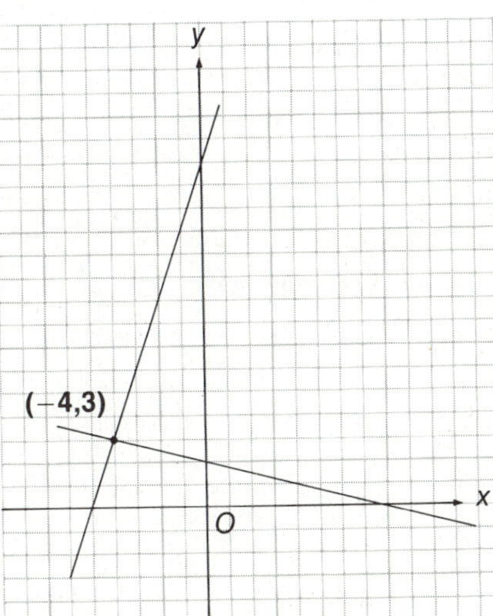

8. (5, 8)
9. (-4, 7)
10. (2, 3)
11. Ø; lines parallel
12. (4, -4); intersecting lines
13. $\{(x, y)|3y = 2x - 2\}$; coincident lines
14. eggs. 65¢ a dozen; milk, 40¢ a quart
15. 6 m.p.h.

Exercises 12.1 A

1. intersecting
2. intersecting
3. parallel
4. intersecting
5. coincident
6. parallel
7. intersecting
8. coincident

Exercises 12.1 B

1. intersecting
2. parallel
3. intersecting
4. intersecting
5. intersecting
6. intersecting
7. parallel
8. coincident

Exercises 12.2 A

1. yes, no, no
2. no, yes, no
3. no, no, yes
4. no, no, yes
5. no, yes, no

Exercises 12.2 B

1. no, yes, no
2. no, yes, no
3. no, no, no
4. yes, yes, no
5. no, no, no

Exercises 12.3 A

1. (7, 5)
2. (6, 2)
3. (5, -1)
4. (3, -5)
5. (-3, -7)

Exercises 12.3 B

1. (-4, -2)
2. (2, 0)
3. (7, 4)
4. (3, 3)
5. (0, 5)

Exercises 12.4 A

1. (5, 4)
2. (8, 5)
3. (2, 1)
4. (3, 6)
5. $\left(-2, \frac{1}{2}\right)$
6. (-3, -9)
7. (2, 0)
8. (4, -10)
9. (5, 1)
10. (9, -1)
11. (-3, -8)
12. (8, 5)

Exercises 12.4 B

1. (4, 20)
2. (7, 4)
3. (9, 4)
4. (-3, 2)
5. (10, -6)
6. (-5, -8)
7. (3, 2)
8. (-10, 0)
9. (30, 5)
10. (-6, 4)

Exercises 12.5 A

1. (6, 4)
2. (9, -1)
3. (12, 2)
4. (5, 2)
5. $\left(\frac{32}{5}, \frac{1}{5}\right)$
6. (7, -4)
7. (5, 7)
8. (8, -5)
9. (-3, -8)
10. $\left(\frac{-1}{2}, \frac{7}{2}\right)$

Exercises 12.5 B

1. (30, 12)
2. (4, 5)
3. (7, −4)
4. (−6, 9)
5. (8, 6)
6. (−2, −3)
7. (0, 12)
8. $\left(\frac{3}{2}, \frac{5}{2}\right)$
9. (2, 0)
10. (4, −7)

Exercises 12.6 A

1. (5, 8), intersecting
2. $\{(x, y)|x = 2y + 3\}$, coincident
3. Ø, parallel
4. Ø, parallel
5. (4, 0), intersecting
6. (4, 0), intersecting
7. $\{(x, y)|y = x + 6\}$, coincident
8. (3, −3), intersecting

Exercises 12.6 B

1. Ø, parallel
2. $\{(x, y)|2x - 5y = 10\}$, coincident
3. (35, 50), intersecting
4. (75, 25), intersecting
5. $\{(x, y)|y = 15 - 5x\}$, coincident
6. Ø, parallel
7. $\{0, 0\}$, intersecting
8. $\left\{\frac{3}{2}, \frac{3}{2}\right\}$, intersecting

Exercises 12.7 A

1. let x = price of lace per yard
   let y = price of seam binding per yard
   $2x + 3y = 96$, $5x + 6y = 222$; $x = 30$¢, $y = 12$¢

2. let x = price of oil per quart
   let y = price of gasoline per gallon
   $x + 5y = 245$, $2x + 7y = 376$; $x = 55$¢, $y = 38$¢

3. let a = percentage of copper in ore A
   let b = percentage of copper in ore B
   $\frac{36a}{100} + \frac{28b}{100} = 39$, $\frac{50a}{100} + \frac{40b}{100} = 55$; $a = 50\%$, $b = 75\%$

4. let x = percentage for stocks
   let y = percentage for bonds
   $\frac{5000x}{100} + \frac{3000y}{100} = 450$, $\frac{2000x}{100} + \frac{1500y}{100} = 195$; $x = 6\%$, $y = 5\%$

5. let x = hourly wage of John
let y = hourly wage of Jill
$52x + 40y = 210$, $40x + 25y = 150$, $x = \$2.50$, $y = \$2.00$

Exercises 12.7 B

1. let x = number of hours carpenter worked first week
let y = number of hours helper worked first week
$8x + 5y = 550$, $4x + 5(y + 10) = 400$
$x = 50$ hr., $y = 30$ hr.
2. let x = number of adults
let y = number of children
$280x + 120y = 69600$, $280(2x) + 120\left(\frac{y}{2}\right) = 85200$; $x = 120$ adults,
$y = 300$ children
3. let x = fixed fee per car
let y = additional fee per passenger
$8x + 30y = 42.5$, $10x + 40y = 55$
$x = \$2.50$ (car), $y = 75¢$ (passenger)
4. let x = price of wine per cask
let y = duty per cask
$5x + 40 = (64 - 5)y$, $2x - 40 = (20 - 2)y$; $x = 110$ francs (wine),
$y = 10$ francs (duty)
5. let x = price of each citron
let y = price of each wood-apple
$9x + 7y = 107$, $7x + 9y = 101$; $x = 8$ (citron), $y = 5$ (wood-apples)

Exercises 12.8 A

1. $3(a + w) = 630$; $\frac{7}{2}(a - w) = 630$
$w = 15$ m.p.h. (speed of wind)
2. $t(20 + r) = 12$; $t(20 - r) = 8$
$r = 4$ m.p.h.
3. let x = number of miles on mountain road
let y = number of miles on level road
$x + y = 200$; $\frac{x}{25} = \frac{y}{50} - 1$; $x = 50$ mi.
4. let r = speed of wind; let t = time to fly 1560 mi. $t(408 - r) = 1560$,
$\frac{t}{2}(408 + r) = 852$; $r = 18$ m.p.h.

5. let x = full speed of helicopter
let y = speed of wind
$\frac{1}{6}(x+y) = 31, \frac{1}{3}\left(\frac{x}{2} - y\right) = 18$
$x = 160$ m.p.h., $y = 26$ m.p.h.

Exercises 12.8 B

1. let r = speed of plane in still air
let t = time to travel 384 mi.
$t(r+18) = 456, t(r-18) = 384$
$t = 2$ hr., $r = 210$ m.p.h.

2. let x = rate of boat; let y = rate of current
$\frac{1}{2}(x+y) = \frac{15}{2}, \frac{3}{4}(x-y) = \frac{15}{2}$
$x = 12\frac{1}{2}$ m.p.h., $y = 2\frac{1}{2}$ m.p.h.

3. let r = rate rowing; let y = rate walking
$x + 2y = 12, \frac{3x}{2} + y = 11$
$x = 5$ m.p.h., $y = 3\frac{1}{2}$ m.p.h.

4. let x = time to go downstream
let y = time to go upstream
$x + y = \frac{1}{2}, (5+3)x = (15-3)y$
$x = \frac{1}{10}$, distance $= 8\left(\frac{1}{10}\right) = 0.8$ or $\frac{4}{5}$ mi.

5. let x = rate uphill; let y = rate downhill
$\frac{1}{x} + \frac{2}{y} = \frac{24}{60} = \frac{2}{5}; \frac{2}{x} + \frac{1}{y} = \frac{30}{60} = \frac{1}{2}$
let $u = \frac{1}{x}$ and $v = \frac{1}{y}$
then $u + 2v = \frac{2}{5}$ and $2u + v = \frac{1}{2}$
$u = \frac{1}{5}$ and $v = \frac{1}{10}$
$x = 5$ m.p.h. and $y = 10$ m.p.h.

Exercises 12.9

1. $x = y + 2.5, 40x + 40y = 420$
$x = \$6.50$ (electrician), $y = \$4.00$ (assistant)

2. $l = w + 15, (l+5)(w+5) = lw + 550$
45 ft. by 60 ft.

3. $\frac{1}{2}(x+y) = 14, \frac{7}{10}(x-y) = 14$
$x = 24$ m.p.h. (boat), $y = 4$ m.p.h. (current)

4. $y = x + 20, (8-5)x + 8(20) = 550$
$x = 130$ min. (time both pipes open)
$y = 150$ min. (time to fill tank)

5. $\frac{6000a}{100} + \frac{9000b}{100} = 690, \frac{6000(2a)}{100} + \frac{9000(b-1)}{100} = 1020$
$a = 7\%$ (company A), $b = 3\%$ (company B)

6. $2.5x + 4y = 1820, x + y = 500$
$x = 120$ sheets (seconds)

7. $6a + 3b = 51, 3a + b = 22$
$a = 5$ cards size A, $b = 7$ cards size B

8. $500x + 40y = 1810, x = y + 2$
   $x = 3\frac{1}{2}$ hr. (on plane), $y = 1\frac{1}{2}$ hr. (on bus)

9. $80x + 60y = 3400, 25x + 10y = 800$
   $x = 20$¢ (each glass), $y = 30$¢ (each mug)

10. $4.4\,(v + w) = 2640, 4.8\,(v - w) = 2640$
    $w = 25$ m.p.h. (speed of wind)

UNIT 12 POSTTEST

1. intersect
2. coincident
3. intersect
4. parallel

5a. yes

5b. no

6a. no

6b. no

7. $\left(\frac{-5}{2}, 3\right)$

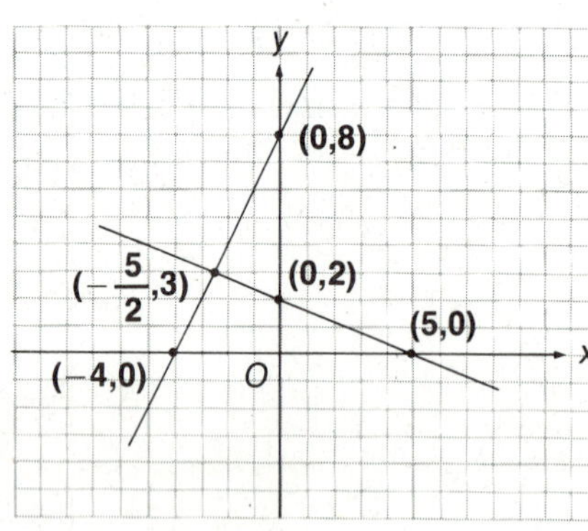

8. $\left(\frac{-5}{2}, 3\right)$
9. $(-6, 4)$
10. $(4, 2)$
11. $\{(x, y) | 3y = x + 2\}$; coincident lines
12. $(0, 4)$, intersecting lines
13. $\emptyset$, parallel lines
14. 5 stoves and 8 refrigerators to Mexico
    8 stoves and 5 refrigerators to Canada
15. 25 m.p.h.

UNIT 13 PRETEST

1. $\{-8, 3\}$
2. $\left\{\frac{-4}{3}, \frac{4}{3}\right\}$
3. 30
4. $\frac{-1}{2}$
5. 6.78
6. 5.37
7. $\left\{\frac{\sqrt{21}}{2}, -\frac{\sqrt{21}}{2}\right\}$
8. $\{6.1, -6.1\}$

9. 30

10. $6\sqrt{5}$

11. $\frac{\sqrt{15}}{5}$

12. $\frac{3\sqrt{2}}{2}$

13. $7\sqrt{3} - 7\sqrt{2}$

14. $5 + \sqrt{6}$

15. $\{7 + \sqrt{5}, 7 - \sqrt{5}\}$

16. $\left\{\frac{-2+\sqrt{14}}{5}, \frac{-2-\sqrt{14}}{5}\right\}$

17. $\left\{\frac{-2+\sqrt{19}}{3}, \frac{-2-\sqrt{19}}{3}\right\}$

18. $\left\{\frac{5+\sqrt{3}}{2}, \frac{5-\sqrt{3}}{2}\right\}$

19. $(45 - 2x)(28 - 2x) = \frac{1}{2}(28)(45)$; $x = 5$ meters

20. $\frac{2(21)}{x} + \frac{21}{x+10} = 1$; $x = 60$ min.

Exercises 13.1 A

1. $\{4, -2\}$
2. $\{-7, 2\}$
3. $\{5, -5\}$
4. $\left\{\frac{7}{6}, \frac{-7}{6}\right\}$
5. $\{0, 6\}$
6. $\left\{\frac{1}{2}, -3\right\}$
7. $\{4\}$
8. $\{-6, 5\}$
9. $\{5, -5\}$
10. $\{-2, -3\}$
11. $\{0, 3\}$
12. $\left\{-1, \frac{3}{5}\right\}$
13. $\{0, 4\}$
14. $\{-1, -4\}$
15. $\left\{\frac{-2}{5}, 3\right\}$
16. $\{0, -10\}$
17. $\left\{\frac{3}{8}, -1\right\}$
18. $\left\{\frac{-3}{2}, \frac{1}{5}\right\}$
19. $\{0, 16\}$
20. $\{0\}$

Exercises 13.1 B

1. $\{-1, -5\}$
2. $\{6, -3\}$
3. $\{3, -3\}$
4. $\left\{\frac{8}{9}, \frac{-8}{9}\right\}$
5. $\{0, -7\}$
6. $\left\{2, \frac{-4}{3}\right\}$
7. $\{-7\}$
8. $\{2, -2\}$
9. $\{8, -6\}$
10. $\{1, -1\}$
11. $\left\{\frac{1}{2}, -5\right\}$
12. $\{0, 9\}$
13. $\{-1\}$
14. $\{6, -7\}$
15. $\{0, 4\}$
16. $\left\{\frac{3}{2}, \frac{-1}{3}\right\}$
17. $\{8, -5\}$
18. $\left\{\frac{5}{2}, \frac{1}{3}\right\}$
19. $\left\{0, \frac{-9}{2}\right\}$
20. $\left\{0, \frac{12}{5}\right\}$

Exercises 13.2 A

1. 8
2. 9
3. 11
4. 10
5. 8
6. 7
7. not real
8. -4
9. not real
10. 0
11. 10
12. 5

13. 8
14. -4
15. 5
16. $\frac{-5}{3}$
17. 20
18. 4.47
19. 9.75
20. 4.47
21. 1.58
22. -0.55
23. 3.41
24. $\{\sqrt{5}, -\sqrt{5}\}$
25. $\left\{\frac{\sqrt{40}}{3}, -\frac{\sqrt{40}}{3}\right\}$
26. $\{3, -3\}$
27. $\{70, -70\}$
28. $\{7.5, -7.5\}$
29. $\left\{\frac{\sqrt{69}}{10}, -\frac{\sqrt{69}}{10}\right\}$

Exercises 13.2 B

1. 3
2. -2
3. 13
4. 15
5. 15
6. not real
7. -6
8. 36
9. 16
10. 3
11. 0
12. 0
13. 3
14. 9
15. $\frac{-1}{4}$
16. 5
17. 16
18. 7.48
19. 9.33
20. 7.94
21. 1.29
22. 4.35
23. 0.63
24. $\{\sqrt{7}, -\sqrt{7}\}$
25. $\left\{\frac{\sqrt{73}}{5}, -\frac{\sqrt{73}}{5}\right\}$
26. $\{40, -40\}$
27. $\{20, -20\}$
28. $\{2.3, -2.3\}$
29. $\{0.82, -0.82\}$

Exercises 13.3 A

1. $6\sqrt{5}$
2. $3\sqrt{6}$
3. $5\sqrt{7}$
4. $7\sqrt{3}$
5. $20\sqrt{10}$
6. 14
7. 12
8. 10
9. 14
10. 36
11. 85
12. $2\sqrt{3}$
13. $3\sqrt{5}$
14. $10\sqrt{6}$
15. $7\sqrt{2}$
16. $2\sqrt{2}$
17. $4\sqrt{10}$
18. $6\sqrt{5}$
19. 30
20. $6\sqrt{5}$

Exercises 13.3 B

1. $7\sqrt{17}$
2. $8\sqrt{3}$
3. $9\sqrt{10}$
4. $4\sqrt{2}$
5. $4\sqrt{5}$
6. 6
7. 12
8. 15
9. 18
10. 44
11. 99
12. $2\sqrt{7}$
13. $3\sqrt{6}$
14. $5\sqrt{3}$
15. $6\sqrt{5}$
16. $3\sqrt{3}$
17. $8\sqrt{3}$
18. $7\sqrt{2}$
19. $80\sqrt{5}$
20. $15\sqrt{7}$

Exercises 13.4 A

1. $\frac{\sqrt{2}}{2}$
2. $\frac{\sqrt{14}}{7}$
3. $\frac{\sqrt{3}}{9}$
4. $\frac{2\sqrt{3}}{15}$
5. $\frac{\sqrt{6}}{6}$
6. $\frac{\sqrt{7}}{7}$
7. $\frac{\sqrt{6}}{12}$
8. $\frac{\sqrt{14}}{14}$
9. $3\sqrt{3}$
10. $\frac{9}{10}$
11. $\frac{\sqrt{14}}{5}$
12. $\frac{\sqrt{30}}{50}$
13. $\frac{\sqrt{2}}{2}$
14. $\frac{\sqrt{42}}{2}$
15. $\frac{5\sqrt{3}}{2}$

Exercises 13.4 B

1. $\frac{\sqrt{3}}{3}$
2. $\frac{\sqrt{30}}{6}$
3. $\frac{\sqrt{2}}{10}$
4. $\frac{3\sqrt{2}}{8}$
5. $\frac{\sqrt{10}}{5}$
6. $\frac{\sqrt{14}}{2}$
7. $\frac{\sqrt{14}}{28}$
8. $\frac{\sqrt{2}}{6}$
9. $\frac{7\sqrt{10}}{100}$
10. $\frac{\sqrt{105}}{50}$
11. $\frac{\sqrt{10}}{4}$
12. $\frac{3}{2}$
13. $\frac{\sqrt{6}}{3}$
14. $\frac{\sqrt{10}}{2}$
15. $\frac{4\sqrt{2}}{3}$

Exercises 13.5 A

1. $7\sqrt{5}$
2. $3\sqrt{2}$
3. $\sqrt{10}$
4. $6\sqrt{2} + 5\sqrt{3}$
5. $7 + 2\sqrt{7}$
6. $\sqrt{6} - 6$
7. $\frac{3 + \sqrt{2}}{2}$
8. $\frac{10 - 2\sqrt{3}}{5}$
9. $\frac{2 - \sqrt{14}}{2}$
10. $\frac{2}{3}$
11. $\frac{6 - \sqrt{2}}{3}$
12. $\frac{3 - \sqrt{5}}{2}$

Exercises 13.5 B

1. $7\sqrt{3}$
2. $\sqrt{6}$
3. $2\sqrt{10} - 5\sqrt{2}$
4. $2\sqrt{3} + 4\sqrt{2}$
5. $6\sqrt{3} - 9$
6. $5 + 2\sqrt{5}$
7. $\frac{1 + \sqrt{5}}{2}$
8. $\frac{2 - \sqrt{6}}{5}$
9. $7 + \sqrt{15}$
10. $-\frac{1}{2}$
11. $\frac{1 - \sqrt{3}}{2}$
12. $\frac{14 - 5\sqrt{3}}{7}$

Exercises 13.6 A

1. $\{4+\sqrt{5}, 4-\sqrt{5}\}$
2. $\{-2+\sqrt{3}, -2-\sqrt{3}\}$
3. $\{8+2\sqrt{3}, 8-2\sqrt{3}\}$
4. $\left\{\frac{-3+\sqrt{10}}{2}, \frac{-3-\sqrt{10}}{2}\right\}$
5. $\left\{\frac{1+2\sqrt{5}}{5}, \frac{1-2\sqrt{5}}{5}\right\}$
6. $\{-1+\sqrt{2}, -1-\sqrt{2}\}$
7. $\{3+\sqrt{6}, 3-\sqrt{6}\}$
8. $\{4+5\sqrt{2}, 4-5\sqrt{2}\}$
9. $\left\{\frac{-7+\sqrt{2}}{2}, \frac{-7-\sqrt{2}}{2}\right\}$
10. $\left\{\frac{1+3\sqrt{3}}{5}, \frac{1-3\sqrt{3}}{5}\right\}$
11. $\left\{\frac{-1+\sqrt{7}}{3}, \frac{-1-\sqrt{7}}{3}\right\}$
12. $\left\{\frac{1}{3}, -1\right\}$
13. $\left\{\frac{-5+3\sqrt{3}}{2}, \frac{-5-3\sqrt{3}}{2}\right\}$
14. $\{9+3\sqrt{5}, 9-3\sqrt{5}\}$

Exercises 13.6 B

1. $\{-7+\sqrt{6}, -7-\sqrt{6}\}$
2. $\{5+\sqrt{7}, 5-\sqrt{7}\}$
3. $\{-9+3\sqrt{2}, -9-3\sqrt{2}\}$
4. $\left\{\frac{4+\sqrt{14}}{3}, \frac{4-\sqrt{14}}{3}\right\}$
5. $\left\{\frac{-1+4\sqrt{2}}{6}, \frac{-1-4\sqrt{2}}{6}\right\}$
6. $\{2+\sqrt{5}, 2-\sqrt{5}\}$
7. $\{-5+\sqrt{10}, -5-\sqrt{10}\}$
8. $\{-6+2\sqrt{7}, -6-2\sqrt{7}\}$
9. $\left\{\frac{8+\sqrt{5}}{3}, \frac{8-\sqrt{5}}{3}\right\}$
10. $\left\{\frac{-3+2\sqrt{3}}{4}, \frac{-3-2\sqrt{3}}{4}\right\}$
11. $\left\{\frac{-3+\sqrt{14}}{5}, \frac{-3-\sqrt{14}}{5}\right\}$
12. $\left\{\frac{-1}{5}, -1\right\}$
13. $\left\{\frac{5+\sqrt{7}}{6}, \frac{5-\sqrt{7}}{6}\right\}$
14. $\{7+2\sqrt{6}, 7-2\sqrt{6}\}$

Exercises 13.7 A

1. $\frac{-5 \pm \sqrt{13}}{6}$
2. $\frac{3 \pm \sqrt{13}}{2}$
3. $3 \pm \sqrt{5}$
4. $\frac{-1 \pm \sqrt{7}}{3}$
5. $-3, \frac{6}{5}$
6. 60, 20
7. $\frac{-1 \pm \sqrt{5}}{2}$
8. $-2 \pm \sqrt{6}$
9. $1 \pm \sqrt{2}$
10. $\frac{5}{4}$
11. $\frac{7}{2}, \frac{-5}{2}$
12. $\frac{-1}{2}, 1$
13. $-b \pm \sqrt{b^2 - c}$
14. $x = 2y$ or $x = \frac{1}{2}y$

Exercises 13.7 B

1. $\frac{-9 \pm \sqrt{21}}{6}$
2. $\frac{2 \pm \sqrt{5}}{3}$
3. $\frac{-5 \pm \sqrt{57}}{4}$
4. $3, \frac{1}{2}$
5. $2, \frac{-9}{10}$
6. 30, −200
7. $\frac{1 \pm \sqrt{5}}{2}$
8. $5 \pm 2\sqrt{7}$
9. $6 \pm 4\sqrt{3}$
10. $\frac{-7}{6}$
11. $\frac{1 \pm \sqrt{2}}{3}$
12. 80, −60
13. $r \pm \sqrt{r^2 + s}$
14. $y = -5x$, or $y = 3x$

Exercises 13.8 A

1. $x^2 = 80^2 + 150^2$; 170 ft.
2. $x^2 = (1450)^2 - (1000)^2$; 1050 ft.
3. $(2x + 4)(2x + 3) = 24$; 0.71 ft. (8.5 in.)
4. $10(x - 20)(2x - 20) = 17{,}500$; 45 cm. by 90 cm.
5. $\frac{4}{x} + \frac{4}{x - 15} = 1$; 20 hr.
6. $\frac{3}{x} + \frac{5}{x + 4} = 1$; 6 hr.
7. $\frac{24}{12 + x} + \frac{24}{12 - x} = \frac{9}{2}$; 4 m.p.h.
8. $9600 = 800t - 16t^2$; 20 sec. (going up), 30 sec. (going down)

9a. $750 = 80x - 2x^2$; 15 or 25

9b. $800 = 80x - 2x^2$; 20

10. $\frac{8 - \sqrt{28}}{3} \approx 0.90$ mole

Exercises 13.8 B

1. $x^2 = 12^2 + 16^2$; 20 ft.
2. $x^2 = 30^2 - 24^2$; 18 ft.
3. $24x + x(18 - x) = 41$; 1 cm.
4. $450x^2 = 50(x + 2)^2$; 1 in.
5. $\frac{7}{x} + \frac{7}{x - 48} = 1$; 56 min.
6. $\frac{20}{x} + \frac{20}{x - 9} = 1$; 45 hr.
7. $\frac{1140}{x + 20} + \frac{870}{x + 30} = \frac{7}{2}$; 550 m.p.h.
8. $2064 = 640t + 16t^2$; 3 sec.
9. $\frac{2100}{x} + 10 = \frac{2100}{x - 5}$; 1 franc = 35¢
10. $27\frac{1}{2}$ amps

UNIT 13 POSTTEST

1. $\left\{\frac{2}{5}, \frac{-2}{5}\right\}$
2. $\left\{\frac{1}{2}, -4\right\}$
3. 6
4. $\frac{16}{5}$
5. 41.23
6. -0.22
7. $\left\{\frac{\sqrt{35}}{3}, -\frac{\sqrt{35}}{3}\right\}$
8. $\{5.4, -5.4\}$
9. 42
10. $8\sqrt{3}$
11. $\frac{\sqrt{6}}{6}$
12. $\frac{\sqrt{21}}{6}$
13. $9\sqrt{6}$
14. $\frac{6-\sqrt{10}}{2}$
15. $\{-4+2\sqrt{3}, -4-2\sqrt{3}\}$
16. $\left\{\frac{2+2\sqrt{7}}{3}, \frac{2-2\sqrt{7}}{3}\right\}$
17. $\left\{\frac{2+\sqrt{6}}{3}, \frac{2-\sqrt{6}}{3}\right\}$
18. $\left\{\frac{-1+\sqrt{3}}{2}, \frac{-1-\sqrt{3}}{2}\right\}$
19. $x(x+3)+\frac{x^2}{2}=72$; 6 ft. by 9 ft.
20. $\frac{36}{20-x}=\frac{36}{20+x}+\frac{3}{4}$; 4 m.p.h.

INDEX